第二部（The Second Book）

Illustrations of the New Camellia Hybrids that Bloom Year-round

四季茶花杂交新品种彩色图集

主　编　刘信凯　钟乃盛　柯欢　武艳芳
Chief Editors　Liu Xinkai　Zhong Naisheng　Ke Huan　Wu Yanfang

浙江科学技术出版社
ZHEJIANG SCIENCE AND TECHNOLOGY PUBLISHING HOUSE

版权所有　侵权必究

图书在版编目（CIP）数据

四季茶花杂交新品种彩色图集. 第二部 / 刘信凯等主编. —杭州：浙江科学技术出版社，2023.2
ISBN 978-7-5739-0484-3

Ⅰ.①四… Ⅱ.①刘… Ⅲ.①山茶花—杂交育种—图集 Ⅳ.① S685.140.36-64

中国国家版本馆 CIP 数据核字（2023）第 024067 号

书　　名	四季茶花杂交新品种彩色图集（第二部）
主　　编	刘信凯　钟乃盛　柯　欢　武艳芳
出　　版	浙江科学技术出版社 杭州市体育场路 347 号　邮政编码：310006 办公室电话：0571-85176593 销售部电话：0571-85176040 网址：www.zkpress.com E-mail：zkpress@zkpress.com
排　　版	杭州兴邦电子印务有限公司
印　　刷	浙江新华数码印务有限公司
经　　销	全国各地新华书店
开　　本	880 mm × 1230 mm　1/16　印　张　38.75
字　　数	980 000
版　　次	2023 年 2 月第 1 版　　印　次　2023 年 2 月第 1 次印刷
书　　号	ISBN 978-7-5739-0484-3　定　价　495.00 元

责任编辑　李宁宁　　　　责任美编　金　晖
责任校对　张　宁　　　　责任印务　崔文红

茶花名人题词

山茶源华夏，四海吐芳华。
品种千千万，舶来占大半。
创新不停息，后浪推前浪。
四季茶花开，启航新时代。

理学博士，研究员，博士生导师
前国际山茶协会主席　　管开云

An Inscription of Famous Camellia People

Camellias originate in China,

Blooming all over the world.

Thousands of the cultivars,

Many of them in China were imported.

Innovation continuously,

Making progress steadily.

Year round blooming camellias,

Started the new era of ornamental camellias.

Guan Kaiyun
Professor & Dr.
Immediate Past President of the ICS

An Inscription of Famous Camellia People

For a thousand years, camellias brightened cold winter days.
The blossoms also gave happiness to autumn and spring.
Now camellias also bloom in the warmth of summer.
A new gift to the world brings joy all year long!

Patricia Short
Herbert Short

Written by the past president of the ICS, Mrs. Patricia Short and the past editor of *International Camellia Journal*, Mr. Herbert Short together.

茶花千年寒冬开，

春秋吐艳幸福来。

而今炎夏亦绽放，

厚礼天下悦整载。

前国际山茶协会主席　帕特丽夏·肯特

前国际山茶协会主办的《国际山茶杂志》编辑　赫伯特·肯特

《四季茶花杂交新品种彩色图集（第二部）》编委会

Editorial Committee of *Illustrations of the New Camellia Hybrids that Bloom Year-round* (*The Second Book*)

总顾问（Chief Advisor）
吴桂昌（Wu Guichang）

编委会主任（Director of the Editorial Committee）
黄万坚（Huang Wanjian）

技术首席专家（Chief Technical Expert）
高继银（Gao Jiyin）

主编（Chief Editors）
刘信凯（Liu Xinkai）　　钟乃盛（Zhong Naisheng）　　柯　欢（Ke Huan）
武艳芳（Wu Yanfang）

副主编（Subeditors）
侯文卿（Hou Wenqing）　　黎艳玲（Li Yanling）　　符秀玉（Fu Xiuyu）
沈　剑（Shen Jian）　　张学平（Zhang Xueping）　　周和达（Zhou Heda）
叶琦君（Ye Qijun）

编委会（Members of the Editorial Committee）
以下按姓氏笔画排列（Under the arrangement by last name stroke order）

王　晶（Wang Jing）　　邓小琴（Deng Xiaoqin）　　叶琦君（Ye Qijun）
冯　骥（Feng Ji）　　刘信凯（Liu Xinkai）　　严丹峰（Yan Danfeng）
李维全（Li Weiquan）　　沈　剑（Shen Jian）　　张　斌（Zhang Bin）
张学平（Zhang Xueping）　　邵太宗（Shao Taizong）　　武艳芳（Wu Yanfang）
周明顺（Zhou Mingshun）　　周和达（Zhou Heda）　　周敏中（Zhou Mingzhong）
赵珊珊（Zhao Shanshan）　　赵鸿杰（Zhao Hongjie）　　胡羡聪（Hu Xiancong）
柯　欢（Ke Huan）　　钟乃盛（Zhong Naisheng）　　侯文卿（Hou Wenqing）
殷广湖（Yin Guanghu）　　殷爱华（Yin Aihua）　　高继银（Gao Jiyin）
黄万坚（Huang Wanjian）　　符秀玉（Fu Xiuyu）　　章丹峰（Zhang Danfeng）
彭逢惠（Peng Fenghui）　　黎艳玲（Li Yanling）　　薛克娜（Xue Kena）

英文翻译（English Translators）
高继银（Gao Jiyin）　　刘信凯（Liu Xinkai）

编委会主要成员简介
Profiles of Major Members of the Editorial Committee

总顾问（Chief Advisor）：
吴桂昌（Wu Guichang）

教授级高级工程师，棕榈生态城镇发展股份有限公司名誉董事长。长期从事园林植物研发产销、园林绿化和园林建筑设计施工，是四季茶花育种项目的主要倡导者和支持者。2009年荣获广东省科技进步奖一等奖，2016年获得国际山茶协会"主席勋章奖"。

A professorate senior engineer Honorary chairman of Palm Eco-town Development Co., Ltd., he has engaged in garden plant research and development, production and marketing, landscaping and garden architecture design and construction for a long time. He is the main advocate and supporter of the ever-blooming camellia breeding project. He won the first prize of Science and Technology Progress of Guangdong Province in 2009 and won the President's Medal Award of ICS in 2016.

编委会主任（Director of the Editorial Committee）：
黄万坚（Huang Wanjian）

园林工程师，肇庆棕榈谷花园有限公司总经理，原棕榈生态城镇发展股份有限公司苗木事业部总经理。长期从事新优植物引种和应用推广，是香港特别行政区和上海两地迪士尼项目核心区苗木供应服务主要负责人，也是四季茶花育种项目的主要倡导者和支持者。参编专著2部，获广东省科技进步奖一等奖1项、中国风景园林学会科技二等奖1项。

A landscape engineer, general manager of Zhaoqing Palm Valley Garden Co., Ltd., and general manager of nursery stock division of original Palm Eco-town development Co., Ltd. He has engaged in new plant introduction and application promotion for a long time. He is the head of nursery stock supple service in the core areas of both Hong Kong and Shanghai Disneyland projects and a leading advocate and supporter of the ever-blooming camellia breeding project. He co-edited two monographs, won the first prize of Science and Technology Progress of Guangdong Province and the second prize of Science and Technology of Chinese Society of Landscape Architecture.

技术首席专家（Chief Technical Expert）：
高继银（Gao Jiyin）

中国林业科学研究院亚热带林业研究所研究员，肇庆棕榈谷花园有限公司茶花育种团队的首席专家。曾获得原国家林业部有关油茶研究的科学进步奖三等奖2项。20世纪80年代开始整理国内外茶花品种，成功参与营建金华国际山茶物种园，在此期间，出版了3部茶花专著。2006年开始，专心研究四季茶花新品种的育种，出版了世界上第一部有关四季茶花新品种的专著。邮箱：y25006@163.com

A researcher of the Research Institute of Subtropical Forestry, CAF., chief expert of the Camellia Breeding Team of Zhaoqing Palm Valley Garden Co., Ltd. He won two scientific progress third level prizes from the former China Ministry of Forestry in oil camellia researches. He began to classify camellia outstanding varieties at home and abroad in the 1980s, and successfully participated in the construction of Jinhua International Camellia Species Garden. During the period, he published three camellia monographs. Since 2006, he has concentrated on the breeding of the new camellia hybrids that bloom year-round and published the world's first monograph on the new camellia hybrids.

第一主编（First Chief Editor）：
刘信凯（Liu Xinkai）

高级工程师，肇庆棕榈谷花园有限公司植物研究所所长，中国花卉协会茶花分会副秘书长，四季茶花育种项目核心成员之一。长期致力于山茶属植物园林开发和应用及四季茶花的产业化推广。主持或参与国家、省、市级科研项目十余项，获广东省科技进步奖一等奖、三等奖各1项，广东省农业推广二等奖1项，梁希林业科技二等奖1项，中国风景园林学会科技二等奖1项。主编或参编专著2部。邮箱：lxk1000@163.com

A senior engineer, director of the Plant Research Institute of Zhaoqing Palm Valley Garden Co., Ltd., deputy secretary general of the Camellia Branch of China Flower Association and one of the core members of the ever-blooming camellia breeding project. He has been devoting to the development and application of Genus *Camellia* and the industrialization promotion of the four season camellia hybrids for a long period. He has presided ever or participated in more than ten national, provincial and municipal scientific research projects. He won the first prize and the third prize of Science and Technology Progress of Guangdong Province, the second prize of Agricultural Promotion of Guangdong Province, the second prize of Liang Xi Forestry Science and Technology, and the second prize of Science and Technology of Chinese Society of Landscape Architecture. He edited or co-edited two monographs.

第二主编（Second Chief Editor）:
钟乃盛（Zhong Naisheng）

园林工程师，肇庆棕榈谷花园有限公司植物研究所研发主管，四季茶花育种项目核心成员之一。自加入育种团队以来，一直致力于四季茶花的育种研究，贡献突出。主持或参与国家、省、市级科研项目十余项，获广东省农业推广一、二等奖各1项，梁希林业科技二等奖1项，中国风景园林学会科技二等奖1项。参编茶花专著2部。邮箱：345466584@qq.com

A landscape engineer, research and development director of the Plant Research Institute of Zhaoqing Palm Valley Garden Co., Ltd., and one of the core members of the ever-blooming camellia breeding project. Since joining the breeding team, he has been devoting to the camellia breeding research and made outstanding contributions. He has presided over or participated in more than ten national, provincial and municipal scientific research projects. He won the first prize and second prize of Agricultural Promotion of Guangdong Province, the second prize of Liang Xi Forestry Science and Technology, and the second prize of Science and Technology of Chinese Society of Landscape Architecture. He co-edited two monographs on camellia.

第三主编（Third Chief Editor）:
柯欢（Ke Huan）

园林正高级工程师，植物学硕士，佛山市林业科学研究所（佛山植物园）园区管理中心副主任，长期从事山茶等木本花卉的育种与应用研究。主持省级科研项目2项，参与国家、省、市级科技项目十余项，获省级农业推广二等奖4项、市级科技进步奖二等奖1项。主编与参编专著14部，发表科技论文52篇，制定省、市行业标准3项，参与培育茶花新品种4个，入选多个省、市级专家库。邮箱：350017258@qq.com

A senior landscape engineer, master of botany, deputy director of Park Management Center of Forestry Research Institute of Foshan City (Foshan Botanical Garden). He has engaged in the breeding and application research of camellia and other woody flower trees for a long time. He has presided over two provincial scientific research projects and participated in more than ten national, provincial and municipal science and technology projects and won four second prizes of provincial agricultural promotion and one second prize of municipal science and technology progress. He edited or co-edited 14 monographs, published 52 science and technology papers, established three industrial standards at the provincial and municipal levels, was involved in breeding four new camellia varieties, and was selected into many provincial and municipal expert libraries.

第四主编(Fourth Chief Editor):
武艳芳(Wu Yanfang)

硕士学位,棕榈生态城镇发展股份有限公司研发副总监,国家花卉产业技术创新战略联盟理事,河南省风景园林学会标准化委员会委员,河南省园林绿化协会理事。长期从事新优园林植物的引种驯化、选育及推广应用研究,在山茶种质资源收集、新品种选育及推广应用方面贡献突出。邮箱:wuyanfang@palm-la.com

Master's degree, Vice-director of Palm Eco-town Development Co., Ltd, director of the National Flower Industry Technology Innovation Strategic Alliance, member of Standardization Committee of Henan Landscape Architecture Association and director of Henan Landscape Architecture Association. She has been engaged in the research of introduction, domestication selection and application of new excellent garden plants for a long time, especially, she has made outstanding contributions to the collection of Camellia germplasm resources, selection, breeding, popularization and application of the new varieties.

序一

让四季茶花开遍城乡大地

茶花是中国的十大传统名花之一，花大、色艳、花期长且四季常绿，戴雪而荣，具松柏之骨，携桃李之姿，是城乡园林绿化的理想植物和优秀观赏花木，深受国人喜爱。

经中国首个茶花育种团队——广东肇庆棕榈谷花园有限公司多年不懈的努力，茶花中的极品——四季茶花于21世纪的头20年里隆重出世，至今已有545个四季茶花新品种、104个越南抱茎茶杂交新品种、39个常规茶花杂交新品种和76个机遇苗新品种，总计764个茶花新品种问世。

四季茶花一经问世就惊艳世界。它们不仅四季开花，而且适应性强，品种丰富，花型多样，花色艳丽，可在公园、花园、居住区、风景区、森林公园及城乡道路等多种场合广泛使用，极大地丰富了城乡园林绿化景观，美化了城乡人居环境，从而使茶花在城乡园林绿化中的应用进入新的时代。

四季茶花叶片革质光亮，植株紧凑，枝叶浓密，开花量大，花期长，春夏秋冬四季开花，弥补了一般茶花只在冬春开花、其他季节无花的不足。四季茶花既具有玫瑰般多样的花型、长的开花期，又具有四季常绿、终年不凋、易于栽培、病虫害少等优点。目前，全国各地的城镇和乡村都在广泛应用四季茶花新品种。四季茶花已成为城乡绿化的一颗新星，它璀璨耀眼，备受人们喜爱。

随着中国生态文明建设的深入发展，城市化进程的加快，以及美丽中国建设的持续推进，美好人居、美丽家园、优美庭院成为中国未来的发展目标和方向，常规茶花和四季茶花将走入千家万户，开遍城乡大地，茶花产业将迎来五彩缤纷的发展时代。愿中国茶花事业更加繁荣昌盛！

中国花卉协会茶花分会会长
浙江省风景园林学会理事长
（施德法）

Let the Four-season Camellia Hybrids Bloom All Over Urban and Rural Areas

Camellia is one of the top ten traditional famous flowers in China. It has the characteristics of big flowers, bright colors, a long blooming period, evergreen all the year round. It is glorious by covering snow with the bones of pine and cypress and the appearance of peach and plum. Camellia is a kind of fine urban and rural landscaping plants and excellent ornamental trees, which is deeply loved by Chinese.

After years of effort by the Camellia Breeding Team of Zhaoqing Palm Valley Garden Co., Ltd., Guangdong province which is the first camellia breeding team in China, the best camellia—four-season camellia hybrids were ceremoniously born in the first 20 years of the 21st century. So far, there are 545 new camellia hybrids, 104 *C. amplexicaulis'* hybrids, 39 normal camellia hybrids, and 76 chance seedling varieties, which a total of 764 new camellia varieties have been bred out.

The four-season camellia hybrids amazed the world as soon as they came out. They are not only blooming year-round, but also have strong adaptability and a lot of varieties with different flower forms and bright colors. Moreover, and they can be widely used in parks, flower gardens, residential areas, scenic areas, forest park and urban and rural roads, etc., which greatly enrich the urban and rural afforestation landscape, beautify the urban and rural living environment, thus making the applications of camellias in urban and rural landscaping enter a new era.

The four-season camellia hybrids' leaves are leathery and shiny, plants are compact, branches and leaves and flowers are dense, the blooming period is long which is from spring and summer to autumn and winter, blooming in four seasons, making up for the defect of normal camellias that only bloom in winter and spring and no flowers in other seasons. Moreover, the four-season camellia hybrids have diversified flower forms and a long blooming period which are the same as rose, and have the advantages of being evergreen in four seasons, not withering all year round, and being easy to cultivate and having fewer pests and diseases. At present, cities, towns and villages of all over China are widely using the new camellia hybrids, making them become a new star of urban and rural greening. The star is bright, dazzling and deeply loved.

With the further development of China's ecological civilization, the speeding up of urbanization and the sustainable development of Beautiful China Construction, good living, beautiful home and beautiful courtyard are becoming China's future development goals and directions. Normal camellias and the four-season camellia hybrids will enter innumerable homes and bloom all over the urban and rural land, camellia industry will usher in a colorful development era. May the cause of Chinese camellia be thriving and prosperous!

President of the Camellia Branch of China Flower Association
Director of Zhejiang Society of Landscape Architecture

Foreword Two

Everblooming Camellia Hybrids, a Milestone in Ornamental Camellias
Written by Prof. Gianmario Motta

    I am proud of writing a foreword for this book, realized by the team led by my old friend and mentor, Professor Gao Jiyin.

    In 2008, Professor Gao Jiyin published a key paper on the *Camellia azalea / Camellia changii* in *International Camellia Journal*. That species, discovered in 1985, has revolutionized our ideas on ornamental camellias because, as we know, it is almost everblooming. Furthermore, it has some nice habits, light red blooms, and a nice foliage. But that species is hardy only in tropical and sub-tropical areas, and it does not set seeds easily. Hence, combining everblooming with hardiness becomes a critical challenge, and requires a large-scale hybridization program, with *C. azalea* as pollen parent. This program has been carried on an industrial scale, by a Chinese company, namely Palm Eco-town Development Co., Ltd., which owns the patents of hybrids. In 2016, a book, namely *Illustrations of New Camellia Hybrids that Bloom Year-round*, was published, with 217 of descriptions of these new hybrids. Now, after 6 years, this second book adds further 328 new hybrids.

    In my opinion, the hybridization of *Camellia azalea* is a milestone in the development of ornamental camellias. First, it can add several months to the blooming period of camellias, not only in tropical areas but also in temperate zones across the world—in China, Japan, Europe, United States, Australia, and New Zealand. Second, the blooming season of these hybrids makes them practically free from the flower blight, thus offering a real complement and alternative to traditional ornamental camellias. Third, these hybrids offer a wide variety of flower forms, since they result from crossing *C. azalea* with various cultivars of *C. japonica* and other camellia species, including the striking *C. amplexicaulis*. Fourth, these hybrid cultivars already are really numerous, with a rather impressive total of 545 cultivars (328 in the second book and 217 in the first one). It is likely that number will grow in the next future.

    Thanks to these superior characteristics, everblooming hybrids are creating great perspectives in the camellia world, by renovating the market for balcony planting, private gardens, and, of course, municipal planting. The related increased demand of camellia will be beneficial for nurseries and the whole camellia world.

Professor
President of the International Camellia Society

序二

四季茶花杂交种，观赏性茶花的里程碑

我能为我的老朋友、导师高继银教授带领的团队所完成的这部书写序言，感到很荣幸。

2008年，高继银教授在《国际山茶杂志》上发表了一篇关于杜鹃红山茶/张氏红山茶的重要论文。该物种发现于1985年，它彻底改变了我们对观赏性山茶花的看法，因为正如我们所知，它几乎全年开花不断。另外，该物种有一些很好的性状，就是花朵淡红色，叶片漂亮。可是，该物种只适应热带和亚热带地区，而且不容易结籽。因此，把四季开花与耐寒性结合起来，是一个极为重要的挑战，而且需要一个以杜鹃红山茶为亲本的规模化的杂交计划。该计划已经由一家拥有杂交新品种专利、名叫棕榈生态城镇发展有限公司的中国公司，大规模地进行了实施。2016年，《四季茶花杂交新品种彩色图集》问世，书中详细介绍了217个杂交新品种。6年后，这部《四季茶花杂交新品种彩色图集（第二部）》将杂交新品种的数量进一步增加了328个。

我认为，杜鹃红山茶的杂交育种是观赏性茶花发展的一个里程碑。第一，它影响范围很广，不仅在热带地区，而且在世界各地的温带地区——中国、日本、欧洲、美国、澳大利亚和新西兰，使茶花的花期延长了几个月。第二，这些杂交种在开花季节几乎没有花枯萎病，从而可以真正地对传统观赏性茶花进行补充和替代。第三，由于这些杂交种是通过杜鹃红山茶与各种各样的红山茶品种或其他的山茶物种——包括引人注目的越南抱茎茶，杂交得到的，因此，它们的花型各式各样。第四，这些杂交品种数量众多，总计已达545个（第二部书中有328个，第一部书中有217个）。这个数字未来很可能还会增加。

由于这些优势性状，四季茶花杂交种通过阳台种植、私人花园营造、市政种植等方面的市场应用，正在山茶世界中展现出广阔的远景。由此而引起的对山茶需求量的增加，将有利于苗圃和整个山茶世界的发展。

国际山茶协会主席　**甘马瑞奥·莫塔教授**

编者的话

自从2016年棕榈生态城镇发展股份有限公司（以下简称"棕榈股份"）出版世界上第一部以介绍四季茶花为主的《四季茶花杂交新品种彩色图集》专著以来，距今已有6年了。在这6年中，肇庆棕榈谷花园有限公司和棕榈股份，联合佛山市林业科学研究所（佛山植物园）、广东阿婆六生态农业发展有限公司、浙江彩园居生态农业发展有限公司、上海星源农业实验场以及浙江省金华市林业技术推广站等在茶花新品种创制和园林应用方面颇具实力的单位，齐心合力，迎难而上，在前期四季茶花新品种培育成果的基础上，继续加大育种力度，全力向此领域的巅峰冲刺。

正如我们在第一部书中所述，四季茶花新品种是对现有常规茶花品种（系指不具杜鹃红山茶基因的全部茶花栽培品种和物种，以下全书同）的一个有力补充。通过多年对四季茶花新品种的栽培和推广，新品种的优势已经被国内外同行和园林界认可。这标志着四季茶花新品种已经从梦想变为现实，颠覆了千百年来人们对茶花只能冬春开花的认知。四季茶花首先成功地应用于棕榈股份营建的广西桂林阳朔三千漓旅游观光项目，继而在"国际杰出茶花园"广东阿婆六茶花谷和佛山植物园大获成功，开创了茶花用于环境美化的新篇章，推动和振兴了我国的茶花业发展。

本书主要详述的四季茶花新品种共328个，其中145个是杜鹃红山茶与常规茶花之间获得的F_1代新品种，该类新品种涉及59个杂交组合（见本书HA）；183个是杜鹃红山茶与杜鹃红山茶F_1代或者常规茶花品种与杜鹃红山茶F_1代之间获得的回交新品种，该类新品种涉及38个回交组合（见本书HAR）。另外，本书还详述了49个越南抱茎茶杂交新品种，它们涉及27个杂交组合（见本书HB）；19个常规茶花杂交新品种，它们涉及12个杂交组合（见本书HC）；还有从25个常规茶花品种中选育出的45个机遇苗新品种（见本书HD）。综上所述，本书详述的新品种数达441个，涉及的杂交组合数为161个。

如果加上第一部书介绍的新品种，到目前为止，我们所获得的新品种数量分别为545个四季茶花新品种，104个越南抱茎茶杂交新品种，39个常规茶花杂交新品种，76个机遇苗新品种，这样，茶花新品种总数已经达到764个。从2006年至2022年这16年间，能培育出如此多的茶花新品种，特别是四季茶花新品种，不得不说这是一个奇迹。

这里应该说明的是，对于第一部书中曾经出现过的杂交组合，本书将保持原杂交组合的编号，而对于这些组合获得的新品种，它们的编号将接续第一部著作中的品种序列编号。因此，本书有些杂交组合的编号和品种的编号是不连续的。如果您需要第一部书，可直接联系作者或出版社。

我们的茶花育种仍在路上，世界各地的茶花育种也在火热进行中，一个更加广泛应用茶花装点环境、绚丽多彩的春天已经到来了。

茶花世界，艳丽无比，展望未来，前景无限！让我们一起继续努力吧！

全体编者

Editors' Words

It has been 6 years since Palm Eco-town Development Co., Ltd. (Hereinafter referred to Palm Shares) published the world's first monograph, *Illustrations of the New Camellia Hybrids that Bloom Year-round*, which mainly introduces the ever-blooming camellia hybrids in 2016. During the past 6 years, Zhaoqing Palm Valley Garden Co., Ltd. and Palm Shares have worked with Forestry Research Institute of Foshan City (Foshan Botanical Garden), Guangdong Apoliu Ecological Agriculture Development Co., Ltd., Zhejiang Caiyuanju Ecological Agriculture Development Co., Ltd., Shanghai Xingyuan Agricilturual Experimental Farm, and Jinhua Forestry Technology Extension Station of Zhejiang Province etc. which are the strong units in new camellia hybrids breeding and garden application. We all pulled together and met difficulties head-on to tackle the difficulties head-on. On the base of the previous achievements of breeding new camellia hybrids that bloom year-round, we continue to put more effort on breeding, and are sprinting to the peak of this field.

As what we stated in our first book, the new camellia hybrids that bloom year-round are a powerful complement to the existing normal camellia cultivars (refers to all camellia cultivars and species without genes of *C. azalea*, all same in the book). Through cultivating and promoting the new hybrids for many years, the advantages of the new hybrids have been recognized by the domestic and foreign counterparts and landscape circles. It marks that the new camellia hybrids have made the dream a reality and have overturned people's knowledge that camellia can only bloom in winter and spring for thousands of years. The four-season camellias were firstly and successfully applied in Sanqianli tourism and sightseeing project that set up by Palm Shares at Yangsuo, Guilin, Guangxi, and then, in International Camellia Gardens of Excellence Guangdong Apoliu Camellia Valley and Foshan Botanical Garden, which opened a new chapter of the camellia application in environmental beautification and promoted the development of the camellia industry in China.

This book mainly describes 328 new hybrids that bloom year-round in detail. Of them, 145 F_1 new hybrids of *C. azalea* are obtained from the crosses between *C. azalea* and normal camellia cultivars, which involving 59 cross-combinations (See HA, this book); 183 backcross new hybrids of *C. azalea* F_1 are obtained from the crosses between *C. azalea* and F_1 new hybrids of *C. azalea* or from the crosses between normal camellia cultivar and F_1 new hybrids of *C. azalea*, which involving in 38 backcross combinations (See HAR, this book). In addition, 49 *C. amplexicaulis* new hybrids in 27 cross-combinations (See HB, this book), 19 new hybrids of normal camellia cultivars in 12 cross-combinations (See HC, this book) and 45 chance new varieties that selected from 25 normal camellia cultivars (See HD, this book) are all described in detail in this book. To sum up, 441 new varieties are described in this book and 161 cross-combinations are involved.

If the new hybrids introduced in our first book are added, there are 545 new camellia hybrids that bloom year-round, 104 *C. amplexicaulis* hybrids, 39 new hybrids obtained from normal camellia cultivars and 76 new varieties selected from chance seedlings. Thus, we have obtained 764 new camellia varieties in total now. It has to be said that it is a miracle that we have been able to breed so many new camellia varieties, especially the new camellia hybrids with its ever-blooming characteristics during the 16 years from 2006 to 2022.

It should be noted that, for the cross-combinations that had already appeared in our first book, their original serial numbers will be maintained in this book, while for the new hybrids obtained from the cross-combinations, their numbers will follow the serial numbers of our first book. Therefore, the numbering of the cross-combinations and their new hybrids are discontinuous in this book. If you need our first book, please contact the authors or the publishing house.

Our camellia breeding is still on the way, the camellia breeding around the world is also in full swing, and a colorful spring that extensively utilizing camellias to decorate environments has arrived.

The gorgeous camellia world will have a nice future. Let's make continuing efforts together!

All editors of the book

目 录

再论四季茶花杂交新品种的六大性状 ... 1

杜鹃红山茶 F_1 代新品种育种概况 ... 8

杜鹃红山茶 F_1 代新品种详解 ... 13

    杂交组合 HA-01. 杜鹃红山茶 × 红山茶品种'克瑞墨大牡丹' ... 13

    杂交组合 HA-02. 杜鹃红山茶 × 红山茶品种'佛蕾德' ... 25

    杂交组合 HA-03. 红山茶品种'媚丽' × 杜鹃红山茶 ... 28

    杂交组合 HA-05. 杜鹃红山茶 × 红山茶品种'超级南天武士' ... 31

    杂交组合 HA-06. 杜鹃红山茶 × 红山茶品种'大菲丽丝' ... 34

    杂交组合 HA-10. 杜鹃红山茶 × 红山茶品种'达婷' ... 37

    杂交组合 HA-11. 杜鹃红山茶 × 红山茶品种'闪烁' ... 51

    杂交组合 HA-12. 红山茶品种'锯叶椿' × 杜鹃红山茶 ... 54

    杂交组合 HA-13. 杜鹃红山茶 × 红山茶品种'红露珍' ... 56

    杂交组合 HA-14. 杜鹃红山茶 × 红山茶品种'花牡丹' ... 60

    杂交组合 HA-16. 杜鹃红山茶 × 红山茶品种'皇家天鹅绒' ... 63

    杂交组合 HA-21. 红山茶品种'都鸟' × 杜鹃红山茶 ... 67

    杂交组合 HA-22. 杜鹃红山茶 × 红山茶品种'毛缘大黑红' ... 70

    杂交组合 HA-24. 红山茶品种'客来邸' × 杜鹃红山茶 ... 72

    杂交组合 HA-26. 红山茶品种'白斑康乃馨' × 杜鹃红山茶 ... 75

    杂交组合 HA-32. 杜鹃红山茶 × 多齿红山茶 ... 77

    杂交组合 HA-34. 杜鹃红山茶 × 浙江红山茶 ... 79

    杂交组合 HA-35. 杜鹃红山茶 × 红山茶品种'霍伯' ... 83

    杂交组合 HA-36. 杜鹃红山茶 × 越南抱茎茶 ... 86

    杂交组合 HA-41. 红山茶品种'金盘荔枝' × 杜鹃红山茶 ... 88

    杂交组合 HA-43. 杜鹃红山茶 × 云南山茶杂交种'蜂露' ... 90

    杂交组合 HA-44. 杜鹃红山茶 × 肥后茶品种'王冠' ... 93

    杂交组合 HA-47. 全缘红山茶 × 杜鹃红山茶 ... 95

    杂交组合 HA-48. 非云南山茶杂交种'珊瑚乐' × 杜鹃红山茶 ... 100

    杂交组合 HA-53. 杜鹃红山茶 × 红山茶品种'黄埔之浪' ... 104

    杂交组合 HA-54. 杜鹃红山茶 × 红山茶品种'莫顿州长' ... 108

    杂交组合 HA-55. 杜鹃红山茶 × 白花抱茎茶 ... 110

    杂交组合 HA-56. 杜鹃红山茶 × 非云南山茶杂交种'甜凯特' ... 112

    杂交组合 HA-57. 杜鹃红山茶 × 红山茶品种'波特' ... 116

    杂交组合 HA-58. 杜鹃红山茶 × 红山茶品种'道温的曙光' ... 119

杂交组合 HA-59. 杜鹃红山茶 × 红山茶品种'金博士'	125
杂交组合 HA-60. 杜鹃红山茶 × 红山茶类型'耐冬'	130
杂交组合 HA-61. 杜鹃红山茶 × 红山茶品种'羽衣'	132
杂交组合 HA-62. 杜鹃红山茶 × 红山茶品种'阿兰'	134
杂交组合 HA-63. 杜鹃红山茶 × 红山茶品种'桑迪玛斯'	137
杂交组合 HA-64. 杜鹃红山茶 × 红山茶品种'琼克莱尔'	140
杂交组合 HA-65. 杜鹃红山茶 × 红山茶品种'香神'	142
杂交组合 HA-66. 杜鹃红山茶 × 威廉姆斯杂交种'詹米'	146
杂交组合 HA-67. 杜鹃红山茶 × 非云南山茶杂交种'龙火珠'	148
杂交组合 HA-68. 非云南山茶杂交种'龙火珠' × 杜鹃红山茶	159
杂交组合 HA-69. 杜鹃红山茶 × 红山茶品种'火把'	164
杂交组合 HA-70. 杜鹃红山茶 × 红山茶品种'艳口红'	166
杂交组合 HA-71. 红山茶品种'艳口红' × 杜鹃红山茶	169
杂交组合 HA-72. 杜鹃红山茶 × 非云南山茶杂交种'克拉丽'	171
杂交组合 HA-73. 杜鹃红山茶 × 红山茶品种'红绒贝蒂'	173
杂交组合 HA-74. 杜鹃红山茶 × 红山茶品种'锦鱼叶椿'	175
杂交组合 HA-75. 杜鹃红山茶 × 红山茶品种'迪士尼乐园'	177
杂交组合 HA-76. 杜鹃红山茶 × 红山茶品种'夸特'	179
杂交组合 HA-77. 杜鹃红山茶 × 非云南山茶杂交种'香四射'	181
杂交组合 HA-78. 杜鹃红山茶 × 云南山茶品种'山茶之都'	185
杂交组合 HA-79. 红山茶品种'牛西奥转马' × 杜鹃红山茶	187
杂交组合 HA-80. 红山茶品种'忠实' × 杜鹃红山茶	192
杂交组合 HA-81. 红山茶品种'孔雀玉浦' × 杜鹃红山茶	195
杂交组合 HA-82. 红山茶品种'玉之浦' × 杜鹃红山茶	204
杂交组合 HA-83. 红山茶品种'黑魔法' × 杜鹃红山茶	206
杂交组合 HA-84. 红山茶品种'咖啡杯' × 杜鹃红山茶	209
杂交组合 HA-85. 杜鹃红山茶 × 雪山茶品种'万代'	211
杂交组合 HA-86. 雪山茶品种'万代' × 杜鹃红山茶	213
杂交组合 HA-87. 云南山茶品种'豪斯' × 杜鹃红山茶	215

杜鹃红山茶 F_1 代回交新品种育种概况 ············ 217

杜鹃红山茶 F_1 代回交新品种详解 ············ 223

回交组合 HAR-01. 杜鹃红山茶 × 杜鹃红山茶 F_1 代新品种'吉利牡丹'	223
回交组合 HAR-02. 杜鹃红山茶 × 杜鹃红山茶 F_1 代新品种'夏风热浪'	225
回交组合 HAR-03. 杜鹃红山茶 × 杜鹃红山茶 F_1 代新品种'夏日粉黛'	253
回交组合 HAR-04. 杜鹃红山茶 × 杜鹃红山茶 F_1 代新品种'夏日红霞'	262
回交组合 HAR-05. 杜鹃红山茶 × 杜鹃红山茶 F_1 代新品种'夏日七心'	269
回交组合 HAR-06. 杜鹃红山茶 × 杜鹃红山茶 F_1 代新品种'夏日台阁'	272
回交组合 HAR-07. 杜鹃红山茶 × 杜鹃红山茶 F_1 代新品种'夏日探戈'	282
回交组合 HAR-08. 杜鹃红山茶 × 杜鹃红山茶 F_1 代新品种'夏咏国色'	285
回交组合 HAR-09. 杜鹃红山茶 × 杜鹃红山茶 F_1 代新品种'茶香飘逸'	294
回交组合 HAR-10. 杜鹃红山茶 × 杜鹃红山茶 F_1 代新品种'红屋积香'	297

回交组合 HAR-11. 杜鹃红山茶 × 杜鹃红山茶 F_1 代新品种'香夏红娇' 308

回交组合 HAR-12. 杜鹃红山茶 × 杜鹃红山茶 F_1 代新品种'夏梦华林' 312

回交组合 HAR-13. 杜鹃红山茶 × 杜鹃红山茶 F_1 代新品种'夏梦小旋' 321

回交组合 HAR-14. 杜鹃红山茶 × 杜鹃红山茶 F_1 代新品种'夏日广场' 326

回交组合 HAR-15. 杜鹃红山茶 × 杜鹃红山茶 F_1 代新品种'夏日粉丽' 351

回交组合 HAR-16. 杜鹃红山茶 × 杜鹃红山茶 F_1 代新品种'满天红星' 354

回交组合 HAR-17. 杜鹃红山茶 × 杜鹃红山茶 F_1 代新品种'红波涌金' 356

回交组合 HAR-18. 杜鹃红山茶 × 杜鹃红山茶 F_1 代新品种'夏梦文清' 360

回交组合 HAR-19. 杜鹃红山茶 × 杜鹃红山茶 F_1 代新品种'不知寒暑' 376

回交组合 HAR-20. 杜鹃红山茶 × 杜鹃红山茶 F_1 代新品种'夏梦春陵' 378

回交组合 HAR-21. 杜鹃红山茶 × 杜鹃红山茶 F_1 代新品种'夏蝶群舞' 381

回交组合 HAR-22. 杜鹃红山茶 × 杜鹃红山茶 F_1 代新品种'夏日红绢' 383

回交组合 HAR-23. 杜鹃红山茶 × 杜鹃红山茶 F_1 代新品种'夏日叠星' 389

回交组合 HAR-24. 杜鹃红山茶 × 杜鹃红山茶 F_1 代新品种'夏梦可娟' 393

回交组合 HAR-25. 杜鹃红山茶 × 杜鹃红山茶 F_1 代新品种'夏梦玉兰' 396

回交组合 HAR-26. 杜鹃红山茶 × 杜鹃红山茶 F_1 代新品种'浪漫粉娘' 398

回交组合 HAR-27. 杜鹃红山茶 × 杜鹃红山茶 F_1 代新品种'火红牡丹' 400

回交组合 HAR-28. 杜鹃红山茶 F_1 代新品种'满天红星' × 杜鹃红山茶 F_1 代新品种'吉利牡丹' 402

回交组合 HAR-29. 杜鹃红山茶 F_1 代新品种'水月紫鹃' × 杜鹃红山茶 F_1 代新品种'夏梦文清' 404

回交组合 HAR-30. 杜鹃红山茶 × 杜鹃红山茶 F_1 代新品种'夏日粉裙' 406

回交组合 HAR-31. 杜鹃红山茶 × 杜鹃红山茶 F_1 代新品种'书香之家' 408

回交组合 HAR-32. 杜鹃红山茶 × 杜鹃红山茶 F_1 代新品种'粤桂大嫂' 410

回交组合 HAR-33. 杜鹃红山茶 × 杜鹃红山茶 F_1 代新品种'夏日红杯' 412

回交组合 HAR-34. 红山茶品种'孔雀玉浦' × 杜鹃红山茶 F_1 代新品种'红屋积香' 414

回交组合 HAR-35. 红山茶品种'玉之浦' × 杜鹃红山茶 F_1 代新品种'红屋积香' 416

回交组合 HAR-36. 红山茶品种'白斑康乃馨' × 杜鹃红山茶 F_1 代新品种'红屋积香' 435

回交组合 HAR-37. 红山茶品种'聚香' × 杜鹃红山茶 F_1 代新品种'夏梦文清' 437

回交组合 HAR-38. 越南抱茎茶 × 杜鹃红山茶 F_1 代新品种'夏日红霞' 442

越南抱茎茶杂交新品种育种概况 444

越南抱茎茶杂交新品种详解 448

常规茶花杂交新品种育种概况 497

常规茶花杂交新品种详解 499

茶花机遇苗新品种选育概况 518

茶花机遇苗新品种详解 520

中国主要茶花苗圃选登 565

参考文献 574

茶花新品种索引 576

Contents

Re-discussion on the Six Major Characteristics of the New Camellia Hybrids that Bloom Year-round ⋯ 1
The Breeding Outline on the *C. azalea* F₁ New Hybrids ⋯ 8
The Detailed Descriptions of the F₁ New Hybrids of *C. azalea* ⋯ 13

    HA-01. *C. azalea* × *C. japonica* cultivar 'Kramer's Supreme' ⋯ 13
    HA-02. *C. azalea* × *C. japonica* cultivar 'Fred Sander' ⋯ 25
    HA-03. *C. japonica* cultivar 'Tama Beauty' × *C. azalea* ⋯ 28
    HA-05. *C. azalea* × *C. japonica* cultivar 'Dixie Knight Supreme' ⋯ 31
    HA-06. *C. azalea* × *C. japonica* cultivar 'Francis Eugene Phillis' ⋯ 34
    HA-10. *C. azalea* × *C. japonica* cultivar 'Mary Agnes Patin' ⋯ 37
    HA-11. *C. azalea* × *C. japonica* cultivar 'Tama Glitters' ⋯ 51
    HA-12. *C. japonica* cultivar 'Nokogiriba-tsubaki' × *C. azalea* ⋯ 54
    HA-13. *C. azalea* × *C. japonica* cultivar 'Hongluzhen' ⋯ 56
    HA-14. *C. azalea* × *C. japonica* cultivar 'Daikagura' ⋯ 60
    HA-16. *C. azalea* × *C. japonica* cultivar 'Royal Velvet' ⋯ 63
    HA-21. *C. japonica* cultivar 'Miyakodori' × *C. azalea* ⋯ 67
    HA-22. *C. azalea* × *C. japonica* cultivar 'Clark Hubbs' ⋯ 70
    HA-24. *C. japonica* cultivar 'Collettii' × *C. azalea* ⋯ 72
    HA-26. *C. japonica* cultivar 'Ville de Nantes' × *C. azalea* ⋯ 75
    HA-32. *C. azalea* × *C. polyodonta* ⋯ 77
    HA-34. *C. azalea* × *C. chekiangoleosa* ⋯ 79
    HA-35. *C. azalea* × *C. japonica* cultivar 'Bob Hope' ⋯ 83
    HA-36. *C. azalea* × *C. amplexicaulis* ⋯ 86
    HA-41. *C. japonica* cultivar 'Jinpan Lizhi' × *C. azalea* ⋯ 88
    HA-43. *C. azalea* × *C. reticulata* hybrid 'Fenglu' ⋯ 90
    HA-44. *C. azalea* × Higo cultivar 'Okan' ⋯ 93
    HA-47. *C. subintegra* × *C. azalea* ⋯ 95
    HA-48. Non-*reticulata* hybrid 'Coral Delight' × *C. azalea* ⋯ 100
    HA-53. *C. azalea* × *C. japonica* cultivar 'Huangpu Zhilang' ⋯ 104
    HA-54. *C. azalea* × *C. japonica* cultivar 'Governor Mouton' ⋯ 108
    HA-55. *C. azalea* × *C. lucii* Orel & Curry (2015) ⋯ 110
    HA-56. *C. azalea* × Non-*reticulata* hybrid 'Sweet Emily Kate' ⋯ 112
    HA-57. *C. azalea* × *C. japonica* cultivar 'Peter Pan' ⋯ 116
    HA-58. *C. azalea* × *C. japonica* cultivar 'Dawn's Early Light' ⋯ 119

HA-59. *C. azalea* × *C. japonica* cultivar 'Dr King' ·········· 125

HA-60. *C. azalea* × *C. japonica* form 'Naidong' ·········· 130

HA-61. *C. azalea* × *C. japonica* cultivar 'Hagoromo' ·········· 132

HA-62. *C. azalea* × *C. japonica* cultivar 'Mark Alan' ·········· 134

HA-63. *C. azalea* × *C. japonica* cultivar 'San Dimas' ·········· 137

HA-64. *C. azalea* × *C. japonica* cultivar 'Jean Clere' ·········· 140

HA-65. *C. azalea* × *C. japonica* cultivar 'Scentsation' ·········· 142

HA-66. *C. azalea* × *C. x williamsii* hybrid 'Jamie' ·········· 146

HA-67. *C. azalea* × Non-*reticulata* hybrid 'Dragon Fireball' ·········· 148

HA-68. Non-*reticulata* hybrid 'Dragon Fireball' × *C. azalea* ·········· 159

HA-69. *C. azalea* × *C. japonica* cultivar 'Firebrand' ·········· 164

HA-70. *C. azalea* × *C. japonica* cultivar 'Lipstick' ·········· 166

HA-71. *C. japonica* cultivar 'Lipstick' × *C. azalea* ·········· 169

HA-72. *C. azalea* × Non-*reticulata* hybrid 'Clarrie Fawcett' ·········· 171

HA-73. *C. azalea* × *C. japonica* cultivar 'Betty Foy Sanders' ·········· 173

HA-74. *C. azalea* × *C. japonica* cultivar 'Kingyo-tsubaki' ·········· 175

HA-75. *C. azalea* × *C. japonica* cultivar 'Disneyland' ·········· 177

HA-76. *C. azalea* × *C. japonica* cultivar 'Bill Quattlebaum' ·········· 179

HA-77. *C. azalea* × Non-*reticulata* hybrid 'Xiangsishe' ·········· 181

HA-78. *C. azalea* × *C. reticulata* cultivar 'Massee Lane' ·········· 185

HA-79. *C. japonica* cultivar 'Nuccio's Carousel' × *C. azalea* ·········· 187

HA-80. *C. japonica* cultivar 'Yours Truly' × *C. azalea* ·········· 192

HA-81. *C. japonica* cultivar 'Tama Peacock' × *C. azalea* ·········· 195

HA-82. *C. japonica* cultivar 'Tama-no-ura' × *C. azalea* ·········· 204

HA-83. *C. japonica* cultivar 'Black Magic' × *C. azalea* ·········· 206

HA-84. *C. japonica* cultivar 'Demi-Tasse' × *C. azalea* ·········· 209

HA-85. *C. azalea* × *C. rusticana* cultivar 'Bandai' ·········· 211

HA-86. *C. rusticana* cultivar 'Bandai' × *C. azalea* ·········· 213

HA-87. *C. reticulata* cultivar 'Frank Houser' × *C. azalea* ·········· 215

The Breeding Outline on the Backcross New Hybrids of *C. azalea* F_1 ·········· 217

The Detailed Descriptions for the Backcross New Hybrids of *C. azalea* F_1 ·········· 223

HAR-01. *C. azalea* × *C. azalea* F_1 new hybrid 'Jili Mudan' ·········· 223

HAR-02. *C. azalea* × *C. azalea* F_1 new hybrid 'Xiafeng Relang' ·········· 225

HAR-03. *C. azalea* × *C. azalea* F_1 new hybrid 'Xiari Fendai' ·········· 253

HAR-04. *C. azalea* × *C. azalea* F_1 new hybrid 'Xiari Hongxia' ·········· 262

HAR-05. *C. azalea* × *C. azalea* F_1 new hybrid 'Xiari Qixin' ·········· 269

HAR-06. *C. azalea* × *C. azalea* F_1 new hybrid 'Xiari Taige' ·········· 272

HAR-07. *C. azalea* × *C. azalea* F_1 new hybrid 'Xiari Tange' ·········· 282

HAR-08. *C. azalea* × *C. azalea* F_1 new hybrid 'Xiayong Guose' ·········· 285

HAR-09. *C. azalea* × *C. azalea* F_1 new hybrid 'Chaxiang Piaoyi' ·········· 294

HAR-10. *C. azalea* × *C. azalea* F_1 new hybrid 'Hongwu Jixiang' ·········· 297

HAR-11. *C. azalea* × *C. azalea* F_1 new hybrid 'Xiangxia Hongjiao' ·········· 308

HAR-12. *C. azalea* × *C. azalea* F$_1$ new hybrid 'Xiameng Hualin' ······ 312

HAR-13. *C. azalea* × *C. azalea* F$_1$ new hybrid 'Xiameng Xiaoxuan' ······ 321

HAR-14. *C. azalea* × *C. azalea* F$_1$ new hybrid 'Xiari Guangchang' ······ 326

HAR-15. *C. azalea* × *C. azalea* F$_1$ new hybrid 'Xiari Fenli' ······ 351

HAR-16. *C. azalea* × *C. azalea* F$_1$ new hybrid 'Mantian Hongxing' ······ 354

HAR-17. *C. azalea* × *C. azalea* F$_1$ new hybrid 'Hongbo Yongjin' ······ 356

HAR-18. *C. azalea* × *C. azalea* F$_1$ new hybrid 'Xiameng Wenqing' ······ 360

HAR-19. *C. azalea* × *C. azalea* F$_1$ new hybrid 'Buzhi Hanshu' ······ 376

HAR-20. *C. azalea* × *C. azalea* F$_1$ new hybrid 'Xiameng Chunling' ······ 378

HAR-21. *C. azalea* × *C. azalea* F$_1$ new hybrid 'Xiadie Qunwu' ······ 381

HAR-22. *C. azalea* × *C. azalea* F$_1$ new hybrid 'Xiari Hongjuan' ······ 383

HAR-23. *C. azalea* × *C. azalea* F$_1$ new hybrid 'Xiari Diexing' ······ 389

HAR-24. *C. azalea* × *C. azalea* F$_1$ new hybrid 'Xiameng Kejuan' ······ 393

HAR-25. *C. azalea* × *C. azalea* F$_1$ new hybrid 'Xiameng Yulan' ······ 396

HAR-26. *C. azalea* × *C. azalea* F$_1$ new hybrid 'Langman Fenniang' ······ 398

HAR-27. *C. azalea* × *C. azalea* F$_1$ new hybrid 'Huohong Mudan' ······ 400

HAR-28. *C. azalea* F$_1$ new hybrid 'Mantian Hongxing' × *C. azalea* F$_1$ new hybrid 'Jili Mudan' ······ 402

HAR-29. *C. azalea* F$_1$ new hybrid 'Shuiyue Zijuan' × *C. azalea* F$_1$ new hybrid 'Xiameng Wenqing' ······ 404

HAR-30. *C. azalea* × *C. azalea* F$_1$ new hybrid 'Xiari Fenqun' ······ 406

HAR-31. *C. azalea* × *C. azalea* F$_1$ new hybrid 'Shuxiang Zhijia' ······ 408

HAR-32. *C. azalea* × *C. azalea* F$_1$ new hybrid 'Yuegui Dasao' ······ 410

HAR-33. *C. azalea* × *C. azalea* F$_1$ new hybrid 'Xiari Hongbei' ······ 412

HAR-34. *C. japonica* cultivar 'Tama Peacock' × *C. azalea* F$_1$ new hybrid 'Hongwu Jixiang' ······ 414

HAR-35. *C. japonica* cultivar 'Tama-no-ura' × *C. azalea* F$_1$ new hybrid 'Hongwu Jixiang' ······ 416

HAR-36. *C. japonica* cultivar 'Ville de Nantes' × *C. azalea* F$_1$ new hybrid 'Hongwu Jixiang' ······ 435

HAR-37. *C. japonica* cultivar 'Juxiang' × *C. azalea* F$_1$ new hybrid 'Xiameng Wenqing' ······ 437

HAR-38. *C. amplexicaulis* × *C. azalea* F$_1$ new hybrid 'Xiari Hongxia' ······ 442

The Breeding Outline on the *C. amplexicaulis*' New Hybrids ······ 444

The Detailed Descriptions for the *C. amplexicaulis*' New Hybrids ······ 448

The Breeding Outline on the Normal Camellia New Hybrids ······ 497

The Detailed Descriptions for the Normal Camellia New Hybrids ······ 499

The Outline on the New Camellia Varieties Selected from Chance Seedlings ······ 518

The Detailed Descriptions for the New Camellia Varieties Selected from Chance Seedlings ······ 520

The Introductions of Selected Major Camellia Nurseries in China ······ 565

Bibliography ······ 574

Index of the New Camellia Hybrids ······ 576

再论四季茶花杂交新品种的六大性状

Re-discussion on the Six Major Characteristics of the New Camellia Hybrids that Bloom Year-round

茶花迎冬傲霜，冬、春开花，这已成为茶花界和园林界千百年来的共识。可是，广东肇庆棕榈谷花园有限公司的茶花育种团队在棕榈股份的大力支持下，在 2006 年至 2022 年这 16 年中，通过将我国特有的杜鹃红山茶（*C. azalea*）/ 张氏红山茶（*C. changii*）与国内外常规茶花品种杂交，获得了一个拥有 500 多个品种的四季茶花新品系。四季茶花这个新品系夏季始花，秋、冬季盛花，翌年春季仍可零星开花，颠覆了千百年来人们对茶花开花期的认知。四季茶花品种的发展和推广，让世界茶花界大开眼界，也进一步证明了四季茶花新品种的培育是茶花领域的一场革命和挑战。可以预见，四季茶花的出现将使茶花世界更加五彩缤纷，这些新品种也必将在今后的环境美化中发挥重要作用。

Camellias welcome winter, resist frost and bloom in winter and spring, which has become the consensus in both camellia circle and garden circle for thousands of years. During 16 years from 2006 to 2022, the Camellia Breeding Team, Zhaoqing Palm Valley Flower Garden Co., Ltd., Guangdong province, China, however, obtained a new strain of more than 500 new camellia hybrids that bloom year-round, under the strong support of Palm Shares, through the crosses between *C. azalea*/*C. changii* which is unique to China and the world's normal camellia cultivars. The new hybrids start to bloom in summer, fully bloom in autumn and winter and sporadically bloom in the spring of the following year, which overturns people's understanding to the camellia blooming period for thousands years. The development and promotion of the new camellia hybrids has opened the eyes of world's camellia circles. It further proves that the cultivation of the new camellia hybrids are a revolution and challenge in the camellia field. Thus, it can be concluded that the camellia world will be more colorful and the new camellia hybrids will play an important role in the future environmental beautification.

为了加深对四季茶花新品种的认知，这里再次对该类茶花的六大性状讨论如下：

In order to deepen the understanding of the new camellia hybrids, the six major characteristics of them will be rediscussed as follows:

性状一

杂交新品种开花期长，而且能重复开花。它们多数夏初始花，秋、冬季盛花，春季零星开花，彻底克服了常规茶花夏季和早秋不能开花的缺点。

Characteristic No.1

The new hybrids have a long blooming period and can ever-bloom. Most of them start to bloom in summer, fully bloom in autumn and winter and sporadically bloom in the spring of the following year, which have completely overcome the shortcomings that normal camellias cannot bloom in summer and early-autumn.

'夏风热浪'夏季开花情况
A blooming case of 'Xiafeng Relang' in summer

'夏风热浪'秋、冬季开花情况
A blooming case of 'Xiafeng Relang' in autumn and winter

'夏风热浪'春季开花情况
A blooming case of 'Xiafeng Relang' in spring

'瑰丽迎夏'夏季开花情况
A blooming case of 'Guili Yingxia' in summer

'瑰丽迎夏'秋、冬季开花情况
A blooming case of 'Guili Yingxia' in autumn and winter

'瑰丽迎夏'春季开花情况
A blooming case of 'Guili Yingxia' in spring

野外栽培的杂交新品种夏季始花
The new hybrids cultivated outside are just blooming in summer

野外栽培的杂交新品种秋、冬季盛花
The new hybrids cultivated outside are fully blooming in autumn and winter

野外栽培的杂交新品种春季零星开花
The new hybrids cultivated outside are sporadically blooming in spring

性状二

杂交新品种花大色艳，花色、花型和花径大小日趋多样化，在一定程度上可以媲美常规茶花品种。

Characteristic No.2

The new hybrids have large and bright flowers and become diversified in color, form and size, which can be, in a certain degree, comparable to normal camellias.

杂交新品种花大
A large flower of a new hybrid

杂交新品种花朵非常大
How large the flowers of the new hybrids are

杂交新品种多样化的花朵
Diversified flowers of the new hybrids

杂交新品种花色鲜艳
Bright color of the new hybrids

花朵多么大呀
How large the flower is

杂交新品种开花多么艳丽呀
How beautiful the new hybrids' flowers are

性状三

杂交新品种开花稠密，花朵盛开时非常壮观，彻底克服了常规茶花花蕾太密而不能正常开花和植株容易衰败死亡的缺点。

Characteristic No.3

The new hybrids bloom densely and the blooming scenes are spectacular, which have completely overcome the shortcomings that normal camellias cannot bloom normally and the plants become easily weak or die owing to too many buds.

| 开花稠密的'夏梦小旋' 'Xiameng Xiaoxuan' with dense flowers | 开花稠密的'夏梦春陵' 'Xiameng Chunling' with dense flowers | 开花稠密的'彤海咏秋' 'Tonghai Yongqiu' with dense flowers | 盛开的'夏日粉裙' 'Xiari Fenqun' in its peak blooming period |

盛开的'夏咏国色' 'Xiayong Guose' in its peak blooming period　　开花稠密的'夏梦华林' 'Xiameng Hualin' with dense flowers　　盛开的'夏日粉黛' 'Xiari Fendai' in its peak blooming period

开花稠密的'曲院风荷' 'Quyuan Fenghe' with dense flowers　　开花稠密的'夏梦可娟' 'Xiameng Kejuan' with dense flowers　　开花极盛的盆栽'怀金拖紫' Potted 'Huaijin Tuozi' with extremely blooming

性状四

　　杂交新品种叶片浓绿、稠密，冬季不畏严寒（－6～－3℃），夏季耐热抗晒，无须遮阳，彻底克服了常规茶花叶片黄绿，夏季易灼伤，且繁殖栽培必须遮阳的缺点。

Characteristic No.4

　　The new hybrids with dark green and dense leaves, are cold-resistant (-6--3℃) in winter, and heat-and sunburn-resistant in summer, which their plants do not need to be shaded, which have completely overcome the shortcomings that the leaves of normal camellias are yellow-green, easy to be burned in summer and the plants must be shaded during their propagations and cultivations in summer.

大面积露天盆栽杂交新品种植株的叶片非常浓绿
The leaves of the potted new hybrid plants which cultivated in a large area without any shade are dark green

高温季节山地栽培的杂交新品种开花正常
The new hybrid plants that cultivated in a hill are blooming normally in hot season

大面积无遮阳条件下栽培的杂交新品种生长健壮
The new hybrids cultivated in a large area without sunshade grow vigorously

杂交新品种无遮阳条件下夏季开花情况
The blooming situation of the new hybrids without sunshade in summer

盆栽的杂交新品种无遮阳条件下秋、冬季开花情况
The blooming situation of the new potted hybrids in autumn and winter without sunshade

盆栽的杂交新品种无遮阳条件下冬季开花情况
The blooming situation of the new potted hybrids in winter without sunshade

嫁接的杂交新品种大树夏、秋、冬季开花情况
The blooming situation of a big grafted tree of the new hybrid in summer, autumn and winter

在裸露土地上栽培的杂交新品种生长旺盛、开花正常
The new hybrids are grow vigorously and bloom normally even they are cultivated in bare ground

大面积露天栽培的杂交新品种繁花似锦
The new hybrids cultivated under the open ground in a large area were fully blooming

在冬季有霜雪的地区,杂交新品种生长开花正常
The new hybrids grow and bloom normally in the areas where have snow or frost in winter

性状五

杂交新品种具有明显的杂种生长优势。在温暖地区,它们每年抽萌新梢3～4次,开花和新梢生长同步进行,而且生长旺盛,彻底克服了常规茶花生长慢、难栽培的缺点。

Characteristic No.5

The new hybrids have obvious heterosis in growth. In the warm regions, their new shoots can grow 3-4 times each year, blooming and new shoots growing are happened at the same time, and growth vigorous, which have completely overcome the shortcomings that normal camellias grow slowly and are difficult to cultivate.

| 杂交新品种边开花边抽新梢 The new hybrids bloom and spread new shoots at the same time | 顶端的嫩梢是杂交新品种的第三次梢 The tender shoots at the top are the new hybrids' third time to sprout | 杂交新品种在秋季同样会萌生很多新梢 The new hybrids also produce many new shoots in autumn | 开花和抽梢同时进行的四季茶花大树受到领导们的称赞 A big tree of the new hybrid, which is blooming and shooting at the same time, was praised by leaders |

性状六

杂交新品种对一直困扰世界茶花界的茶花花腐病具有极强的抗性。经对国内茶花苗圃 50 多万株四季茶花品种，历经 5 年的观测，四季茶花新品种即便栽培在有花腐病的常规茶花栽培区，也没有出现花朵感染花腐病的情况。

Characteristic No.6

The new hybrids are highly resistant to camellia blight disease, which has been plagued the camellia world. In five years of observation on more than 500, 000 plants of the four-season camellia hybrids in camellia nurseries of China, even in the normal camellia growing area with camellia blight disease that the new hybrids planted, no flower was found infected by camellia blight disease.

四季茶花没有花腐病，花朵鲜艳漂亮
Four-season camellia hybrids never have flower blight and their flowers are fresh and beautiful

四季茶花不会受到花腐病威胁，花朵保持亮丽多彩
Four-season camellia hybrids are not threatened by flower blight, keeping their flowers beautiful and colorful

普通茶花红花品种感染花腐病，完全失去了观赏价值
The red flowers of normal camellia cultivars infected with flower blight completely lost their ornamental value

普通茶花白花品种感染花腐病，花朵格外难看
The white flowers of normal camellia cultivars infected with flower blight look especially ugly

最后，用中国的顺口溜方式，总结四季茶花杂交新品种的优点：
四季开花不断，叶片碧绿好看，花朵艳丽无限，花型花色多变；
生长旺盛强健，耐热抗旱不蔫，冬季无畏严寒，栽培管理方便；
庭院栽培首选，园林美化惊艳，茶花家族新篇，美丽中国贡献。

Finally, the advantages of the new hybrids are summarized in a Chinese doggerel as follows:

The blooming is unceasing in four seasons, the leaves are dark-green and good-looking, the flowers are infinitely gorgeous, and the flower form and color are diverse;

The growth is vigorous and strong, the heat and drought resistances make the plants never withering, the cold-resistance is strong in winter, and the cultivation and management are easy;

The first choice for cultivating in yard, the landscaping is stunning, a new chapter has opened in camellia family and a contribution to the beautifying China process.

这个四季茶花新品种花朵多么漂亮啊
How beautiful are the flowers of the new hybrid

一个四季茶花新品种盆栽
A potted plant of the new hybrid

四季茶花新品种用于庭院美化
The new hybrids are used for landscaping

四季茶花新品种种在公园中
The new hybrids grow in a park

四季茶花种在花槽中也很漂亮
The new hybrids to be planted in a flower bed are also very beautiful

这个四季茶花品种开花稠密，生长旺盛
This new hybrid has a dense flowering and vigorous growth

四季茶花嫁接的大树在8月至10月份已经开花了
Big grafted trees of the new hybrids have fully bloomed from August to October

几株四季茶花嫁接的大树在6月份开花
Some big grafted trees of new hybrids were blooming in June

杜鹃红山茶 F_1 代新品种育种概况

The Breeding Outline on the *C. azalea* F_1 New Hybrids

杜鹃红山茶杂交组合及获得的杂交新品种数量

杜鹃红山茶 F_1 代系指杜鹃红山茶与常规茶花品种或物种之间进行杂交所获得的杂交新品种。在我们的第一部书《四季茶花杂交新品种彩色图集》(以下简称"第一部书")中，曾详细介绍了 217 个杜鹃红山茶 F_1 代杂交新品种（见第一部书 HA-01 ～ HA-50 和 HA-52）。这些 F_1 代新品种涉及的杂交组合达 51 个。

从 2016 年到 2022 年，在以前杂交育种的基础上，我们又进行了 59 个杂交组合的探索，其中有 24 个杂交组合是与第一部书相同的，有 35 个组合是新的。本书总计获得了 145 个杜鹃红山茶 F_1 代新品种。加上第一部书中的那些新品种，到目前为止，杜鹃红山茶 F_1 代杂交组合总数已达到 86 个，获得的 F_1 代新品种总数达到 362 个。

The Quantities of Cross-combinations and Obtained New Hybrids from *C. azalea*

C. azalea F_1 generation refers to the new hybrids obtained from the crosses between *C. azalea* and normal camellia cultivars or between camellia species. Two hundred and seventeen *C. azalea*'s F_1 new hybrids (See HA-01 to HA-50 and HA-52, our first book) had been introduced in detail in our first book *Illustrations of the New Camellia Hybrids that Bloom Year-round* (hereinafter referred to our first book). These F_1 new hybrids are involved in 51 cross-combinations.

During the 6 years from 2016 to 2022, we have carried out 59 cross-combinations on the basis of the previous cross breeding, 24 cross-combinations of them are the same as those in our first book and 35 combinations are new. In total, 145 F_1 new hybrids from *C. azalea* have been obtained in this book. If those new hybrids of our first book are added, we have carried out 86 cross-combinations and obtained 362 F_1 new hybrids in total now.

杜鹃红山茶与常规茶花品种之间杂交的两种方式

在此类杂交组合中，我们所进行的有两种杂交方式。一种是以杜鹃红山茶为母本，常规茶花品种为父本；另一种则反之，以常规茶花品种为母本，杜鹃红山茶为父本。这里要特别指出，作为母本的必须要能坐果，作为父本的必须要有花粉，并且两亲本之间应该具有一定的杂交亲和性。

Two Patterns of the Crosses Between *C. azalea* and Normal Camellia Cultivars

In this kind of cross combinations, we have carried out two cross patterns. One is *C. azalea* as female parent and normal camellia cultivars as male parent, the other is just contrary, that is normal camellia cultivars as female parent and *C. azalea* as male parent. It should be pointed out that the female parent must be able to bear fruit, the male parent must have pollens, and these two parents should have a certain cross compatibility.

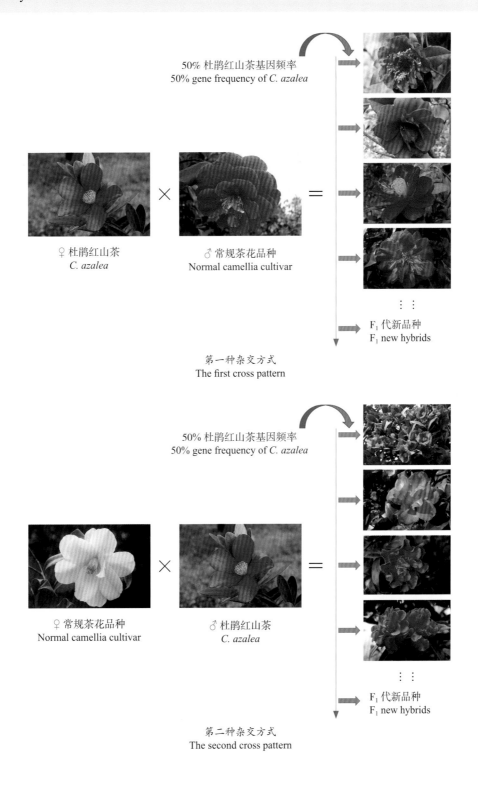

杜鹃红山茶 F_1 代新品种的性状表达

很明显，在这些杂交组合中，无论杜鹃红山茶作母本还是作父本，在理论上这些杂交新品种都有50%的基因来自杜鹃红山茶，有50%的基因来自常规茶花品种或者山茶原种。因此，在性状上，杂交种才表现出双亲的特性。

因为杂交新品种具有这样的基因型，所以它们在叶性状、株形、开花期等方面的表达，主要倾向于杜鹃红山茶的性状，而在花色、花型、花径大小以及抗寒性方面的表达，则主要倾向于常规茶花品种的性状。另外，在生长方面，杂交新品种表现出明显的杂种优势（见第一部书第67～69页）。

与常规茶花品种相比，杜鹃红山茶 F_1 代新品种不仅开花期比常规茶花要长，而且开花季节也与常规茶花相反。据多年观测，杜鹃红山茶 F_1 代新品种通常夏季始花（6月至7月份），秋季至冬季盛花（8月至12月份），翌年春季零星开花（1月至2月份），整个花期可达9个月，比常规茶花多2个月，比杜鹃红山茶仅少1个月。详情请看表1：

The Characteristic Expression of *C. azalea* F_1 New Hybrids

Obviously, in these cross-combinations, no matter *C. azalea* is as female parent or as male parent, in theory, these hybrids have 50% genes that came from *C. azalea* and 50% genes that came from normal camellia cultivars or camellia species. Therefore, the hybrids can show the characteristics of their two parents.

Because the new hybrids have such genotypes, they mainly tend to have the characteristics of *C. azalea* in the expressions of leaf trait, plant shape and the blooming period, while they mainly tend to have the characteristics of normal camellia cultivars in the expressions of flower color, flower form, flower size and cold resistance. In addition, the new hybrids also can express obvious heterosis in growth (See p.67-69, our first book).

Compared with normal camellia cultivars, the *C. azalea* F_1 new hybrids not only have a longer blooming period, but also have the opposite blooming seasons. According to years of observation, the F_1 new hybrids usually start to bloom in summer (from June to July), fully bloom from autumn to winter (from August to December) and bloom sporadically in the spring of the following year (from January to February). Its entire blooming period can reach to nine months, two months longer than normal camellia cultivars, and only one month shorter than *C. azalea*'s. Please see the table 1 below for details:

表 1　杜鹃红山茶 F_1 代与其杂交亲本开花期的比较
Table 1　Comparison between the Blooming Period of the *C. azalea* F_1 New Hybrids and Their Two Cross-parents

杂交亲本或杂交新品种 Cross-parents or new hybrids	杜鹃红山茶基因频率 / % Gene frequency of *C. azalea* / %	春季 3—5 月份 Spring Mar.–May	夏季 6—8 月份 Summer Jun.–Aug.	秋季 9—11 月份 Autumn Sept.–Nov.	冬季 12—2 月份 Winter Dec.–Feb.	全花期 / 月数 Entire blooming / Mon.	盛花期 / 月数 Fully blooming / Mon.	无花期 / 月数 No blooming / Mon.
常规茶花品种 Normal camellia cultivar	0					7	3	5
杜鹃红山茶 *C. azalea*	100					10	8	2
杜鹃红山茶 F_1 代新品种 F_1 new hybrids of *C. azalea*	50					9	6	3

无花期　No blooming period　　始花期　Started blooming period
盛花期　Fully blooming period　　末花期　Sparsely blooming period

杜鹃红山茶 F_1 代新品种的编号与排序

应该指出的是，对与第一部书相同的杂交组合，本书将保持原杂交组合编号，其杂交新品种编号也将接续第一部书中的品种序列编号（见 HA-01 ～ HA-48）。因为本书中只有少数杂交组合与第一部书相同，因此这些杂交组合编号是不连贯的。对于第一部书中没有的杂交组合，本书将接续第一部书中的杂交组合编号（见本书 HA-53 ～ HA-87）。

Numbering and Arranging for *C. azalea* F_1 New Hybrids

It should be noted that for the cross-combinations introduced in our first book, their numbers in this book will be the same in our first book and their new hybrids numbers will also follow the number of our first book (See HA-01 to HA-48). Since only several cross-combinations in this book have introduced in our first book, the numbers of them are discontinuous. For the cross-combinations that are absent in our first book, their numbers in this book will follow the cross-combination numbers of our first book (See HA-53 to HA-87 in this book).

四季茶花杂交新品种彩色图集（第二部）

梦幻世界
(Menghuan Shijie) HA-10-19

满堂喝彩
(Mantang Hecai) HA-10-8

夏日红霞
(Xiari Hongxia) HA-35-5

后起之秀
(Houqi Zhixiu) HA-64-1

杜鹃红山茶 F_1 代新品种详解

The Detailed Descriptions of the F_1 New Hybrids of *C. azalea*

◉ 杂交组合 HA-01. 杜鹃红山茶 × 红山茶品种'克瑞墨大牡丹'
HA-01. *C. azalea* × *C. japonica* cultivar 'Kramer's Supreme'

本杂交组合将介绍11个新品种,其编号将接续第一部书本组的序列编号。

'克瑞墨大牡丹'是一个具有红色、牡丹型花朵和芳香性状的红山茶(*C. japonica*)名种。

第一部书中介绍的本组7个新品种中有6个新品种具芳香性状(见第一部书第71～81页)。这次介绍的11个品种中,有3个具芳香性状。可见,'克瑞墨大牡丹'的芳香性状在一定程度上是可以遗传的。

Eleven new hybrids will be introduced in this cross-combination and their numbers will follow the serial numbers of the cross-combination in our first book.

'Kramer's Supreme' is a famous *C. japonica* cultivar with the characteristics of red peony flower form and fragrance.

Seven new hybrids were introduced in the cross-combination in our first book, six of these new hybrids are with fragrant characteristics (See p.71-81, our first book). Three of the eleven new hybrids introduced this time are with fragrance. It can be seen that the fragrant characteristics of 'Kramer's Supreme', to some degree, can be inherited.

HA-01-8. 心醉神迷
Xinzui Shenmi（Mental Intoxication）

杂交组合：杜鹃红山茶 × 红山茶品种'克瑞墨大牡丹'（*C. azalea* × *C. japonica* cultivar 'Kramer's Supreme'）。

杂交苗编号：DK-No.5。

性状：花朵深桃红色至红色，略带芳香，牡丹型，巨型花，花径 13.2～14.6cm，花朵厚 5.8cm，外轮大花瓣 7～10 枚，呈 1～2 轮排列，花瓣阔倒卵形，外翻，先端凹缺，中部小花瓣很多，松散排列，半直立，波浪状或者扭曲，偶有大花瓣出现，雄蕊簇生，混生于中部花瓣之间。叶片浓绿色，叶背面灰绿色，厚革质，椭圆形，长 8.4～9.5cm，宽 4.0～4.6cm，上部边缘具稀锯齿。植株开张，生长旺盛。夏季始花，秋、冬季盛花，春季零星开花。

Abstract: Flowers deep-peach red to red, slightly with fragrance, peony form, very large size, 13.2-14.6cm across and 5.8cm deep with 7-10 exterior large petals arranged in 1-2 rows, petals broad obovate, rolled outward, apices emarginate, central small petals numerous, semi-erected or wavy, stamens mixed with the petals. Leaves dark green, heavy leathery, elliptic, margins sparsely serrate. Plant spread and growth vigorous. Starts to bloom from summer, fully blooms from autumn to winter and sporadically in spring.

HA-01-9. 热恋季节
Relian Jijie (Lovestruck Season)

杂交组合：杜鹃红山茶 × 红山茶品种'克瑞墨大牡丹'（*C. azalea* × *C. japonica* cultivar 'Kramer's Supreme'）。

杂交苗编号：DK-No.8。

性状：花朵橘粉红色至橘红色，有时花瓣上具少量白色隐斑，半重瓣型，中型花，花径 8.5～10.0cm，花朵厚 4.5cm，花瓣 15～18 枚，呈 2 轮松散排列，倒卵形，波浪状，中部有时出现雄蕊瓣，雄蕊近离生，花丝淡粉红色，花药淡黄色。叶片浓绿色，革质，椭圆形，长 9.5～10.5cm，宽 4.0～5.0cm，边缘浅锯齿。植株立性，紧凑，生长旺盛。夏末始花，秋、冬季盛花，春季零星开花。

Abstract: Flowers orange pink to orange red, sometimes with a few faintly white blotches, semi-double form, medium size, 8.5-10.0cm across and 4.5cm deep with 15-18 petals arranged loosely in 2 rows, petals obovate, wavy, sometimes a few petaloids at the center, stamens nearly free at the base, filaments pale pink, anthers pale yellow. Leaves dark green, leathery, elliptic, margins shallowly serrate. Plant upright, compact and growth vigorous. Starts to bloom from late-summer, fully blooms from autumn to winter and sporadically in spring.

HA-01-10. 绚丽多彩
Xuanli Duocai (Bright & Colorful)

杂交组合： 杜鹃红山茶 × 红山茶品种'克瑞墨大牡丹'（*C. azalea* × *C. japonica* cultivar 'Kramer's Supreme'）。

杂交苗编号： DK-No.12。

性状： 花朵深粉红色至红色，中部花瓣偶有隐约的白斑块，渐向花瓣上部呈粉白色，半重瓣型至玫瑰重瓣型，小型花，花径6.0～7.5cm，花朵厚4.0cm，大花瓣13～15枚，呈3轮有序排列，雄蕊呈短柱状，花丝淡红色，花药黄色。叶片浓绿色，叶背面灰绿色，革质，阔椭圆形，长8.5～9.5cm，宽3.5～4.0cm，基部楔形，边缘具浅齿。植株紧凑，矮生，生长旺盛。夏季始花，秋、冬季盛花，春季零星开花。

Abstract: Flowers deep pink to red, occasionally with faint white patches, fading to pink-white at the upper part, semi-double to rose-double form, small size, 6.0-7.5cm across and 4.0cm deep with 13-15 large petals arranged orderly in 3 rows, stamens shortly columnar, filaments pale red, anthers yellow. Leaves dark green, leathery, broad elliptic, cuneate at the base, margins shallowly serrate. Plant compact, dwarf and growth vigorous. Starts to bloom from summer, fully blooms from autumn to winter and sporadically in spring.

HA-01-11. 繁华街景
Fanhua Jiejing (Bustling Streetscape)

杂交组合：杜鹃红山茶 × 红山茶品种'克瑞墨大牡丹'（*C. azalea* × *C. japonica* cultivar 'Kramer's Supreme'）。

杂交苗编号：DK-No.13。

性状：花朵桃红色至红色，花瓣偶有白条纹或白斑，玫瑰重瓣型至牡丹型，大型至巨型花，花径11.6～13.5cm，花朵厚5.5cm，大花瓣约60枚，呈6轮排列，边缘略内扣或者波浪状，中部小花瓣5枚左右，直立，与簇生雄蕊混生。叶片浓绿色，叶背面灰绿色，厚革质，椭圆形，长9.5～10.5cm，宽3.8～4.2cm，基部楔形，上部边缘具尖齿。植株紧凑，立性，生长旺盛。夏季始花，秋、冬季盛花，春季零星开花。

> **Abstract:** Flowers peach red to red, occasionally with white stripes or patches, rose-double to peony form, large to very large size, 11.6-13.5cm across and 5.5cm deep with about 60 large petals arranged in 6 rows, 5 central small petals erected and mixed with stamens. Leaves dark green, heavy leathery, elliptic, cuneate at the base, margins sharply serrate at the upper part. Plant compact, upright and growth vigorous. Starts to bloom from summer, fully blooms from autumn to winter and sporadically in spring.

HA-01-12. 生命阳光
Shengming Yangguang (Sunny Life)

杂交组合：杜鹃红山茶 × 红山茶品种'克瑞墨大牡丹'（*C. azalea* × *C. japonica* cultivar 'Kramer's Supreme'）。

杂交苗编号：DK-No.14 + 19。

性状：花朵红色至艳红色，单瓣型，中型至大型花，花径9.5～11.5cm，花瓣6～7枚，雄蕊呈柱状，花丝粉红色，花药金黄色，开花稠密。嫩叶泛红色，成熟叶浓绿色，叶背面灰绿色，革质，椭圆形，长8.7～10.0cm，宽4.0～5.0cm，边缘近光滑。植株紧凑，生长旺盛。夏季始花，秋、冬季盛花，春季零星开花。

Abstract: Flowers red to bright red, single form, medium to large size, 9.5-11.5cm across with 6-7 petals, stamens columnar, filaments pink, anthers golden yellow, bloom dense. Young leaves reddish, mature leaves dark green, leathery, elliptic, margins nearly smooth. Plant compact and growth vigorous. Starts to bloom from summer, fully blooms from autumn to winter and sporadically in spring.

HA-01-13. 季节色彩
Jijie Secai (Seasonal Colors)

杂交组合：杜鹃红山茶 × 红山茶品种'克瑞墨大牡丹'（ *C. azalea* × *C. japonica* cultivar 'Kramer's Supreme'）。

杂交苗编号：DK-No.15。

性状：炎热季节花朵为淡红色，冷凉季节花朵为艳红色，花瓣背面偶有白条纹，单瓣型，中型花，花径8.5～10.0cm，花瓣5～6枚，阔倒卵形，波浪状，瓣面略皱褶，先端凹，雄蕊柱状，花丝淡红色，花药黄色。叶片浓绿色，叶背面灰绿色，厚革质，阔椭圆形，中脉凸起，长7.5～11.0cm，宽4.5～5.0cm，上部边缘齿钝。植株紧凑，枝条稠密，生长旺盛。夏末始花，秋、冬季盛花，春季零星开花。

Abstract: Flowers light red in hot season and bright red in cold season, occasionally with white stripes on the back surfaces of petals, single form, medium size, 8.5-10.0cm across with 5-6 petals, petals wavy, the surfaces crinkled, stamens columnar, filaments pale red, anthers yellow. Leaves dark green, heavy leathery, broad elliptic, midrib raised, margins obtusely serrate at the upper part. Plant compact, branches dense and growth vigorous. Starts to bloom from late-summer, fully blooms from autumn to winter and sporadically in spring.

HA-01-14. 九九艳阳
Jiujiu Yanyang (The Bright Sun of Sept. Ninth)

杂交组合： 杜鹃红山茶 × 红山茶品种'克瑞墨大牡丹'（*C. azalea* × *C. japonica* cultivar 'Kramer's Supreme'）。

杂交苗编号： DK-No.20。

性状： 花朵艳红色，偶有白条纹，略具芳香，托桂型至牡丹型，大型至巨型花，花径10.5～14.5cm，最大可达17.0cm，花朵厚5.5cm，外部大花瓣13～15枚，呈2轮排列，阔倒卵形，外翻，中部小花瓣簇拥，波浪状或者扭曲，雄蕊散射。叶片浓绿色，叶背面灰绿色，革质，阔椭圆形，中脉凸起，长10.5～11.0cm，宽5.0～5.5cm，上部边缘具浅锯齿。植株立性，紧凑，生长旺盛。夏季始花，秋、冬季盛花，春季零星开花。

Abstract: Flowers bright red, occasionally with white stripes, slightly fragrant, anemone to peony form, large to very large size, 10.5-14.5cm across, the largest across can reach to 17cm, 5.5cm deep with 13-15 exterior large petals arranged in 2 rows, exterior large petals rolled outward, central small petaloids crowded, wavy or twisted, stamens scattered. Leaves dark green, leathery, broad elliptic, midrib raised, margins shallowly serrate at the upper part. Plant upright, compact and growth vigorous. Starts to bloom from summer, fully blooms from autumn to winter and sporadically in spring.

HA-01-15. 桃红羞面
Taohong Xiumian (Peach Red & Bashful Face)

杂交组合： 杜鹃红山茶 × 红山茶品种'克瑞墨大牡丹'（*C. azalea* × *C. japonica* cultivar 'Kramer's Supreme'）。

杂交苗编号： DK-No.25。

性状： 花朵桃红色，雄蕊瓣具白条纹，略带芳香，半重瓣型至松散的牡丹型，大型花，花径11.0～12.5cm，花朵厚5.0cm，外部大花瓣15～18枚，呈2～3轮排列，花瓣阔倒卵形，边缘内卷，雄蕊散射状，与雄蕊瓣混生。叶片浓绿色，叶背面灰绿色，厚革质，阔椭圆形，长9.0～10.0cm，宽4.5～5.0cm，上部边缘锯齿浅，下部光滑。植株立性，生长旺盛。夏末始花，秋、冬季盛花，春季零星开花。

> **Abstract:** Flowers peach red, petaloids with white stripes, slightly fragrant, semi-double to loose peony form, large size, 11.0-12.5cm across and 5.0cm deep, 15-18 exterior large petals arranged in 2-3 rows, edges rolled inward, stamens scattered and mixed with petaloids. Leaves dark green, heavy leathery, broad elliptic, margins shallowly serrate at the upper part and smooth at the lower part. Plant upright and growth vigorous. Starts to bloom from late-summer, fully blooms from autumn to winter and sporadically in spring.

HA-01-16. 紫绶金章
Zishou Jinzhang (Gold Seal with Purple Ribbon)

杂交组合： 杜鹃红山茶 × 红山茶品种'克瑞墨大牡丹'（*C. azalea* × *C. japonica* cultivar 'Kramer's Supreme'）。

杂交苗编号： DK-No.28。

性状： 花朵紫红色，半重瓣型至牡丹型，大型至巨型花，花径11.0～13.5cm，花朵厚6.0cm，外部大花瓣13～15枚，波浪状，呈2轮排列，中部雄蕊瓣25～30枚，半直立，波浪状或扭曲，与雄蕊混生。叶片浓绿色，叶背面灰绿色，革质，长椭圆形，长10.0～11.5cm，宽3.0～4.0cm，边缘具浅锯齿。植株开张，矮性，生长旺盛。夏末始花，秋、冬季盛花，春季零星开花。

> **Abstract:** Flowers purple red, semi-double to peony form, large to very large size, 11.0-13.5cm across and 6.0cm deep, 13-15 exterior wavy large petals arranged in 2 rows, 25-30 central petaloids semi-erected, wavy or twisted and mixed with stamens. Leaves dark green, leathery, long elliptic, margins shallowly serrate. Plant spread, dwarf and growth vigorous. Starts to bloom from late-summer, fully blooms from autumn to winter and sporadically in spring.

HA-01-17. 变装魔女
Bianzhuang Monü (Magic Girl in Changing Dresses)

杂交组合：杜鹃红山茶 × 红山茶品种'克瑞墨大牡丹'（ *C. azalea* × *C. japonica* cultivar 'Kramer's Supreme'）。

杂交苗编号：DK-No.50。

性状：花朵艳红色，渐渐变为淡橘红色，花瓣上部近白色，玫瑰重瓣型至牡丹型，中型至大型花，花径9.5～11.5cm，花朵厚5.0cm，大花瓣25～30枚，呈5～6轮排列，边缘内卷，偶见雄蕊，开花稠密。叶片浓绿色，叶背面灰绿色，厚革质，长椭圆形，长8.0～9.0cm，宽4.0～4.5cm，基部楔形，边缘齿浅。植株紧凑，枝条稠密，生长旺盛。夏季始花，秋、冬季盛花，春季零星开花。

> **Abstract:** Flowers bright red, gradually turning into light orange red, upper part of the petals nearly white, rose-double to peony form, medium to large size, 9.5-11.5cm across and 5.0cm deep, 25-30 large petals arranged in 5-6 rows, edges rolled inward, stamens occasionally visible, bloom dense. Leaves dark green, heavy leathery, long elliptic, cuneate at the base, margins shallowly serrate. Plant compact, branches dense and growth vigorous. Starts to bloom from summer, fully blooms from autumn to winter and sporadically in spring.

HA-01-18. 佛植华章
Fozhi Huazhang (Foshan Botanical Garden's Brilliant Works)

杂交组合： 杜鹃红山茶 × 红山茶品种'克瑞墨大牡丹'（*C. azalea* × *C. Japonica* cultivar 'Kramer's Supreme'）。

杂交苗编号： Fozhi –No.1。

性状： 花朵艳红色，半重瓣型至牡丹型，中型至大型花，花径8.5～11.7cm，厚5.0cm，大花瓣22～28枚，呈3～4轮排列，花瓣阔倒卵形，先端近圆，瓣面有明显脉纹，先端瓣缘偶有白色条纹，花丝红色，花药黄色。叶片浓绿色，厚革质，阔椭圆形，长8.0～9.0cm，宽4.4～5.5cm，主脉凸起，侧脉明显，近平行，叶缘锯齿浅，先端钝尖。植株紧凑，枝叶稠密，生长旺盛。夏季始花，秋季盛花，冬季零星开花。

注：本品种由广东省佛山市佛山植物园培育。

Abstract: Flowers bright red, semi-double to peony form, medium to large size, 8.5-11.7cm across and 5.0cm deep, 22-28 large petals arranged in 3-4 rows, petals broad obovate, deep veins visible, occasionally with white stripes edged at petal apices, filaments red, anthers yellow. Leaves dark green, heavy leathery, broad elliptic, margins shallowly serrate. Plant compact and growth vigorous. Starts to bloom from summer, fully blooms from autumn and sporadically in winter.
Mark: The hybrid is bred by Foshan Botanical Garden, Foshan City, Guangdong Province.

⦿ 杂交组合 HA-02. 杜鹃红山茶 × 红山茶品种'佛蕾德'
HA-02. *C. azalea* × *C. japonica* cultivar 'Fred Sander'

本杂交组合介绍 2 个新品种，其编号将接续第一部书本组的序列编号。

'佛蕾德'是一个花瓣带锯齿的品种（见第一部书第 82 页）。第一部书曾介绍过本杂交组合的 1 个新品种（见第一部书第 82～83 页）。加上本次介绍的 2 个新品种，到目前为止，本组合共获得了 3 个新品种。

可以看出，本杂交组合所获得的新品种虽然没有表达'佛蕾德'品种花瓣带锯齿的性状，但是花色都是桃红色至深红色的，而且都能够全年重复开花。这证明，杂交新品种已经具有双亲的性状。

Two new hybrids will be introduced in this cross-combination and their numbers will follow the serial numbers of the cross-combination in our first book.

'Fred Sander' is a nice cultivar that petals are serrated (See p.82, our first book). A new hybrid was introduced in this cross-combination in our first book (See p.82-83, our first book). Adding two new hybrids introduced this time, three new hybrids have been obtained in this cross-combination so far.

It can be seen that all the new hybrids obtained from this cross-combination do not express any characteristic of petals' serrated edges, but their flowers are peach red to crimson and can bloom year-round, which shows that these hybrids have the traits from their cross parents.

HA-02-2. 娟好静秀
Juanhao Jingxiu (Beautiful in Appearance & Gentle in Disposition)

杂交组合： 杜鹃红山茶 × 红山茶品种'佛蕾德'（*C. azalea* × *C. japonica* cultivar 'Fred Sander'）。

杂交苗编号： DF-No.2。

性状： 花朵深粉红色至桃红色，半重瓣型，中型花，花径9.5～10.0cm，花朵厚4.5cm，花瓣15～18枚，呈3轮排列，花瓣略皱褶，雄蕊基部略连生，偶有雄蕊瓣，花丝淡红色，花药黄色，开花稠密，花瓣逐片掉落。叶片浓绿色，略扭曲，革质，椭圆形，长8.0～9.5cm，宽3.5～4.0cm，上部边缘具稀疏锯齿。植株紧凑，枝叶稠密，生长旺盛。夏末始花，秋、冬季盛花，春季零星开花。

Abstract: Flowers deep pink to peach red, semi-double form, medium size, 9.5-10.0cm across and 4.5cm deep with 15-18 petals arranged in 3 rows, petals slightly crinkled, stamens slightly united at the base, occasionally a few petaloids appeared, bloom dense, petals fall one by one. Leaves dark green, slightly twisted, leathery, elliptic, margins sparsely serrate at the upper part. Plant compact, branches dense and growth vigorous. Starts to bloom from later-summer, fully blooms from autumn to winter and sporadically in spring.

HA-02-3. 瑶林仙境
Yaolin Xianjing（Yaolin's Fairyland）

杂交组合： 杜鹃红山茶 × 红山茶品种'佛蕾德'（*C. azalea* × *C. japonica* cultivar 'Fred Sander'）。

杂交苗编号： DF-No.3。

性状： 花朵深红色，泛蜡光，雄蕊瓣具白条纹或白斑，半重瓣型至玫瑰重瓣型，大型花，花径11.5～12.5cm，花朵厚5.5cm，花瓣18～22枚，呈3～4轮排列，花瓣阔倒卵形，偶有雄蕊瓣，雄蕊少，花丝淡红色，花瓣逐片掉落，不留残朵。叶片浓绿色，叶背面灰绿色，革质，长椭圆形，长9.5～10.5cm，宽3.0～4.0cm，基部楔形，上部边缘具钝锯齿。植株紧凑，矮灌状，枝叶稠密，生长旺盛。夏末始花，秋、冬季盛花，春季零星开花。

> **Abstract:** Flowers crimson with waxy luster and some petaloids with white stripes or patches, semi-double to rose-double form, large size, 11.5-12.5cm across and 5.5cm deep with 18-22 petals arranged in 3-4 rows, occasionally a few petaloids appeared, stamens a few, petals fall one by one. Leaves dark green, leathery, long elliptic, cuneate at the base, margins obtusely serrate at the upper part. Plant compact, bushy, branches dense and growth vigorous. Starts to bloom from late-summer, fully blooms from autumn to winter and sporadically in spring.

⦿ 杂交组合 HA-03. 红山茶品种'媚丽'× 杜鹃红山茶
HA-03. *C. japonica* cultivar 'Tama Beauty' × *C. azalea*

本杂交组合将介绍 2 个新品种，其编号将接续第一部书本组的序列编号。

第一部书曾介绍过本杂交组合的 45 个新品种（见第一部书第 84～148 页）。加上本次介绍的 2 个新品种，到目前为止，本组合共获得了 47 个新品种。

'媚丽'与杜鹃红山茶杂交组合获得如此多性状各异的杂交新品种，证明了'媚丽'品种的杂合度很高，是创制四季茶花新品种非常理想的亲本材料。

Two new hybrids will be introduced in this cross-combination and their numbers will follow the serial numbers of the cross-combination in our first book.

Forty five new hybrids were introduced in this cross-combination in our first book (See p. 84-148, our first book). Adding two new hybrids introduced this time, 47 new hybrids have been obtained in this cross-combination so far.

This cross-combination can obtain so many new hybrids with different characteristics, which proves that the cultivar 'Tama Beauty' has a high heterozygosity and is an ideal parent material for creating new hybrids that bloom year-round.

HA-03-46. 赤诚红心
Chicheng Hongxin (Sincere Red Heart)

杂交组合：红山茶品种'媚丽'× 杜鹃红山茶（*C. japonica* cultivar 'Tama Beauty' × *C. azalea*）。

杂交苗编号：MD-No.6。

性状：花朵深红色至黑红色，半重瓣型至托桂型，大型花，花径 10.5～12.5cm，花朵厚 5.2cm，外轮大花瓣 8～9 枚，中部小花瓣簇拥成球，大花瓣倒卵形，雄蕊基部连生，花丝红色，花药金黄色。叶片浓绿色，叶背面灰绿色，革质，椭圆形，长 9.0～10.5cm，宽 3.0～4.0cm，基部楔形，上部边缘具钝锯齿。植株立性，枝叶稠密，生长旺盛。夏末始花，秋、冬季盛花，春季零星开花。

Abstract: Flowers crimson to dark red, semi-double to anemone form, large size, 10.5-12.5cm across and 5.2cm deep with 8-9 exterior large petals, central many small petals clustered into a ball, large petals obovate, stamens united at the base. Leaves dark green, leathery, elliptic, cuneate at the base, margins obtusely serrate at the upper part. Plant upright, branches dense and growth vigorous. Starts to bloom from late-summer, fully blooms from autumn to winter and sporadically in spring.

HA-03-47. 蕊珠头饰
Ruizhu Toushi (Miss Ruizhu's Headdress)

杂交组合：红山茶品种'媚丽'× 杜鹃红山茶（*C. japonica* cultivar 'Tama Beauty' × *C. azalea*)。

杂交苗编号：MD–No.70。

性状：花朵橘红色至红色，小花瓣偶有小白斑，托桂型，小型至中型花，花径7.0～9.0cm，花朵厚5.0cm，外轮大花瓣7～12枚，呈1～2轮排列，中部雄蕊瓣簇拥呈紧实球状，并与雄蕊混生，开花稠密。叶片浓绿色，叶背面灰绿色，厚革质，阔椭圆形，长9.5～11.0cm，宽4.0～5.0cm，上部边缘具稀齿。植株立性，枝叶稠密，生长旺盛。夏末始花，秋、冬季盛花，春季零星开花。

注：蕊珠是我国古典小说《红楼梦》中一个丫鬟的名字。

Abstract: Flowers orange red to red, occasionally with tiny white dots, anemone form, small to medium size, 7.0-9.0cm across and 5.0cm deep with 7-12 exterior large petals arranged in 1-2 rows, central petaloids tightly grouped together into a ball and mixed with stamens, bloom dense. Leaves dark green, heavy leathery, broad elliptic, margins sparsely serrate at the upper part. Plant upright, branches dense and growth vigorous. Starts to bloom from late-summer, fully blooms from autumn to winter and sporadically in spring.

Mark: Miss Ruizhu is the name of a servant girl in *The Dream of Red Mansions* which is a classical Chinese novel.

⊙ 杂交组合 HA-05. 杜鹃红山茶 × 红山茶品种'超级南天武士'
HA-05. *C. azalea* × *C. japonica* cultivar 'Dixie Knight Supreme'

×

本杂交组合将介绍2个新品种，其编号将接续第一部书本组的序列编号。

第一部书曾介绍过3个新品种（见第一部书第152～156页）。至此，本组共获得了5个新品种。

这些新品种都具有叶片浓绿、重复开花的性状，颇像杜鹃红山茶。可是在花色上，有黑红色的品种，也有粉红色的品种；在花型上，单瓣型的、半重瓣型的品种都有。这可能是杜鹃红山茶和'超级南天武士'基因共同作用的结果。

Two new hybrids will be introduced in this cross-combination and their numbers will follow the serial numbers of the cross-combination in our first book.

We introduced three new hybrids in our first book (See p.152-156, our first book). There have been a total of five new hybrids introduced in this cross-combination until now.

These new hybrids all have the characteristics of dark green leaves and ever-blooming trait, which are very similar to *C. azalea*. However, their flowers are dark red or pink in color and single or semi-double in form. It would be the results of joint actions from the genes of *C. azalea* and 'Dixie Knight Supreme'.

HA-05-4. 仙女下凡
Xiannü Xiafan (Fairy from Sky Down to the Earth)

杂交组合： 杜鹃红山茶 × 红山茶品种'超级南天武士'（*C. azalea* × *C. japonica* cultivar 'Dixie Knight Supreme'）。

杂交苗编号： DCN-No.10。

性状： 花朵外部花瓣上部桃红色，渐向中部和基部呈淡粉白色，玫瑰重瓣型，微型至小型花，花径5.5～7.0cm，花朵厚4.5cm，花瓣18～21枚，呈3～4轮有序排列，倒卵形，长4.0cm，宽3.0cm，花芯偶现少量雄蕊和雄蕊瓣，开花稠密。叶片浓绿色，叶背面灰绿色，革质，椭圆形，长7.0～8.0cm，宽3.0～3.5cm，边缘叶齿稀。植株紧凑，生长旺盛。夏末始花，秋、冬季盛花，春季零星开花。

Abstract: Flowers peach red at the upper part of outside petals and fading to pink-white at both the center and the base, rose-double form, miniature to small size, 5.5-7.0cm across and 4.5cm deep with 18-21 petals arranged orderly in 3-4 rows, bloom dense. Leaves dark-green, leathery, elliptic, margins sparsely serrate. Plant compact and growth vigorous. Starts to bloom from late-summer, fully blooms from autumn to winter and sporadically in spring.

HA-05-5. 夏秋粉妞
Xiaqiu Fenniu (Pink Girl in Summer & Autumn)

杂交组合： 杜鹃红山茶 × 红山茶品种'超级南天武士'（ C. azalea × C. japonica cultivar 'Dixie Knight Supreme'）。

杂交苗编号： DCN–No.6。

性状： 花朵外部花瓣淡红色，中部花瓣深粉红色，具白斑，半重瓣型，小型至中型花，花径7.0～8.5cm，花朵厚4.3cm，花瓣约23枚，呈3轮排列，倒卵形，长4.3cm，宽3.5cm，大波浪状，雄蕊簇状，开花稠密。叶片浓绿色，叶背面灰绿色，革质，小椭圆形，长7.0～8.0cm，宽3.0～3.5cm，边缘叶齿密。植株紧凑，矮性，枝条匍匐，生长旺盛。夏末始花，秋、冬季盛花，春季零星开花。

Abstract: Flowers light red in outside petals and deep pink in mid-petals, blotched white, semi-double form, small to medium size, 7.0-8.5cm across and 4.3cm deep with about 23 petals arranged in 3 rows, bloom dense. Leaves dark-green, leathery, small elliptic, margins densely serrate. Plant compact, dwarf, branches creeping and growth vigorous. Starts to bloom from late-summer, fully blooms from autumn to winter and sporadically in spring.

杂交组合 HA-06. 杜鹃红山茶 × 红山茶品种'大菲丽丝'
HA-06. *C. azalea* × *C. japonica* cultivar 'Francis Eugene Phillis'

×

本杂交组合将介绍2个新品种，其编号将接续第一部书本组的序列编号。

第一部书曾介绍了本组的一个名为'夏日光辉'的杂交新品种（见第一部书第157～158页）。至此，本组共获得3个新品种。

可以看到，它们在性状上均具有双亲的特性：花期和叶色倾向于母本杜鹃红山茶，而叶片边缘具深尖锯齿则倾向于'大菲丽丝'。因此，可以确定，它们都是真实的杂交品种。

Two new hybrids will be introduced in this cross-combination and their numbers will follow the serial numbers of the cross-combination in our first book.

A hybrid named 'Xiari Guanghui' was introduced in our first book (See p.157-158, our first book). A total of three new hybrids have been obtained in this cross-combination until now.

It can be seen that they all have characteristics of their cross parents: their blooming period and leaves color tend to their female *C. azalea* and their leave margins with sharp serrations are similar to 'Francis Eugene Phillis'. Therefore, it can be confirmed that these two hybrids we obtained are all true.

HA-06-2. 夏秋褶裙
Xiaqiu Zhequn (Wrinkled Skirt of Summer & Autumn)

杂交组合：杜鹃红山茶 × 红山茶品种'大菲丽丝'（ *C. azalea* × *C. japonica* cultivar 'Francis Eugene Phillis'）。

杂交苗编号：DDF-No.4。

性状：花朵深红色至黑红色，具蜡光，偶有少量白斑，渐向基部变为淡红色，半重瓣型至牡丹型或者玫瑰重瓣型，中型花，花径9.5～10.0cm，花朵厚6.0cm，大花瓣40～50枚，3～4轮排列，倒卵形，边缘皱褶，内卷，中部小花瓣半直立，扭曲，雄蕊短，簇生，偶有白色雄蕊瓣，开花稠密，花朵整朵掉落。叶片浓绿色，叶背面灰绿色，革质，长倒卵形，长9.5～10.0cm，宽4.5～5.0cm，略扭曲，基部急楔形，边缘叶齿尖而深。植株紧凑，灌丛状，生长旺盛。夏末始花，秋、冬季盛花，春季偶有开花。

Abstract: Flowers deep red to dark red with waxy luster, occasionally with a few white blotches, fading to pale red at the petal bases, semi-double to peony form or rose-double form, medium size, 9.5-10.0cm across and 6.0cm deep with 40-50 large petals arranged in 3-4 rows, petal edges crinkled, rolled inward, central small petals twisted, stamens short, occasionally with some white petaloids, bloom dense, flowers fall whole. Leaves dark green, leathery, long obovate, urgently cuneate at the base, slightly twisted, margins sharply and deeply serrate. Plant compact, bushy and growth vigorous. Starts to bloom from late-summer, fully blooms from autumn to winter and occasionally in spring.

HA-06-3. 大菲升级
Dafei Shengji ('Francis Eugene Phillis' Upgrade)

杂交组合：杜鹃红山茶 × 红山茶品种'大菲丽丝'（*C. azalea* × *C. japonica* cultivar 'Francis Eugene Phillis'）。

杂交苗编号：DDF-No.5。

性状：花朵深粉红色至桃红色，泛紫色调，偶有隐约的白条纹，半重瓣型至牡丹型，中型至巨型花，花径8.5～13.0cm，花朵厚5.0cm，大花瓣35～40枚，呈3～4轮排列，花瓣边缘波浪状，中部少量小花瓣，半直立，雄蕊簇生。叶片浓绿色，叶背面灰绿色，革质，长倒卵形，长8.5～9.5cm，宽4.0～5.0cm，基部楔形，边缘叶齿细。植株立性，生长旺盛。夏末始花，秋、冬季盛花，春季零星开花。

Abstract: Flowers deep pink to peach red with purple hue, occasionally with faintly white stripes, semi-double to peony form, medium to very large size, 8.5-13.0cm across and 5.0cm deep with 35-40 large petals arranged in 3-4 rows, petal edges wavy, a few central small petals, semi-erected, stamens clustered. Leaves dark green, leathery, long obovate, cuneate at the base, margins thinly serrate. Plant upright, growth vigorous. Starts to bloom from late-summer, fully blooms from autumn to winter and sporadically in spring.

⦿ 杂交组合 HA-10. 杜鹃红山茶 × 红山茶品种'达婷'
HA-10. *C. azalea* × *C. japonica* cultivar 'Mary Agnes Patin'

本杂交组合将介绍 13 个新品种,其编号将接续第一部书本组的序列编号。

第一部书曾介绍了本组的 7 个新品种(见第一部书第 172～181 页)。至此,本组共获得了 20 个新品种。

本杂交组合是杜鹃红山茶杂交组合中获得新品种数最多的杂交组合之一,这证明'达婷'是创制四季茶花的好品种。

Thirteen new hybrids will be introduced in this cross-combination and their numbers will follow the serial numbers of the cross-combination in our first book.

Seven new hybrids were introduced in this cross-combination of our first book (See p.172-181, our first book). A total of 20 new hybrids have been obtained in this cross-combination until now.

This cross-combination is one of most productive *C. azalea* cross-combination, which shows that 'Mary Agnes Patin' is an ideal cultivar of creating new hybrids that bloom year-round.

HA-10-8. 满堂喝彩
Mantang Hecai (Universal Applause)

杂交组合：杜鹃红山茶 × 红山茶品种'达婷'（ *C. azalea* × *C. japonica* cultivar 'Mary Agnes Patin' ）。

杂交苗编号：DDA-No.7。

性状：花朵橘红色，略泛蜡光，中部雄蕊瓣具白条纹，松散的托桂型至牡丹型，中型至大型花，花径9.0～12.5cm，花朵厚5.5cm，大花瓣15～18枚，呈2～3轮排列，上部花色较淡，雄蕊瓣簇拥呈松散的球状，花丝淡红色，花药金黄色。叶片浓绿色，叶背面灰绿色，厚革质，椭圆形，长9.5～10.5cm，宽4.0～5.5cm，上部边缘具稀疏尖齿。植株紧凑，立性，生长旺盛。夏末始花，秋、冬季盛花，春季零星开花。

Abstract: Flowers orange red, slightly with waxy luster, central petaloids with white stripes, loose anemone to peony form, medium to large size, 9.0-12.5cm across and 5.5cm deep, 15-18 large petals arranged in 2-3 rows, petals light color at the upper part, petaloids clustered into a loose globose. Leaves dark green, back surfaces gray-green, heavy leathery, elliptic, margins sparsely and sharply serrate at the upper part. Plant compact, upright and growth vigorous. Starts to bloom from late-summer, fully blooms from autumn to winter and sporadically in spring.

HA-10-9. 欣欣向荣
Xinxin Xiangrong (Flourishment)

杂交组合：杜鹃红山茶 × 红山茶品种'达婷'（ *C. azalea* × *C. japonica* cultivar 'Mary Agnes Patin'）。

杂交苗编号：DDA-No.9。

性状：花朵艳红色，略显橘红色调，具白色条纹，半重瓣型至玫瑰重瓣型，大型至巨型花，花径11.0～13.5cm，花朵厚5.3cm，大花瓣21～24枚，呈3～5轮排列，花瓣外翻，花芯处少量小花瓣半直立，雄蕊簇生。叶片浓绿色，叶背面灰绿色，厚革质，阔椭圆形，长7.5～9.0cm，宽5.0～5.5cm，上部边缘具稀疏钝齿。植株紧凑，立性，生长旺盛。夏末始花，秋、冬季盛花，春季零星开花。

Abstract: Flowers bright red, slightly with tangerine hue and white stripes, semi-double to rose-double form, large to very large size, 11.0-13.5cm across and 5.3cm deep with 21-24 large petals arranged in 3-5 rows, petals rolled outward, a few of small petals semi-erected at the center, stamens clustered. Leaves dark green, heavy leathery, broad elliptic, margins sparsely and obtusely serrate at the upper part. Plant compact, upright and growth vigorous. Starts to bloom from late-summer, fully blooms from autumn to winter and sporadically in spring.

HA-10-10. 广场舞步
Guangchang Wubu (Dance Steps in Square)

杂交组合： 杜鹃红山茶 × 红山茶品种'达婷'（*C. azalea* × *C. japonica* cultivar 'Mary Agnes Patin'）。

杂交苗编号： DDA-No.11。

性状： 花朵艳红色，具蜡光，中部小花瓣有白条纹，半重瓣型至牡丹型，大型花，花径10.5～12.5cm，花朵厚5.0cm，外部大花瓣15～18枚，呈2～3轮松散排列，花瓣厚质，外部花瓣外翻，波浪状，内部花瓣半直立，扭曲，雄蕊近散生。叶片浓绿色，叶背面灰绿色，厚革质，阔椭圆形，长7.5～8.5cm，宽4.5～5.5cm，基部楔形，边缘具稀疏尖齿。植株紧凑，立性，生长旺盛。夏末始花，秋、冬季盛花，春季零星开花。

Abstract: Flowers bright red with waxy luster, central small petals with white stripes, semi-double to peony form, large size, 10.5-12.5cm across and 5.0cm deep with 15-18 exterior large petals arranged loosely in 2-3 rows, petals thick texture, rolled outward, wavy, central small petals semi-erected and twisted, stamens nearly scattered. Leaves dark green, heavy leathery, broad elliptic, cuneate at the base, margins sparsely and sharply serrate. Plant compact, upright and growth vigorous. Starts to bloom from late-summer, fully blooms from autumn to winter and sporadically in spring.

HA-10-11. 女模时装
Nümo Shizhuang (Female Models' Fashionable Dress)

杂交组合：杜鹃红山茶 × 红山茶品种'达婷'(*C. azalea* × *C. japonica* cultivar 'Mary Agnes Patin')。

杂交苗编号：DDA-No.14。

性状：花朵艳红色，有时泛紫色调，单瓣型，大型花，花径11.0～12.0cm，花瓣7枚，宽大，阔倒卵形，厚质，全开放后外翻，雄蕊基部略连生，呈短管状，花丝淡粉红色，开花稠密。叶片浓绿色，叶背面灰绿色，光亮，中脉凸起，革质，长椭圆形，长10.0～10.5cm，宽3.5～4.0cm，基部楔形，叶缘具钝齿。植株开张，枝叶稠密，生长旺盛。夏末始花，秋、冬季盛花，春季零星开花。

Abstract: Flowers bright red, sometimes with purple hue, single form, large size, 11.0-12.0cm across with 7 broad petals, petals thick texture, roll outward when they are in full bloom, stamens united into a short column at the base, bloom very dense. Leaves dark green, leathery, long elliptic, margins obtusely serrate. Plant spread, branches dense and growth vigorous. Starts to bloom from late-summer, fully blooms from autumn to winter and sporadically in spring.

HA-10-12. 灿烂辉煌
Canlan Huihuang (Splendid & Glorious)

杂交组合：杜鹃红山茶 × 红山茶品种'达婷'（*C. azalea* × *C. japonica* cultivar 'Mary Agnes Patin'）。

杂交苗编号：DDA-No.15。

性状：花朵红色，泛橘红色调，半重瓣型至玫瑰重瓣型，大型花，花径10.0～12.5cm，花朵厚5.6cm，花瓣30枚以上，倒卵形，呈3～4轮整齐、松散排列，边缘小波浪状，雄蕊基部连生，呈筒状，偶有簇生雄蕊，雄蕊长约3cm，雌蕊奶白色，柱头5～7深裂，开花稠密。叶片浓绿色，叶背面灰绿色，厚革质，长椭圆形，长11.2～11.6cm，宽4.2～5.0cm，上部边缘具浅锯齿。植株紧凑，枝条稠密，生长旺盛。早秋始花，可持续开花到翌年春季。

> **Abstract:** Flowers red with some tangerine hue, semi-double to rose-double form, large size, 10.0-12.5cm across and 5.6cm deep with over 30 petals arranged orderly and loosely in 3-4 rows, stamens united at the base, filaments milk white, stigma 5-7 deep splits. Leaves dark green, heavy leathery, long elliptic, margins shallowly serrate at the upper part. Plant compact, branches dense and growth vigorous. Starts to bloom from early-autumn and blooms continuously to the following spring.

HA-10-13. 红色畅想
Hongse Changxiang (Red Imagination)

杂交组合：杜鹃红山茶 × 红山茶品种'达婷'（ *C. azalea* × *C. japonica* cultivar 'Mary Agnes Patin'）。

杂交苗编号：DDA-No.16。

性状：花朵桃红色至淡红色，略显橘红色调，花芯处小花瓣偶有白斑，托桂型至牡丹型，大型花，花径 11.0～12.5cm，花朵厚 5.5cm，外部大花瓣 18 枚，呈 2 轮紧实排列，花瓣阔倒卵形，中部外翻，大部分雄蕊瓣化，簇拥成团，间杂有金黄色雄蕊。叶片浓绿色，叶背面灰绿色，厚革质，阔椭圆形，长 8.7～9.5cm，宽 5.0～5.5cm，上部边缘具稀疏尖齿。植株紧凑，立性，生长旺盛。夏末始花，秋、冬季盛花，春季零星开花。

Abstract: Flowers peach red to light red slightly with tangerine hue, some white blotches on the central small petals, anemone to peony form, large size, 11.0-12.5cm across and 5.5cm deep with 18 large petals arranged tightly in 2 rows, central small petals clustered into a ball with the stamens together. Leaves dark green, heavy leathery, broad elliptic, margins sparsely and sharply serrate at the upper part. Plant compact, upright and growth vigorous. Starts to bloom from late-summer, fully blooms from autumn to winter and sporadically in spring.

HA-10-14. 红绸舞浪
Hongchou Wulang (Red Silk Dancing Waves)

杂交组合：杜鹃红山茶 × 红山茶品种'达婷'（ *C. azalea* × *C. japonica* cultivar 'Mary Agnes Patin'）。

杂交苗编号：DDA-No.17。

性状：花朵艳红色，渐向花瓣基部呈淡红色，单瓣型，中型至大型花，花径8.5～10.5cm，花瓣6～7枚，阔倒卵形，边缘皱褶，外翻，波浪状，雄蕊柱状。叶片浓绿色，叶背面灰绿色，厚革质，长椭圆形，长9.5～10.5cm，宽4.0～4.5cm，基部楔形，上部边缘具浅齿。植株紧凑，矮性，生长旺盛。夏末始花，秋、冬季盛花，春季零星开花。

Abstract: Flowers bright red, gradually fading to light red at the petal bases, single form, medium to large size, 8.5-10.5cm across with 6-7 petals, petals broad obovate, rolled outward, wavy, edges crinkled, stamens united in a column. Leaves dark green, heavy leathery, long elliptic, cuneate at the base, margins shallowly serrate at the upper part. Plant compact, dwarf and growth vigorous. Starts to bloom from late-summer, fully blooms from autumn to winter and sporadically in spring.

HA-10-15. 粉娇醉秋
Fenjiao Zuiqiu (Pink Girls Making Autumn Intoxicate)

杂交组合：杜鹃红山茶 × 红山茶品种'达婷'（*C. azalea* × *C. japonica* cultivar 'Mary Agnes Patin'）。

杂交苗编号：DDA-No.21。

性状：花朵初开放时多呈蜡烛状，粉红色至淡桃红色，偶有花瓣顶部出现模糊白斑，单瓣型，中型花，花径 8.5～9.5cm，花瓣 8～9 枚，瓣面可见深红色脉纹，花瓣呈勺状，雄蕊基部连生，呈短管状，花丝近白色，花朵整朵掉落，开花稠密。叶片浓绿色，叶背面灰绿色，厚革质，长椭圆形，长 9.6～10.0cm，宽 3.8～4.2cm，边缘具尖齿。植株开张，枝条软，略下垂，生长旺盛。夏末始花，秋、冬季盛花，春季零星开花。

Abstract: Flower buds always are candle shaped when they just open. Flowers pink to pale peach red, occasionally with a fuzzy white patch, single form, medium size, 8.5-9.5cm across with 8-9 petals, deep red veins visible, petals spoon-shaped, stamens united at the base, filaments nearly white, flowers fall whole, bloom dense. Leaves dark green, heavy leathery, long elliptic, margins sharply serrate. Plant spread, branches soft and slightly hanging down, and growth vigorous. Starts to bloom from late-summer, fully blooms from autumn to winter and sporadically in spring.

HA-10-16. 万家灯火
Wanjia Denghuo (Myriad Families' Lights)

杂交组合： 杜鹃红山茶 × 红山茶品种'达婷'（ *C. azalea* × *C. japonica* cultivar 'Mary Agnes Patin'）。

杂交苗编号： DDA-No.25。

性状： 花朵桃红色至红色，单瓣型，小型至中型花，花径7.0～8.0cm，花瓣5～6枚，夏季多呈半开放状态，但秋季花朵开放正常，雄蕊基部略连生，开花稠密。叶片浓绿色，叶背面灰绿色，革质，长椭圆形，长9.5～10.0cm，宽3.5～4.0cm，叶缘具钝齿。植株开张，枝条稠密，生长旺盛。夏末始花，秋、冬季盛花，翌年春季零星开花。

Abstract: Flowers peach red to red, single form, small to medium size, 7.0-8.0cm across with 5-6 petals, most of the flowers are semi-opened in summer, but the flowers open normally in autumn, stamens united at the base, bloom very dense. Leaves dark green, leathery, long elliptic, margins obtusely serrate. Plant spread, branches dense and growth vigorous. Starts to bloom from late-summer, fully blooms from autumn to winter and sporadically in spring.

HA-10-17. 夕阳余晖
Xiyang Yuhui（Afterglow of the Sunset）

杂交组合：杜鹃红山茶 × 红山茶品种'达婷'（*C. azalea* × *C. japonica* cultivar 'Mary Agnes Patin'）。

杂交苗编号：DDA-No.26。

性状：花朵橙红色，单瓣型，中型花，花径9.5～10.0cm，花瓣6～7枚，阔倒卵形，瓣面可见深红色脉纹，边缘略皱，波浪状，雄蕊柱状，开花稠密。叶片浓绿色，叶背面灰绿色，厚革质，长椭圆形，长9.0～10.0cm，宽4.5～5.0cm，边缘具浅齿。植株立性，生长旺盛。夏末始花，秋、冬季盛花，春季零星开花。

Abstract: Flowers orange red, single form, medium size, 9.5-10.0cm across with 6-7 petals, petals broad obovate with deep red veins visible, slightly wavy, stamens united into a column, bloom dense. Leaves dark green, heavy leathery, long elliptic, margins shallowly serrate. Plant upright and growth vigorous. Starts to bloom from late-summer, fully blooms from autumn to winter and sporadically in spring.

HA-10-18. 天伦之乐
Tianlun Zhile（Enjoying Family's Happiness）

杂交组合：杜鹃红山茶 × 红山茶品种'达婷'（*C. azalea* × *C. japonica* cultivar 'Mary Agnes Patin'）。

杂交苗编号：DDA-No.27。

性状：花朵红色至艳红色，阳光下泛橘红色调，中部花瓣有少量白色条纹，玫瑰重瓣型至牡丹型，大型至巨型花，花径11.5～13.5cm，花朵厚5.0～6.0cm，外部大花瓣24枚，呈3轮排列，花瓣外翻，中部半直立，花芯处有散生雄蕊。叶片浓绿色，叶背面灰绿色，厚革质，阔椭圆形，长8.5～9.6cm，宽5.2～5.7cm，基部楔形，上部边缘具浅齿。植株紧凑，立性，生长旺盛。夏末始花，秋、冬季盛花，春季零星开花。

Abstract: Flowers red to bright red with tangerine hue under sunlight, a few white stripes appear on the central petals, rose-double to peony form, large to very large size, 11.5-13.5cm across and 5.0-6.0cm deep with 24 exterior large petals arranged in 3 rows, petals rolled outward, central petals semi-erected. Leaves dark green, heavy leathery, broad elliptic, cuneate at the base, margins shallowly serrate at the upper part. Plant compact, upright and growth vigorous. Starts to bloom from late-summer, fully blooms from autumn to winter and sporadically in spring.

HA-10-19. 梦幻世界
Menghuan Shijie (Dream-like World)

杂交组合： 杜鹃红山茶 × 红山茶品种'达婷'（*C. azalea* × *C. japonica* cultivar 'Mary Agnes Patin'）。

杂交苗编号： DDA-No.30。

性状： 花朵顶生，深红色，夏季和秋季花瓣上部具无数隐约的白条纹，秋末至翌年春季花瓣逐渐变为红色，半重瓣型至玫瑰重瓣型，偶然会出现松散的牡丹型，大型至巨型花，花径 11.5～14.0cm，花朵厚 5.5～6.0cm，花瓣 60 枚以上，呈 7～8 轮有序排列，中部花瓣半直立，先端近圆，雄蕊基部连生。叶片浓绿色，叶背面灰绿色，厚革质，椭圆形，长 9.0～9.5cm，宽 4.2～4.6cm，边缘齿浅。植株开张，生长旺盛。夏末始花，秋、冬季盛花，春季零星开花。

> **Abstract:** Flower solitary at the tips of shoots and deep red, with a lot of hazy white stripes at petal upper part in summer and autumn, flower color gradually changes into red from late-autumn to the following spring, semi-double to rose-double form, occasionally loose peony form, large to very large size, 11.5-14.0cm across and 5.5-6.0cm deep with over 60 petals arranged orderly in 7-8 rows, stamens united at the base. Leaves dark green, heavy leathery, elliptic, margins shallowly serrate. Plant spread and growth vigorous. Starts to bloom from late-summer, fully blooms from autumn to winter and sporadically in spring.

HA-10-20. 夏秋盛典
Xiaqiu Shengdian (Great Ceremony in Summer & Autumn)

杂交组合：杜鹃红山茶 × 红山茶品种'达婷'（*C. azalea* × *C. japonica* cultivar 'Mary Agnes Patin'）。

杂交苗编号：DDA-No.40。

性状：花朵桃红色至红色，有时泛紫色调，中部雄蕊瓣具白色斑块，托桂型至牡丹型，大型至巨型花，花径11.5～13.7cm，花朵厚5.8cm，外轮大花瓣24～26枚，呈3轮有序排列，平铺或波浪状，中部偶有几枚半直立的较大花瓣，簇拥成团，少量黄色雄蕊外露，花丝红色，开花与抽梢同时进行。叶片浓绿色，叶背面灰绿色，厚革质，椭圆形，长8.5～10.0cm，宽3.5～4.0cm，边缘齿浅。植株立性，枝条稠密，生长非常旺盛。夏末始花，秋、冬季盛花，春季零星开花。

Abstract: Flowers peach red to red, sometimes with purple hue, central petaloids blotched white, anemone form to peony form, large to very large size, 11.5-13.7cm across and 5.8cm deep with 24-26 exterior large petals arranged orderly in 3 rows, petals flat or wavy, some semi-erected larger petals clustered into a ball at the center, filaments red, bloom and sprout at the same time. Leaves dark green, heavy leathery, elliptic, margins shallowly serrate. Plant upright, branches dense and growth very vigorous. Starts to bloom from late-summer, fully blooms from autumn to winter and sporadically in spring.

◉ 杂交组合 HA-11. 杜鹃红山茶 × 红山茶品种'闪烁'
HA-11. *C. azalea* × *C. japonica* cultivar 'Tama Glitters'

本杂交组合将介绍2个新品种，其编号将接续第一部书本组的序列编号。

第一部书曾介绍了该组的2个新品种（见第一部书第182～184页）。至此，本组已经获得了4个新品种。

'闪烁'是一个花瓣具白色宽边的品种。虽然本次获得的新品种花瓣不具白边，但其花型和花色颇具魅力，而且开花期大大提前了。

Two new hybrids will be introduced in this cross-combination and their numbers will follow the serial numbers of the cross-combination in our first book.

We introduced two new hybrids in this combination in our first book (See p.182-184, our first book). A total of four new hybrids have been obtained in this cross-combination until now.

'Tama Glitters' is a cultivar of petals with white borders. Although the new hybrids do not have white borders, its shape and color are attractive, and its blooming period is much earlier.

HA-11-3. 秋冬野美
Qiudong Yemei (Wild Beauty in Autumn & Winter)

杂交组合：杜鹃红山茶 × 红山茶品种'闪烁'（*C. azalea* × *C. japonica* cultivar 'Tama Glitters'）。

杂交苗编号：DSS-No.3。

性状：花朵红色，瓣面可见模糊的细白脉纹，半重瓣型，中型至大型花，花径 8.0～11.5cm，花朵厚 5.0cm，花瓣 21 枚，呈 3 轮排列，阔倒卵形，厚质，外翻，雄蕊基部连生，呈短柱状，偶有小花瓣，呈白色。叶片浓绿色，叶背面灰绿色，革质，椭圆形，长 8.0～9.5cm，宽 4.0～5.0cm，中脉明显，边缘叶齿浅钝。植株开张，枝叶稠密，生长旺盛。夏末始花，秋、冬季盛花，春季零星开花。

> **Abstract:** Flowers red with misty and tiny white veins on the petal surfaces, semi-double form, medium to large size, 8.0-11.5cm across and 5.0cm deep with 21 petals arranged in 3 rows, petals broad obovate, thick texture, rolled outward. Leaves dark green, leathery, elliptic, midrib visible, margins shallowly and obtusely serrate. Plant spread, branches dense and growth vigorous. Starts to bloom from late-summer, fully blooms from autumn to winter and sporadically in spring.

HA-11-4. 火红双节
Huohong Shuangjie (Fire Red Double Festivals)

杂交组合： 杜鹃红山茶 × 红山茶品种'闪烁'（*C. azalea* × *C. japonica* cultivar 'Tama Glitters'）。

杂交苗编号： DSS-No.4。

性状： 花朵艳红色，牡丹型，大型至巨型花，花径11.0～13.0cm，花朵厚4.5～5.0cm，大花瓣16～21枚，呈3轮排列，花瓣阔倒卵形，中部雄蕊瓣很多，半直立，与雄蕊混生。叶片浓绿色，叶背面灰绿色，革质，椭圆形，长9.0～10.0cm，宽4.0～4.5cm，侧脉模糊，基部楔形，边缘叶齿浅钝。植株立性，枝叶稠密，生长旺盛。夏末始花，秋、冬季盛花，春季零星开花。

Abstract: Flowers bright red, peony form, large to very large size, 11.0-13.0cm across and 4.5-5.0cm deep with 16-21 large petals arranged in 3 rows, petals broad obovate, central petaloids numerous, semi-erected and mixed with stamens. Leaves dark green, leathery, elliptic, lateral veins blurry, cuneate at the base, margins shallowly and obtusely serrate. Plant upright, branches dense and growth vigorous. Starts to bloom from late-summer, fully blooms from autumn to winter and sporadically in spring.

杂交组合 HA-12. 红山茶品种'锯叶椿' × 杜鹃红山茶
HA-12. *C. japonica* cultivar 'Nokogiriba-tsubaki' × *C. azalea*

本杂交组合将介绍1个新品种,其编号将接续第一部书本组的序列编号。

第一部书曾介绍了本组的1个杂交新品种(见第一部书第185～186页)。至此,本杂交组合共获得了2个新品种。

'锯叶椿'品种的叶片具明显且规则的粗锯齿,花朵艳红,单瓣型,很容易坐果。可是,获得的杂交新品种叶形没有变化,因此,有待尝试更多的杂交。

One new hybrid will be introduced in this cross-combination and its number will follow the serial number of the cross-combination in our first book.

One new hybrid was introduced in this combination of our first book (See p.185-186, our first book), so two new hybrids have been obtained in this cross-combination until now.

'Nokogiriba-tsubaki' has leaves with obviously and regularly deep serrations, opens bright flowers with single form and can set capsules easily. However, there is no change in leaf shape of the new hybrids obtained, so more crosses should be done.

HA-12-2. 彤海咏秋
Tonghai Yongqiu (Red Sea Singing Autumn)

杂交组合：红山茶品种'锯叶椿' × 杜鹃红山茶（ *C. japonica* cultivar 'Nokogiriba-tsubaki' × *C. azalea* ）。

杂交苗编号：JYD-No.2。

性状：花朵粉红色，泛橘红色调，单瓣型，中型花，花径8.5～9.0cm，花瓣5～7枚，阔倒卵形，长6.0cm，宽5.0cm，瓣面皱褶，先端深缺刻，雄蕊柱状，基部略连生，花丝淡粉红色，花药黄色，开花极稠密。叶片绿色，革质，椭圆形，长6.5～7.8cm，宽3.5～4.5cm，边缘叶齿浅。植株紧凑，立性，枝叶稠密，生长旺盛。夏末始花，秋、冬季盛花，春季零星开花。

Abstract: Flowers pink with some tangerine hue, single form, medium size, 8.5-9.0cm across, petals broad obovate, crinkled, apices with a deep notch, stamens united into a column, bloom very dense. Leaves green, leathery, elliptic, margins shallowly serrate. Plant compact, upright, branches dense and growth vigorous. Starts to bloom from late-summer, fully blooms from autumn to winter and sporadically in spring.

◉ 杂交组合 HA-13. 杜鹃红山茶 × 红山茶品种 '红露珍'
HA-13. *C. azalea* × *C. japonica* cultivar 'Hongluzhen'

本杂交组合将介绍 3 个新品种，其编号将接续第一部书本组的序列编号。

在第一部书中，本组曾介绍过 4 个新品种（见第一部书第 187～192 页）。至此，本组获得的新品种已达 7 个。

如第一部书所述，本组获得的杂交新品种既遗传了杜鹃红山茶叶片浓绿、生长旺盛、多季开花的性状，也遗传了'红露珍'品种大花、多瓣、花色鲜艳的性状。

Three new hybrids will be introduced in this cross-combination and their numbers will follow the serial numbers of the cross-combination in our first book.

Four new hybrids were introduced in this cross-combination in our first book (See p.187-192, our first book). A total of seven new hybrids have been obtained in this cross-combination until now.

As what we wrote in our first book, the new hybrids obtained in this cross-combination have inherited both the characteristics of dark green leaves, strong growth and year-round bloom from *C. azalea* and the characteristics of large flowers, multi-petals and bright color from 'Hongluzhen'.

HA-13-5. 熙攘庙会
Xirang Miaohui (Bustling Fairs)

杂交组合： 杜鹃红山茶 × 红山茶品种'红露珍'(*C. azalea* × *C. japonica* cultivar 'Hongluzhen')。

杂交苗编号： DH-No.15。

性 状： 花朵深酒红色至红色，中部小花瓣具白条纹，松散的牡丹型，大型花，花径11.0～12.0cm，花朵厚5.0cm，外轮大花瓣12～13枚，呈2～3轮排列，花瓣阔倒卵形，雄蕊大部分瓣化，散射状，花丝酒红色，花药淡黄色，开花稠密。叶片浓绿色，叶背面灰绿色，厚革质，阔倒卵形，长9.5～10.5cm，宽5.5～6.5cm，边缘叶齿稀尖。植株立性，紧凑，生长旺盛。夏季始花，秋、冬季盛花，春季零星开花。

Abstract: Flowers deep wine red to red, white stripes appeared on the central petaloids, loose peony form, large size, 11.0-12.0cm across and 5.0cm deep with 12-13 exterior large petals arranged in 2-3 rows, petals broad obovate, most of stamens are petaloids, which are scattered, filaments wine red, bloom dense. Leaves dark green, heavy leathery, broad obovate, margins sparsely and sharply serrate. Plant upright, compact and growth vigorous. Starts to bloom from summer, fully blooms from autumn to winter and sporadically in spring.

HA-13-6. 国色天姿
Guose Tianzi (Possess Surpassing Beauty)

杂交组合：杜鹃红山茶 × 红山茶品种'红露珍'(*C. azalea* × *C. japonica* cultivar 'Hongluzhen')。

杂交苗编号：DH-No.17。

性状：花朵深红色，渐向花瓣中部呈淡红色或白色，半重瓣型至牡丹型，大型花，花径 10.5～12.5cm，花朵厚5.5cm，花瓣20～23枚，呈3轮松散排列，阔倒卵形，略波浪状，瓣面深红色脉纹清晰可见，雄蕊簇生，少量雄蕊瓣与雄蕊混生。叶片浓绿色，叶背面灰绿色，革质，阔椭圆形，长9.0～10.5cm，宽5.5～6.3cm。植株立性，生长旺盛。夏季始花，秋、冬季盛花，春季零星开花。

Abstract: Flowers deep red, fading to light red or white at the center of petals, semi-double to peony form, large size, 10.5-12.5cm across and 5.5cm deep with 20-23 petals arranged loosely in 3 rows, petals broad obovate, slightly wavy, deep red veins visible on the petal surfaces, stamens clustered, a few of petaloids mixed with the stamens. Leaves dark green, leathery, broad elliptic. Plant upright and growth vigorous. Starts to bloom from summer, fully blooms from autumn to winter and sporadically in spring.

HA-13-7. 恋人约会
Lianren Yuehui (Lovers Dating)

杂交组合：杜鹃红山茶 × 红山茶品种'红露珍'(*C. azalea* × *C. japonica* cultivar 'Hongluzhen')。

杂交苗编号：DH-No.30。

性状：花朵深粉红色至红色，中部花瓣具隐约的白条纹，半重瓣型，中型花，花径8.5～10.0cm，花朵厚5.0cm，花瓣25～30枚，呈3轮紧实排列，倒卵形，略皱褶，雄蕊基部略连生。叶片浓绿色，叶背面灰绿色，革质，椭圆形，长8.0～9.0cm，宽4.0～4.5cm。植株立性，紧凑，生长旺盛。仲夏始花，秋、冬季盛花，春季零星开花。

Abstract: Flowers deep pink to red with faint white stripes at the central petals, semi-double form, medium size, 8.5-10.0cm across and 5.0cm deep with 25-30 petals arranged tightly in 3 rows, petals obovate, slightly crinkled, stamens slightly united at the base. Leaves dark green, leathery, elliptic. Plant upright, compact and growth vigorous. Starts to bloom from mid-summer, fully blooms from autumn to winter and sporadically in spring.

⊙ 杂交组合 HA-14. 杜鹃红山茶 × 红山茶品种'花牡丹'
HA-14. *C. azalea* × *C. japonica* cultivar 'Daikagura'

本杂交组合将介绍2个新品种，其编号将接续第一部书本组的序列编号。

第一部书曾介绍了本组的4个新品种（见第一部书第193～199页）。至此，本组获得的新品种数已达到6个。

可以看到，本杂交组合非常完美，获得的6个杂交新品种中，有5个是多瓣类型的。

Two new hybrids will be introduced in this cross-combination and their numbers will follow the serial numbers of the cross-combination in our first book.

Four hybrids were introduced in this combination of the first book (See p.193-199, our first book). A total of six new hybrids have been obtained in this cross-combination until now.

It can be seen that this cross-combination is very perfect which five of six new hybrids obtained have multi-petals.

HA-14-5. 处处欢腾
Chuchu Huanteng (Exult Everywhere)

杂交组合： 杜鹃红山茶 × 红山茶品种'花牡丹'（ *C. azalea* × *C. japonica* cultivar 'Daikagura'）。

杂交苗编号： DHUA-No.6。

性状： 花朵深桃红色至红色，花芯处小花瓣偶具白条纹，花开6天后，中部花瓣渐渐转淡，牡丹型至完全重瓣型，巨型花，花径13.0～14.7cm，花朵厚5.0～5.5cm，大花瓣30枚以上，呈4～5轮有序排列，花瓣阔倒卵形，边缘皱褶，少量雄蕊簇生，与大花瓣混生。叶片浓绿色，革质，阔椭圆形，长9.0～10.0cm，宽4.0～4.5cm，叶缘有钝齿。植株开张，矮性，生长旺盛。夏末始花，秋、冬季盛花，翌年春季零星开花。

Abstract: Flowers deep peach red to red, occasionally with white stripes on the central small petals, after 6 days' bloom, fading to light color at the central petals, peony to formal double form, very large size, 13.0-14.7cm across and 5.0-5.5cm deep with more than 30 large petals arranged orderly in 4-5 rows, petals broad obovate, edges crinkled, a few stamens clustered and mixed with large petals. Leaves dark green, leathery, broad elliptic, margins obtusely serrate. Plant spread, dwarf and growth vigorous. Starts to bloom from late-summer, fully blooms from autumn to winter and sporadically in spring.

HA-14-6. 花衣小旋
Huayi Xiaoxuan (Miss Xiaoxuan Dressed Colorful Clothes)

杂交组合：杜鹃红山茶 × 红山茶品种'花牡丹'（*C. azalea* × *C. japonica* cultivar 'Daikagura'）。

杂交苗编号：DHUA-No.2Mu。

性状：本品种是'夏梦小旋'（见第一部书第 194～196 页）的一个复色突变。其大部分性状与'夏梦小旋'品种相似，但是花瓣上具大量云状白斑，花型半重瓣型至玫瑰重瓣型，花径大型至巨型，花朵厚可达 6.0cm，叶片偶然出现少量黄斑，这些性状与其母本'夏梦小旋'有明显不同。

Abstract: It is a variegated mutation obtained from the hybrid 'Xiameng Xiaoxuan' (See p.194-196, our first book). Most of its traits are similar to 'Xiameng Xiaoxuan', but there are a lot of white markings on the petal surfaces, flowers semi-double to rose-double form, large to very large size, 6.0cm deep and a few yellow patches on the leaves occasionally, which are obviously different from its female parent 'Xiameng Xiaoxuan'.

◉ 杂交组合 HA-16. 杜鹃红山茶 × 红山茶品种'皇家天鹅绒'
HA-16. *C. azalea* × *C. japonica* cultivar 'Royal Velvet'

本杂交组合将介绍3个新品种，其编号将接续第一部书本组的序列编号。

第一部书曾介绍了本杂交组合的2个新品种（见第一部书第202～204页）。至此，本杂交组合共获得了5个新品种。

可以看出，本杂交组合的新品种都是艳红色至黑红色的，而且都能够从夏季开始开花。这证明，杂交新品种已经具有双亲的性状。

Three new hybrids will be introduced in this cross-combination and their numbers will follow the serial numbers of the cross-combination in our first book.

Two new hybrids were introduced in this cross-combination in our first book (See p.202-204, our first book). A total of five new hybrids have been obtained in this cross-combination until now.

It can be seen that the new hybrids gotten from this cross-combination all can open bright red to dark red flowers and all can start to bloom from summer, which shows that these hybrids have the traits of their cross parents.

HA-16-3. 炎夏红浪
Yanxia Honglang (Red Waves in Hot Summer)

杂交组合：杜鹃红山茶 × 红山茶品种'皇家天鹅绒'（ *C. azalea* × *C. japonica* cultivar 'Royal Velvet'）。

杂交苗编号：DHJ-No.4。

性状：花朵深红色至黑红色，具蜡光，单瓣型，中型至大型花，花径9.5～11.0cm，花瓣6～7枚，阔倒卵形，波浪状，边缘皱褶，雄蕊基部连生，呈柱状，花丝红色，开花稠密，花朵整朵掉落。叶片浓绿色，叶背面灰绿色，厚革质，椭圆形，长9.0～10.5cm，宽4.0～5.0cm，上部边缘具稀疏锯齿。植株立性，枝叶稠密，生长旺盛。夏末始花，秋、冬季盛花，春季零星开花。

Abstract: Flowers deep red to dark red with some waxy luster, single form, medium to large size, 9.5-11.0cm across with 6-7 petals, petals broad obovate, wavy, edges crinkled, stamens united into a column at the base, filaments red, bloom dense, flowers fall whole. Leaves dark green, thick leathery, elliptic, margins sparsely serrate at the upper part. Plant upright, branches dense and growth vigorous. Starts to bloom from late-summer, fully blooms from autumn to winter and sporadically in spring.

HA-16-4. 中秋红喜
Zhongqiu Hongxi (Mid-Autumn Festival's Red Delight)

杂交组合： 杜鹃红山茶 × 红山茶品种'皇家天鹅绒'（ *C. azalea* × *C. japonica* cultivar 'Royal Velvet'）。

杂交苗编号： DHJ-No.6。

性状： 花朵深红色至黑红色，泛蜡光，偶有白斑，单瓣型，大喇叭状，大型至巨型花，花径10.5～13.5cm，花瓣5～6枚，长8.0cm，宽5.0cm，阔倒卵形，厚质，勺状，雄蕊连生呈柱状，乳白色，花朵整朵掉落。叶片浓绿色，叶背面灰绿色，厚革质，长椭圆形，长10.5～11.5cm，宽3.5～4.0cm，基部急楔形，叶缘齿稀钝。植株立性，枝条软，生长旺盛。秋初始花，冬季盛花，春季零星开花。

Abstract: Flowers deep red to dark red with waxy luster, occasionally with a faint white patch on a petal, single form which is trumpet-shaped, large to very large size, 10.5-13.5cm across with 5-6 petals, petals broad obovate, thick texture, spoon shaped, stamens united into a column, flowers fall whole. Leaves dark green, thick leathery, long elliptic, urgently cuneate at the base, margins sparsely and obtusely serrate. Plant upright, branches soft and growth vigorous. Starts to bloom from early-autumn, fully blooms in winter and sporadically in spring.

HA-16-5. 余霞成绮
Yuxia Chengqi (The Sunset Likely Beautiful Silk)

杂交组合：杜鹃红山茶 × 红山茶品种'皇家天鹅绒'（*C. azalea* × *C. japonica* cultivar 'Royal Velvet'）。

杂交苗编号：DHJ-No.7。

性状：花朵深红色至黑红色，花瓣边缘略黑，中部雄蕊瓣偶有白斑，半重瓣型至牡丹型，大型花，花径11.5～12.6cm，花朵厚5.0cm，花瓣18～20枚，呈2～3轮排列，阔倒卵形，外翻，雄蕊近散生，花瓣逐片掉落。嫩叶淡红色，成熟叶浓绿色，叶背面灰绿色，厚革质，椭圆形，长9.5～10.9cm，宽4.6～5.1cm，边缘具浅锯齿。植株立性，枝叶稠密，生长旺盛。夏末始花，秋、冬季盛花，春季偶尔开花。

Abstract: Flowers deep red to dark red, petal edges slightly dark, occasionally with white patches on central petaloids, semi-double to peony form, large size, 11.5-12.6cm across and 5.0cm deep with 18-20 petaloids arranged in 2-3 rows, petals broad obovate, stamens nearly separated, petals fall one by one. Tender leaves light red, mature leaves dark green, thick leathery, elliptic, margins shallowly serrate. Plant upright, branches dense and growth vigorous. Starts to bloom from late-summer, fully blooms from autumn to winter and occasionally in spring.

◉ 杂交组合 HA-21. 红山茶品种'都鸟' × 杜鹃红山茶
HA-21. *C. japonica* cultivar 'Miyakodori' × *C. azalea*

本杂交组合将介绍2个新品种，其编号将接续第一部书本组的序列编号。

第一部书曾介绍了本组7个新品种（见第一部书第218～228页）。至此，本组已获得的新品种达9个。

可以看到，本组获得的9个新品种花朵都是红色的，证明母本的白色花性状在遗传上对于父本的红色花性状是隐性的。

Two new hybrids will be introduced in this cross-combination and their numbers will follow the serial numbers of the cross-combination in our first book.

There were seven new hybrids introduced in this cross-combination of our first book (See p.218-228, our first book). So far nine new hybrids have been obtained in this combination.

It can be seen that these nine new hybrids obtained all have red flowers, which have shown that the white flowers of the female parent are genetically recessive to the red flowers of the male parent.

HA-21-8. 四季风情
Siji Fengqing (Four-season Amorous Feelings)

杂交组合：红山茶品种'都鸟' × 杜鹃红山茶（*C. japonica* cultivar 'Miyakodori' × *C. azalea*）。

杂交苗编号：HED-No.3。

性状：花朵深粉红色，边缘花色较淡，半重瓣型至牡丹型，中型至大型花，花径9.5～11.0cm，花朵厚4.8cm，大花瓣18～23枚，呈2～3轮排列，花瓣阔倒卵形，边缘波浪状，雄蕊基部略连生，呈簇状，开花极稠密。叶片浓绿色，叶背面灰绿色，革质，椭圆形，长8.0～9.0cm，宽3.0～3.5cm，上部叶缘具钝齿。植株紧凑，立性，生长旺盛。夏末始花，秋、冬季盛花，春季零星开花。

Abstract: Flowers deep pink, fading to light pink at the petal edges, semi-double to peony form, medium to large size, 9.5-11.0cm across and 4.8cm deep with 18-23 petals arranged in 2-3 rows, petals wavy, stamens clustered, bloom very dense. Leaves dark green, leathery, elliptic, margins obtusely serrate at the upper part. Plant compact, upright and growth vigorous. Starts to bloom from late-summer, fully blooms from autumn to winter and sporadically in spring.

HA-21-9. 争奇斗艳
Zhengqi Douyan (Contend in Beauty & Fascination)

杂交组合： 红山茶品种'都鸟'× 杜鹃红山茶（*C. japonica* cultivar 'Miyakodori'× *C. azalea*)。

杂交苗编号： HED-No.7。

性状： 花朵红色至深红色，略具蜡光，单瓣型，中型花，花径8.5～9.5cm，花瓣5～6枚，阔倒卵圆形，上部外翻，皱褶，先端略凹，雄蕊基部短柱状，花丝和花药淡黄色，开花极稠密。叶片浓绿色，叶背面灰绿色，革质，椭圆形，长8.5～9.5cm，宽3.5～4.0cm，叶缘上部具浅齿，下部光滑。植株立性，枝条稠密，生长旺盛。夏季始花，秋、冬季盛花，翌年春季零星开花，极适合于美化环境。

Abstract: Flowers red to deep red, slightly with waxy luster, single form, medium size, 8.5-9.5cm across with 5-6 petals, petals broad obovate, rolled outward at the top part, apices emarginate, surfaces crinkled, stamens united into a short column at the base, filaments and anthers yellow, bloom very dense. Leaves dark-green, leathery, elliptic, margins shallowly serrate at the upper part, smooth at the lower part. Plant upright, branches dense and growth vigorous. Starts to bloom from summer, fully blooms from autumn to winter and sporadically in the spring of following year. It is very suitable for beautifying environments.

⦿ 杂交组合 HA-22. 杜鹃红山茶 × 红山茶品种'毛缘大黑红'
HA-22. *C. azalea* × *C. japonica* cultivar 'Clark Hubbs'

本杂交组合将介绍1个新品种，其编号将接续第一部书本组的序列编号。

第一部书曾介绍了本组的1个名为'清纯红裳'的新品种（见第一部书第229～230页）。至此，本组获得的新品种总数已达2个。

可以看到，获得的杂交新品种都较好地具备了其两个杂交亲本的性状，既能开出类似于'毛缘大黑红'品种的黑红色花朵，也能像杜鹃红山茶一样叶片浓绿，重复开花。

One new hybrid will be introduced in this cross-combination and its number will follow the serial number of the cross-combination in our first book.

One new hybrid named as 'Qingchun Hongshang' was introduced in our first book (See p.229-230, our first book). A total of two hybrids have been obtained in this cross-combination until now.

It can be seen that the hybrids we got all have the characteristics of the two parents, such as dark red flowers which is similar to 'Clark Hubbs' and dark green leaves and repeated bloom which are the same as *C. azalea*.

HA-22-2. 回龙颂歌
Huilong Songge (Huilong Town's Ode)

杂交组合： 杜鹃红山茶 × 红山茶品种'毛缘大黑红'（*C. azalea* × *C. japonica* cultivar 'Clark Hubbs'）。

杂交苗编号： DMY-No.3。

性状： 花朵深红色至黑红色，中部花瓣可见白条纹，半重瓣型，中型至大型花，花径9.5～12.0cm，花朵厚4.7cm，大花瓣15～18枚，呈2～3轮松散排列，花瓣阔倒卵形，中部偶有具白斑的雄蕊瓣，花芯处雄蕊少。叶片浓绿色，叶背面灰绿色，厚革质，长椭圆形，长9.0～11.4cm，宽3.5～4.0cm，边缘叶齿明显。植株立性，生长旺盛。夏末始花，秋末至冬季盛花，春季零星开花。

Abstract: Flowers deep red to dark red with white stripes visible on the central petals, semi-double form, medium to large size, 9.5-12.0cm across and 4.7cm deep with 15-18 large petals arranged loosely in 2-3 rows, petals broad obovate, occasionally some petaloids with white patches at the center, stamens only few. Leaves dark green, thick leathery, long elliptic, margins obviously serrate. Plant upright and growth vigorous. Starts to bloom from late-summer, fully blooms from late-autumn to winter and sporadically in spring.

杂交组合 HA-24. 红山茶品种'客来邸' × 杜鹃红山茶
HA-24. *C. japonica* cultivar 'Collettii' × *C. azalea*

×

本杂交组合将介绍 2 个新品种，其编号将接续第一部书本组的序列编号。

第一部书曾介绍了本组的 2 个新品种（见第一部书第 234～237 页）。至此，本组获得的新品种总计已达 4 个。

不难发现，杂交种花朵艳红色、偶具云状白斑，开花稠密，叶片浓绿，株型紧凑，这些都与两个亲本相似。花型保持了'客来邸'品种的轮廓，开花期则与杜鹃红山茶非常相似。

Two new hybrids will be introduced in this cross-combination and their numbers will follow the serial numbers of the cross-combination in our first book.

Two new hybrids were introduced in this cross-combination of our first book (See p.234-237, our first book), so a total of four new hybrids have been obtained in this combination until now.

It is not difficult to find that bright red flowers occasionally with cloud-shaped white markings, dense flowers, dark green leaves, compact plants of the hybrids are all similar to their two parents'. Hybrids' flower shapes have maintained the outline of 'Collettii' and the blooming period of the hybrids is very similar to *C. azalea*.

HA-24-3. 星源红霞
Xingyuan Hongxia (Xingyuan's Red Glow)

杂交组合：红山茶品种'客来邸' × 杜鹃红山茶（*C. japonica* cultivar 'Collettii' × *C. azalea*）。

杂交苗编号：KLD-No.3。

性状：花朵红色至深红色，偶尔出现白斑，玫瑰重瓣型至牡丹型，中型至大型花，花径9.5～12.0cm，花朵厚5.5cm，大花瓣21枚，呈3轮排列，花瓣阔倒卵形，中部小花瓣20枚以上，雄蕊簇生，开花稠密。叶片浓绿色，叶背面灰绿色，厚革质，阔椭圆形，长9.0～12.0cm，宽5.0～6.0cm，边缘叶齿浅。植株紧凑，枝条稠密，生长旺盛。夏末始花，秋、冬季盛花，春季零星开花。

注：本品种由上海茶花园培育，已成功获得国家林业和草原局专利授权。

Abstract: Flowers red to deep red, rose-double to peony form, medium to large size, 9.5-12.0cm across and 5.5cm deep with 21 large petals arranged in 3 rows, petals broad obovate, more than 20 small petals at the center, stamens clustered, bloom dense. Leaves dark green, heavy leathery, broad elliptic, margins shallowly serrate. Plant compact, branches dense and growth vigorous. Starts to bloom from late-summer, fully blooms from autumn to winter and sporadically in spring.

Mark: The hybrid is bred by Shanghai Camellia Garden and its patent right has been obtained from National Forestry and Grassland Administration.

HA-24-4. 星源云海
Xingyuan Yunhai (Xingyuan's Cloud Sea)

杂交组合：红山茶品种'客来邸'×杜鹃红山茶（*C. japonica* cultivar 'Collettii' × *C. azalea*）。

杂交苗编号：KLD-No.3Mu。

性状：本品种是'星源红霞'的一个复色突变。其大部分性状与'星源红霞'品种相似，但是花朵更大了，而且具大量云状白斑，叶片偶然出现少量黄斑。

注：本品种于2018年由上海茶花园培育。

> **Abstract:** It is a variegated mutation obtained from the hybrid 'Xingyuan Hongxia'. Most of its characteristics are similar to 'Xingyuan Hongxia', but its flowers are larger with many white cloud-shaped markings, and a few yellow patches appeared on the leaves occasionally.
> Mark: The hybrid was bred by Shanghai Camellia Garden in 2018.

杂交组合 HA-26. 红山茶品种'白斑康乃馨' × 杜鹃红山茶
HA-26. *C. japonica* cultivar 'Ville de Nantes' × *C. azalea*

本杂交组合将介绍 1 个新品种，其编号将接续第一部书本组的序列编号。

第一部书曾介绍了本组的 5 个新品种（见第一部书第 241～247 页）。至此，本组合已获得 6 个新品种。

本次获得的新品种是本组的'羊城之夜'品种（见第一部书第 246～247 页）的一个芽变，其花朵具白色云斑，非常漂亮，而且性状已经稳定。

One new hybrid will be introduced in this combination and its number will follow the serial numbers of the cross-combination in our first book.

Five new hybrids were introduced in the cross-combination in our first book (See p.241-247, our first book). There have been a total of six new hybrids obtained in this cross-combination until now.

The new variety is a bud mutation obtained from the cultivar 'Yangcheng Zhiye' (See p.246-247, our first book) in the combination. Its flowers with white marbled markings are very beautiful and its characteristics are stable.

HA-26-6. 花城闹市
Huacheng Naoshi（Guangzhou's Busy Streets）

杂交组合： 红山茶品种'白斑康乃馨'× 杜鹃红山茶（*C. japonica* cultivar 'Ville de Nantes'× *C. azalea*）。

杂交苗编号： NTD-No.4E-Mu。

性状： 这是'羊城之夜'（见第一部书第246～247页）品种的突变种。花朵黑红色，具大理石白色斑纹，单瓣型，大型花，花径通常为11.0～12.0cm，偶有超过13.0cm的大花朵，花瓣7～9枚，阔倒卵圆形，雄蕊基部连生，呈柱状，花丝红色，开花稠密。叶片浓绿色，叶背面灰绿色，长椭圆形，厚革质，长9.5～11.5cm，宽3.9cm。植株紧凑，立性，生长极旺盛。夏季始花，秋、冬季盛花，春季零星开花。

> **Abstract:** It is a mutation from the hybrid 'Yangcheng Zhiye' (See p. 246-247, our first book). Flowers dark red with white marbled markings, single form, large size, usually 11.0-12.0cm across, sometimes over 13.0cm in size, 7-9 petals, stamens united into a column. Leaves dark green, long elliptic, heavy leathery. Plant compact, upright and growth very vigorous. Starts to bloom in summer, fully blooms from autumn to winter and sporadically in spring.

◉ 杂交组合 HA-32. 杜鹃红山茶 × 多齿红山茶
HA-32. *C. azalea* × *C. polyodonta*

本杂交组合将介绍 1 个新品种，其编号将接续第一部书本组的序列编号。

第一部书曾介绍了本组的 1 个新品种（见第一部书第 284 ～ 285 页）。至此，本组共获得了 2 个新品种。

虽然本组的两个杂交亲本的花都是单瓣型的，但是新品种的花型都是半重瓣型至牡丹型的，而且非常漂亮，这很可能是多齿红山茶的基因杂合度很高的缘故。

One new hybrid will be introduced in this cross-combination and its number will follow the serial number of the cross-combination in our first book.

We introduced one new hybrid in our first book (See p.284-285, our first book). There are a total of two new hybrids obtained from this cross-combination until now.

The flowers of the two parents in this combination are all single form, however, the flowers of their new hybrids are all semi-double to peony form and very beautiful, which was, perhaps, caused by the higher level of gene heterozygosity of *C. polyodonta*.

HA-32-2. 粤桂宝珠
Yuegui Baozhu (Guangdong & Guangxi's Pearl)

杂交组合： 杜鹃红山茶 × 多齿红山茶（*C. azalea* × *C. polyodonta*）。

杂交苗编号： DDC-No.3。

性状： 花朵红色，花瓣具白条纹，半重瓣型至玫瑰重瓣型，中型至大型花，花径 9.0～12.5cm，花朵厚 6.0cm，大花瓣 35～45 枚，呈 4～5 轮排列，厚质，长倒卵形，长 6.5cm，宽 4.5cm，外翻，雄蕊近散生。叶片浓绿色，叶背面灰绿色，厚革质，椭圆形，长 10.5～11.5cm，宽 5.0～5.5cm，边缘叶齿钝。植株立性，生长旺盛。夏末始花，秋、冬季盛花，春季零星开花。

Abstract: Flowers red, petals with white stripes on the surfaces, semi-double to rose-double form, medium to large size, 9.0-12.5cm across and 6.0cm deep with 35-45 petals arranged in 4-5 rows, petals thick texture, long obovate, rolled outward, stamens nearly radial. Leaves dark green, heavy leathery, elliptic, margins obtusely serrate. Plant upright and growth vigorous. Starts to bloom from late-summer, fully blooms from autumn to winter and sporadically in spring.

● 杂交组合 HA-34. 杜鹃红山茶 × 浙江红山茶
HA-34. *C. azalea* × *C. chekiangoleosa*

本杂交组合将介绍 3 个新品种，其编号将接续第一部书本组的序列编号。

第一部书曾介绍了本组的 7 个新品种（见第一部书第 298～305 页）。至此，本组获得的新品种总数已达到 10 个。

本组的新品种虽然都是单瓣型的，但是它们有花径大、抗寒性强、生长旺盛等优点，更适合于园林美化。

Three new hybrids will be introduced in this cross-combination and their numbers will follow the serial numbers of the cross-combination in our first book.

Seven new hybrids were introduced in this cross-combination of our first book (See p.298-305, our first book). There have been a total of ten new hybrids obtained in this cross-combination until now.

The new hybrids of this combination all open single form flowers, but they have the advantages such as large flowers, strong cold resistance and vigorous growth, which are more suitable for gardening.

HA-34-8. 秋艳冬红
Qiuyan Donghong (Gorgeous Autumn & Red Winter)

杂交组合：杜鹃红山茶 × 浙江红山茶（ *C. azalea* × *C. chekiangoleosa*)。

杂交苗编号：DZ-No.35。

性状：萼片被银白色绒毛，宿存。花朵深桃红色至红色，单瓣型，中型至大型花，花径9.5～11.5cm，花瓣5～6枚，厚质，阔倒卵形，雄蕊基部略连生，柱头3深裂。叶片浓绿色，叶背面灰绿色，硬革质，椭圆形，长11.5～12.5cm，宽4.5～5.5cm，边缘叶齿明显。植株立性，生长旺盛。夏末始花，秋、冬季盛花，春季零星开花。

Abstract: Sepals with silver white-tomentose, persistent. Flowers deep peach red to red, single form, medium to large size, 9.5-11.5cm across with 5-6 petals, petals thick texture, stamens slightly united at the base. Leaves dark green, hard leathery, elliptic, margins obviously serrate. Plant upright and growth vigorous. Starts to bloom from late-summer, fully blooms from autumn to winter and sporadically in spring.

HA-34-9. 秋冬桃红
Qiudong Taohong (Peach Red in Autumn & Winter)

杂交组合：杜鹃红山茶 × 浙江红山茶（*C. azalea* × *C. chekiangoleosa*）。

杂交苗编号：DZ-No.42。

性状：萼片被褐色短绒毛，宿存。花朵浅桃红色至橘红色，偶有一枚小花瓣具白斑，单瓣型，大型至巨型花，花径 11.0～14.0cm，花瓣 6 枚，厚质，波浪状，雄蕊基部略连生，子房光滑。叶片浓绿色，叶背面灰绿色，硬革质，椭圆形，长 11.0～12.8cm，宽 5.0～5.5cm，边缘叶齿钝。植株立性，生长旺盛。秋初始花，冬季盛花，春季零星开花。

Abstract: Sepals with brown-tomentose, persistent. Flowers light peach red to orange red, occasionally with a white marking on a small petal, single form, large to very large size, 11.0-14.0cm across with 6 petals, petals thick texture, wavy, stamens slightly united at the base, ovary glabrous. Leaves dark green, hard leathery, elliptic, margins obtusely serrate. Plant upright and growth vigorous. Starts to bloom from early-autumn, fully blooms in winter and sporadically in spring.

HA-34-10. 暮鼓晨钟
Mugu Chenzhong（Evening Drum & Morning Bell）

杂交组合：杜鹃红山茶 × 浙江红山茶（*C. azalea* × *C. chekiangoleosa*）。

杂交苗编号：DZ–No.55。

性状：萼片被褐色短绒毛，宿存。花朵粉红色至深粉红色，单瓣型，长喇叭状，中型花，花径8.5～9.5cm，花瓣5～6枚，长倒卵形，雄蕊长5.0cm，基部略连生。叶片浓绿色，光亮，叶背面灰绿色，硬革质，长椭圆形，长10.5～11.5cm，宽4.0～4.5cm，叶片上部边缘齿尖。植株立性，枝条硬挺，生长旺盛。秋初始花，冬季盛花，春季零星开花。

Abstract: Sepals with brown-tomentose, persistent. Flowers pink to deep pink, single form, long trumpet shaped, medium size, 8.5-9.5cm across with 5-6 petals, petals long obovate, stamens 5.0cm long, slightly united at the base. Leaves dark green, hard leathery, long elliptic, margins sharply serrate at the upper part. Plant upright and growth vigorous. Starts to bloom from early-autumn, fully blooms in winter and sporadically in spring.

⦿ 杂交组合 HA-35. 杜鹃红山茶 × 红山茶品种'霍伯'
HA-35. *C. azalea* × *C. japonica* cultivar 'Bob Hope'

本杂交组合将介绍 2 个新品种，其编号将接续第一部书本组的序列编号。

第一部书曾介绍了本组的 3 个新品种（见第一部书第 306～310 页）。至此，本杂交组合获得的具有观赏价值的品种总计已达 5 个。

这 5 个杂交新品种的花朵多为黑红色，明显倾向于父本'霍伯'品种。而杂交新品种的开花期从夏季开始至翌年春季，叶片浓绿色，叶缘近光滑，基部楔形，明显倾向于母本杜鹃红山茶。

×

Two new hybrids will be introduced in this cross-combination and their numbers will follow the serial numbers of the cross-combination in our first book.

Three new hybrids were introduced in the cross-combination in our first book (See p.306-310, our first book). A total of five hybrids with ornamental value have been obtained in this cross-combination until now.

Most of the five hybrids' flowers are dark red, which obviously tends to their male parent 'Bob Hope' cultivar, and the blooming period of the new hybrids is from summer to the spring of the following year, their leaves are dark green, margins are nearly smooth and bases are cuneate, which obviously tend to their female parent *C. azalea*.

HA-35-4. 叠瓣金蕊
Dieban Jinrui (Overlapped Petals with Golden Stamens)

杂交组合：杜鹃红山茶 × 红山茶品种'霍伯'（*C. azalea* × *C. japonica* cultivar 'Bob Hope'）。

杂交苗编号：DHB-No.2。

性状：花朵红色至黑红色，中部花瓣偶有白斑，半重瓣型至牡丹型，小型花，花径6.5～7.5cm，花朵厚4.0cm，大花瓣18枚，呈2～3轮排列，阔倒卵形，中部花瓣5～8枚，波浪状，雄蕊簇生。叶片浓绿色，叶背面灰绿色，革质，椭圆形，长9.5～10.0cm，宽4.0～5.0cm，边缘具浅齿。植株立性，枝叶稠密，生长旺盛。夏末始花，秋、冬季盛花，春季零星开花。

Abstract: Flowers red to dark red, occasionally with white markings on central petals, semi-double to peony form, small size, 6.5-7.5cm across and 4.0cm deep with 18 petals arranged in 2-3 rows, petals broad obovate, 5-8 central petals wavy, stamens clustered. Leaves dark green, leathery, elliptic, margins shallowly serrate. Plant upright, branches dense and growth vigorous. Starts to bloom from late-summer, fully blooms from autumn to winter and sporadically in spring.

HA-35-5. 夏日红霞
Xiari Hongxia（Summer's Red Glow）

杂交组合： 杜鹃红山茶 × 红山茶品种'霍伯'（*C. azalea* × *C. japonica* cultivar 'Bob Hope'）。

杂交苗编号： DHB-No.5。

性状： 花朵红色至黑红色，边缘显紫红色，偶具白色条纹，半重瓣型至牡丹型，中型至巨型花，花径8.0～13.0cm，有时可达15cm以上，花朵厚6.0～7.0cm，外轮花瓣13～15枚，呈1～2轮排列，平铺，阔倒卵形，中部常有3～5枚较大的花瓣，与雄蕊混生，雄蕊多数，基部略连生，雌蕊乳白色。叶片浓绿色，叶背面灰绿色，革质，阔椭圆形，长11.0～11.3cm，宽4.8～5.3cm，叶缘有钝齿。植株紧凑，立性或半开张，生长旺盛。夏末始花，秋、冬季盛花，春季零星开花。

Abstract: Flowers red to dark red, petal edges purple red, occasionally with white stripes, semi-double to peony form, medium to very large size, 8.0-13.0cm across, sometimes over 15cm, and 6.0-7.0cm deep with 13-15 exterior petals arranged in 1-2 rows, usually 3-5 large petals at the center and mixed with the stamens, pistil milk white. Leaves dark green, leathery, broad elliptic, margins obtusely serrate. Plant compact, upright or semi-spread and growth vigorous. Starts to bloom in late-summer, fully blooms from autumn to winter and sporadically in spring.

⦿ 杂交组合 HA-36. 杜鹃红山茶 × 越南抱茎茶
HA-36. *C. azalea* × *C. amplexicaulis*

本杂交组合将介绍 1 个新品种，其编号将接续第一部书本组的序列编号。

第一部书曾介绍了本组的 10 个新品种（见第一部书第 311～323 页）。至此，本组已总计获得新品种 11 个。

本杂交组合的杂交新品种都具有四季开花、叶色浓绿、生长旺盛的性状，这些性状颇像母本杜鹃红山茶。而它们同时也具有花朵粉红色、花柄长、叶片大、植株高大的性状，这些又颇像父本越南抱茎茶。

One new hybrid will be introduced in this cross-combination and its number will follow the serial numbers of the cross-combination in our first book.

Ten new hybrids were introduced in the cross-combination of our first book (See p.311-323, our first book). A total of eleven new hybrids have been obtained in this combination until now.

The hybrids of this cross-combination all have the characteristics as year-round blooming, dark green leaves and vigorous growth, which are very similar to their female parent, *C. azalea*. Meanwhile, they also have the characteristics as pink flowers, long pedicels, large leaves and tall plants, which are very similar to their male parent, *C. amplexicaulis*.

HA-36-11. 一见钟情
Yijian Zhongqing (Love at First Sight)

杂交组合： 杜鹃红山茶 × 越南抱茎茶（*C. azalea* × *C. amplexicaulis*）。

杂交苗编号： DB-No.52 + 53 + 56。

性状： 花柄长 0.8cm。花朵粉红色至深粉红色，单瓣型，大型至巨型花，11.5～13.5cm，花瓣 8～9 枚，肉质，长倒卵形，边缘内卷，雄蕊柱状，高 3cm。叶片浓绿色，叶背面灰绿色，厚革质，长椭圆形，长 12.0～15.5cm，宽 5.0～6.0cm，中脉凸起，叶齿浅稀。植株立性，枝条立性，高大，生长旺盛。夏季始花，秋、冬季盛花，春季零星开花。

Abstract: Pedicels 0.8cm long. Flowers pink to deep pink, single form, large to very large size, 11.5-13.5cm across with 8-9 petals, petals thick fleshy, long obovate, edges rolled inward, stamens columnar. Leaves dark green, heavy leathery, long elliptic, midrib raised, margins shallowly and sparsely serrate. Plant upright, tall and growth vigorous. Starts to bloom from summer, fully blooms from autumn to winter and sporadically in spring.

杂交组合 HA-41. 红山茶品种'金盘荔枝' × 杜鹃红山茶
HA-41. *C. japonica* cultivar 'Jinpan Lizhi' × *C. azalea*

本杂交组合将介绍1个新品种，其编号将接续第一部书本组的序列编号。

第一部书曾介绍了本组的4个新品种（见第一部书第342～346页）。至此，本杂交组合共获得了5个杂交新品种。

杂交新品种的花色和花型与其母本'金盘荔枝'非常相似，而开花期和叶色则与其父本杜鹃红山茶非常相似。

One new hybrid will be introduced in this cross-combination and its number will follow the serial numbers of the cross-combination in our first book.

Four new hybrids were introduced in this cross-combination of our first book (See p.342-346, our first book) and therefore five new hybrids in total have been obtained in this combination until now.

The flower color and flower form of the hybrids are very similar to those of their female parent, 'Jinpan Lizhi', and the blooming period and leaf color of the hybrids are very similar to those of their male parent, *C. azalea*.

HA-41-5. 婚庆元宝
Hunqing Yuanbao（Gold Ingot for Wedding）

杂交组合： 红山茶品种'金盘荔枝'×杜鹃红山茶（*C. japonica* cultivar 'Jinpan Lizhi'× *C. azalea*）。

杂交苗编号： JD-No.3-15。

性状： 花朵红色至深红色，中部雄蕊瓣偶现小白斑，托桂型，中型花，花径8.5～9.0cm，花朵厚3.5～4.0cm，外轮花瓣6枚，长倒卵形，中部雄蕊瓣簇拥成球，雌蕊高出球面1.0cm。叶片浓绿色，叶背面灰绿色，革质，长椭圆形，叶片略波浪状，长9.5～10.0cm，宽3.0～3.5cm，基部急楔形，略波浪状，边缘齿浅。植株立性，生长旺盛。夏、秋季盛花，冬、春季零星开花。

Abstract: Flowers red to deep red, occasionally with small white spots at the central petaloids, anemone form, medium size, 8.5-9.0cm across and 3.5-4.0cm deep with 6 exterior petals, petals long obovate, petaloids crowded into a ball at the center. Leaves dark green, leathery, long elliptic, slightly wavy, urgently cuneate at the base, margins shallowly serrate. Plant upright and growth vigorous. It blooms fully in summer and autumn and sporadically in winter and spring.

杂交组合 HA-43. 杜鹃红山茶 × 云南山茶杂交种'蜂露'
HA-43. *C. azalea* × *C. reticulata* hybrid 'Fenglu'

×

本杂交组合将介绍 2 个新品种，其编号将接续第一部书本组的序列编号。

'蜂露'品种由红山茶品种'媚丽'和云南山茶品种'山茶之都'杂交而得（见第一部书第 504～505 页）。第一部书曾介绍了本组的 3 个新品种（见第一部书第 349～353 页）。至此，本杂交组合已获得 5 个新品种。

该组新品种具有云南山茶基因，因此其花径较大。

Two new hybrids will be introduced in this cross-combination and their numbers will follow the serial numbers of the cross-combination in our first book.

'Fenglu' was obtained from *C. japonica* cultivar 'Tama Beauty' × *C. reticulata* cultivar 'Massee Lane' (See p.504-505, our first book). Three new hybrids were introduced in the cross-combination of our first book (See p.349-353, our first book). There have been five new hybrids obtained in total in this combination until now.

The hybrids in this cross-combination contain the genes from *C. reticulata*, so, their flower acrosses are large.

HA-43-4. 吉日良辰
Jiri Liangchen (Auspicious Day)

杂交组合： 杜鹃红山茶 × 云南山茶杂交种'蜂露'（*C. azalea* × *C. reticulata* hybrid 'Fenglu'）。

杂交苗编号： DFL-No.13。

性状： 花朵淡桃红色，花瓣上部粉白色，渐渐变为粉红色，半重瓣型至牡丹型，大型花，花径11.5～13.0cm，花朵厚5.5cm，大花瓣20～25枚，呈3～4轮排列，花瓣阔倒卵形，厚质，雄蕊簇生，与花芯处的小花瓣混生。叶片浓绿色，叶背面灰绿色，厚革质，椭圆形，长11.5～12.0cm，宽4.0～4.5cm，边缘上部叶齿钝。植株开张，生长旺盛。夏末始花，秋、冬季盛花，春季零星开花。

Abstract: Flowers lightly peach red with pink-white at the upper part of the petals, and gradually changing into pink, semi-double to peony form, large size, 11.5-13.0cm across and 5.5cm deep with 20-25 large petals arranged in 3-4 rows, petals broad obovate, thick texture, stamens clustered and mixed with small petals at the center. Leaves dark green, heavy leathery, elliptic, margins obtusely serrate at the upper part. Plant spread and growth vigorous. Starts to bloom from late-summer, fully blooms from autumn to winter and sporadically in spring.

HA-43-5. 紫粉舞秋
Zifen Wuqiu (Purple Pink Dancing Autumn)

杂交组合： 杜鹃红山茶 × 云南山茶杂交种'蜂露'（ *C. azalea* × *C. reticulata* hybrid 'Fenglu'）。

杂交苗编号： DFL-No.30。

性状： 花朵淡紫粉红色，渐向上部变为淡红色，偶具纵向的白条纹，半重瓣型，中型至大型花，花径9.0～11.5cm，花朵厚5.0cm，花瓣40～45枚，呈4～5轮排列，花瓣狭长倒卵形，瓣缘两侧外卷，先端"V"形深凹，雄蕊簇生，开花稠密，花瓣逐片掉落。叶片浓绿色，叶背面灰绿色，硬革质，长椭圆形，长8.5～9.0cm，宽2.5～3.5cm，边缘上部叶齿尖，下部光滑，基部楔形。植株立性，生长旺盛。夏末始花，秋、冬季盛花，春季零星开花。

Abstract: Flowers light purple pink, gradually fading to light pink at the upper part of the petals, semi-double form, medium to large size, 9.0-11.5cm across and 5.0cm deep with 40-45 petals arranged in 4-5 rows, petals long and narrow obovate, both edges rolled outward, apices sunken as "V", stamens clustered, bloom dense, petals fall one by one. Leaves dark green, hard leathery, long elliptic, margins sharply serrate at the upper part, cuneate at the base. Plant upright and growth vigorous. Starts to bloom from late-summer, fully blooms from autumn to winter and sporadically in spring.

⦿ 杂交组合 HA-44. 杜鹃红山茶 × 肥后茶品种'王冠'
HA-44. *C. azalea* × Higo cultivar 'Okan'

本杂交组合将介绍 1 个新品种，其编号将接续我们第一部书本组的序列编号。

第一部书曾介绍了一个名为'肥厚首喜'的杂交新品种（见第一部书第 354～355 页）。至此，本组共获得 2 个新品种。

可以看到，杂交种在性状上均具有双亲的特性：花期和叶色倾向于母本杜鹃红山茶，花朵雄蕊散射的性状则与父本'王冠'品种非常相似。因此，可以确定它们都是真实的杂交品种。

One new hybrid will be introduced in this cross-combination and their numbers will follow the serial numbers of the cross-combination in our first book.

A hybrid named 'Feihou Shouxi' was introduced in our first book (See p.354-355, our first book). A total of two new hybrids have been obtained in this cross-combination until now.

It can be seen that the hybrids all have the characteristics of their cross parents: their blooming period and leaves color tend to their female parent *C. azalea* and their scattering stamens are similar to their male parent 'Okan'. Therefore, it can be confirmed that they are all true hybrids.

HA-44-2. 肥后二喜
Feihou Erxi（Higo's Second Pleasure）

杂交组合：杜鹃红山茶 × 肥后茶品种'王冠'（C. azalea × Higo cultivar 'Okan'）。

杂交苗编号：DFH-No.3。

性状：花朵红色至黑红色，单瓣型，中型至大型花，花径9.0～11.7cm，花瓣6～7枚，质地厚，阔倒卵形，长5.0cm，宽4.5cm，上部瓣面略皱褶，雄蕊呈辐射状，基部连生，花丝粉红色，花药黄色，花朵整朵掉落。叶片浓绿色，叶背面灰绿色，厚革质，阔椭圆形，长9.3cm，宽4.5cm，上部边缘具浅齿，下部边缘光滑。植株立性，枝条稠密，生长旺盛。夏、秋季盛花，冬、春季零星开花。

Abstract: Flowers red to dark red, single form, medium to large size, 9.0-11.7cm across with 6-7 broad obovate petals, petals thick, stamens radialized, base united, filaments pink, anthers yellow, flowers fall whole. Leaves dark green, broad elliptic, thick leathery, the upper edges shallow serrated and the lower edges smooth. Plant upright, branches dense and growth vigorous. Blooms fully in summer and autumn and sporadically in winter and spring.

⦿ 杂交组合 HA-47. 全缘红山茶 × 杜鹃红山茶
HA-47. *C. subintegra* × *C. azalea*

本杂交组合将介绍 4 个杂交新品种，其编号将接续第一部书本组的序列编号。

第一部书曾介绍了本组的 1 个新品种（见第一部书第 359～360 页）。至此，本杂交组合所获得的新品种总数已达 5 个。

全缘红山茶是一个高抗寒性、叶片边缘光滑的山茶原生种。尽管所获得新品种的花朵是单瓣型的，但是它们可四季开花，开花也非常稠密，而且具有一定的抗寒性。这些新品种在环境美化方面大有用武之地。

×

Four new hybrids will be introduced in this cross-combination and their numbers will follow the serial number of the cross-combination in our first book.

We introduced one new hybrid in the cross-combination in our first book (See p.359-360, our first book), and therefore five new hybrids have been obtained in this combination until now.

C. subintegra is a camellia species with strong cold resistance and smooth edges of leaves. Although the new hybrids we obtained are single in flower form, they can ever-bloom with dense flowers and have certain cold resistance. These new hybrids will be very useful for beautifying environments.

HA-47-2. 星光争辉
Xingguang Zhenghui (Starlit Contending for Brilliancy)

杂交组合：全缘红山茶 × 杜鹃红山茶（*C. subintegra* × *C. azalea*）。

杂交苗编号：QYD-No.27。

性状：花朵红色至深红色，瓣面上部有隐约白斑，单瓣型，中型花，花径 8.0～9.5cm，花瓣 6 枚，阔倒卵形，瓣面略皱褶，雄蕊基部连生。叶片浓绿色，叶背面灰绿色，硬革质，椭圆形，长 8.0～9.5cm，宽 3.5～4.0cm，边缘光滑。植株紧凑，灌丛状，生长中等。秋初始花，秋末至冬季盛花，春季零星开花。

Abstract: Flowers red to deep red with faint white markings on the top part of petal surfaces, single form, medium size, 8.0-9.5cm across with 6 petals, petals broad obovate, slightly crinkled, stamens united at the base. Leaves dark green, hard leathery, elliptic, margins smooth. Plant compact, bushy, and growth normal. Starts to bloom from early-autumn, fully blooms from late-autumn to winter and sporadically in spring.

HA-47-3. 星光高照
Xingguang Gaozhao (Starlit's Brightly Shining)

杂交组合：全缘红山茶 × 杜鹃红山茶 (*C. subintegra* × *C. azalea*)。

杂交苗编号：QYD-No.29。

性状：花朵深粉红色至红色，瓣面上部具隐约白斑，单瓣型，中型花，花径7.5～9.0cm，花瓣6枚，倒卵形，雄蕊基部连生，开花稠密。叶片浓绿色，叶背面灰绿色，厚革质，长椭圆形，长8.5～11.0cm，宽3.0～4.0cm，略波浪状，边缘光滑，具明显黄边。植株紧凑，枝叶稠密，灌丛状，生长中等。秋初始花，秋末至冬季盛花，春季零星开花。

> **Abstract:** Flowers deep pink to red with faint white markings on the top part of petal surfaces, single form, medium size, 7.5-9.0cm across with 6 petals, petals obovate, stamens united at the base, bloom dense. Leaves dark green, heavy leathery, long elliptic, wavy, margins smooth with an obvious yellow border. Plant compact, branches dense, bushy and growth normal. Starts to bloom from early-autumn, fully blooms from late-autumn to winter and sporadically in spring.

HA-47-4. 星光闪闪
Xingguang Shanshan (Starlit Sparkle)

杂交组合：全缘红山茶 × 杜鹃红山茶（*C. subintegra* × *C. azalea*）。

杂交苗编号：QYD-No.42。

性状：花朵深橘红色，瓣面上部渐呈白色，单瓣型，中型花，花径8.5～10.0cm，花瓣6枚，倒卵形，外翻，雄蕊基部连生，开花稠密。叶片浓绿色，叶背面灰绿色，厚革质，狭长椭圆形，长8.0～9.5cm，宽3.0～4.0cm，波浪状，边缘光滑，具隐约黄边。植株紧凑，枝叶稠密，灌丛状，生长中等。秋初始花，秋末至冬季盛花，春季零星开花。

Abstract: Flowers deep orange red, gradually fading to white on the top part of petal surfaces, single form, medium size, 8.5-10.0cm across with 6 petals, petals obovate, rolled outward, stamens united at the base. Leaves dark green, heavy leathery, narrowly and long elliptic, slightly wavy, margins smooth with a faint yellow border. Plant compact, branches dense, bushy and growth normal. Starts to bloom from early-autumn, fully blooms from late-autumn to winter and sporadically in spring.

HA-47-5. 星光灿烂
Xingguang Canlan（Starlit Splendid）

杂交组合： 全缘红山茶 × 杜鹃红山茶（*C. subintegra* × *C. azalea*）。

杂交苗编号： QYD-No.49。

性状： 花朵淡红色至红色，瓣面上部具许多白色小斑点，单瓣型，中型至大型花，花径8.5～11.0cm，花瓣6枚，花瓣阔倒卵形，雄蕊基部连生，开花极稠密。叶片中等绿色，叶背面灰绿色，厚革质，狭长椭圆形，长8.5～10.5cm，宽3.0～3.5cm，略扭曲或波浪状，边缘光滑，具隐约黄边。植株紧凑，枝叶稠密，灌丛状，生长中等。秋初始花，秋末至冬季盛花，春季零星开花。

> **Abstract:** Flowers light red to red with a lot of tiny white dots on the top part of petal surfaces, single form, medium to large size, 8.5-11.0cm across with 6 petals, petals broad obovate, stamens united at the base, bloom very dense. Leaves normal green, heavy leathery, narrowly and long elliptic, slightly twisted or wavy, margins smooth with a faint yellow border. Plant compact, branches dense, bushy and growth normal. Starts to bloom from early-autumn, fully blooms from late-autumn to winter and sporadically in spring.

杂交组合 HA-48. 非云南山茶杂交种'珊瑚乐' × 杜鹃红山茶
HA-48. Non-*reticulata* hybrid 'Coral Delight' × *C. azalea*

本杂交组合将介绍3个新品种,其编号将接续第一部书本组的序列编号。

第一部书曾介绍了本组1个命名为'四季鸿运'的新品种(见第一部书第360～361页)。至此,本组所获得的新品种总数已达4个。

新品种已经融入了两个杂交亲本的性状。它们既能开出类似于'珊瑚乐'的花朵,也能像杜鹃红山茶一样叶片浓绿、重复开花。

Three new hybrids will be introduced in this cross-combination and their numbers will follow the serial number of the cross-combination in our first book.

One new hybrid named as 'Siji Hongyun' was introduced in our first book (See p.360-361, our first book). A total of four hybrids have been obtained in this cross-combination until now.

The new hybrids have been integrated into the characteristics of their two cross parents. They can open flowers which are similar to 'Coral Delight' and also have dark green leaves and ever-blooming characteristics which are similar to *C. azalea*.

HA-48-2. 珊瑚新貌
Shanhu Xinmao (Coral's New Look)

杂交组合：非云南山茶杂交种'珊瑚乐'× 杜鹃红山茶（Non-*reticulata* hybrid 'Coral Delight' × *C. azalea*）。

杂交苗编号：SHLD-No.1。

性状：花朵艳红色至黑红色，瓣面具少量白条纹或白斑，半重瓣型，中型至大型花，花径 9.5～11.5cm，花朵厚 4.5cm，大花瓣 11～14 枚，呈 2 轮松散排列，长倒卵形，边缘波浪状，质地较硬，中部小花瓣 3～5 枚，雄蕊少。叶片浓绿色，叶背面灰绿色，厚革质，椭圆形，长 8.5～10.0cm，宽 4.0～4.5cm，边缘齿稀。植株紧凑，生长旺盛。夏末始花，秋、冬季盛花，春季零星开花。

Abstract: Flowers bright red to dark red with a few white stripes or white markings on the petal surfaces, semi-double form, medium to large size, 9.5-11.5cm across and 4.5cm deep with 11-14 petals arranged loosely in 2 rows, petals wavy and hard textured, stamens a few. Leaves dark green, heavy leathery, elliptic, margins thinly serrate. Plant compact and growth vigorous. Starts to bloom from late-summer, fully blooms from autumn to winter and sporadically in spring.

HA-48-3. 珊瑚田歌
Shanhu Tiange (Coral's Field Songs)

杂交组合： 非云南山茶杂交种'珊瑚乐'×杜鹃红山茶（Non-*reticulata* hybrid 'Coral Delight'× *C. azalea* ）。

杂交苗编号： SHLD-No.2。

性状： 花朵深粉红色至淡桃红色，瓣面可见白条纹或白斑，单瓣型，中型至大型花，花径 8.5～11.0cm，花瓣6～7枚，阔倒卵形，雄蕊柱状。叶片浓绿色，叶背面灰绿色，革质，椭圆形，长8.0～9.0cm，宽3.5～4.0cm，边缘叶齿明显。植株立性，枝叶稠密，生长旺盛。夏末始花，秋、冬季盛花，春季零星开花。

Abstract: Flowers deep pink to light peach red, white stripes or patches visible on the petal surfaces, single form, medium to large size, 8.5-11.0cm across with 6-7 petals, petals broad obovate, stamens columnar. Leaves dark green, leathery, elliptic, margins obviously serrate. Plant upright, branches dense and growth vigorous. Starts to bloom from late-summer, fully blooms from autumn to winter and sporadically in spring.

HA-48-4. 珊瑚姑娘
Shanhu Guniang (Coral Girl)

杂交组合：非云南山茶杂交种'珊瑚乐'×杜鹃红山茶（Non-*reticulata* hybrid 'Coral Delight' × *C. azalea*）。

杂交苗编号：SHLD-No.4。

性状：花朵淡粉红色至淡橘粉红色，渐向瓣面中部呈模糊的白色，半重瓣型至玫瑰重瓣型，中型至大型花，花径9.0～12.0cm，花朵厚5.0cm，大花瓣15～16枚，呈3轮松散排列，阔倒卵形，边缘内卷，先端近圆，雄蕊簇生，偶有小花瓣与雄蕊混生。叶片浓绿色，叶背面灰绿色，厚革质，椭圆形，长8.0～10.5cm，宽4.0～4.5cm，边缘齿稀。植株紧凑，矮性，生长旺盛。夏末始花，秋、冬季盛花，春季零星开花。

Abstract: Flowers bright pink to light orange-pink, fading to misty white at mid-part of petal surfaces, semi-double to rose-double form, medium to large size, 9.0-12.0cm across and 5.0cm deep with 15-16 petals arranged loosely in 3 rows, petals broad obovate, edges inward curved, stamens clustered, occasionally some petaloids mixed with the stamens. Leaves dark green, heavy leathery, elliptic, margins thinly serrate. Plant compact, dwarf and growth vigorous. Starts to bloom from late-summer, fully blooms from autumn to winter and sporadically in spring.

⦿ 杂交组合 HA-53. 杜鹃红山茶 × 红山茶品种'黄埔之浪'
HA-53. *C. azalea* × *C. japonica* cultivar 'Huangpu Zhilang'

本杂交组合将介绍3个新品种。

红山茶品种'黄埔之浪'是从红山茶品种'皇家天鹅绒'×红山茶品种'丝纱罗'这一杂交组合中获得的。其花朵黑红色，具蜡质感，半重瓣型至牡丹型，大花型（见第一部书第493～494页）。

本组获得的杂交新品种基本上具有了杜鹃红山茶多季开花、叶片浓绿的特性和'黄埔之浪'花朵黑红色、半重瓣型至牡丹型的特性。

Three new hybrids will be introduced in this cross-combination.

'Huangpu Zhilang' was obtained from a cross-combination of *C. japonica* cultivar 'Royal Velvet' × *C. japonica* cultivar 'Tiffany'. Its flowers are dark red with waxy luster in color and large semi-double to peony in form (See p.493-494, our first book).

The new hybrids basically have the characteristics with ever-blooming and dark green leaves of *C. azalea* and the characteristics with dark red and semi-double to peony form in flowers of 'Huangpu Zhilang'.

HA-53-1. 红龙舞天
Honglong Wutian (Red Dragon Dancing in the Sky)

杂交组合：杜鹃红山茶 × 红山茶品种'黄埔之浪'（*C. azalea* × *C. japonica* cultivar 'Huangpu Zhilang'）。

杂交苗编号：DHP-No.1。

性状：花朵深红色至黑红色，具蜡光，中部花瓣偶有白条纹，半重瓣型至牡丹型，中型至大型花，花径9.0～11.0cm，花朵厚5.5cm，外部大花瓣14～16枚，呈2轮排列，阔倒卵形，略扭曲，中部小花瓣直立，波浪状，雄蕊簇生，花朵整朵掉落。嫩叶淡红色，成熟叶浓绿色，叶背面灰绿色，革质，阔椭圆形，长9.5～11.0cm，宽4.5～5.5cm，边缘叶齿浅。植株立性，枝条稠密，生长旺盛。夏末始花，秋、冬季盛花，春初零星开花。

Abstract: Flowers deep red to dark red with some waxy luster, occasionally with white stripes on the central petals, semi-double to peony form, medium to large size, 9.0-11.0cm across and 5.5cm deep with 14-16 exterior large petals arranged in 2 rows, petals broad obovate, slightly twisted, central small petals erected, wavy, stamens clustered, flowers fall whole. Tender leaves light red, mature leaves dark green, leathery, broad elliptic, margins shallowly serrate. Plant upright, branches dense and growth vigorous. Starts to bloom from late-summer, fully blooms from autumn to winter and sporadically in early-spring.

HA-53-2. 广场红艺
Guangchang Hongyi (Square Red Art)

杂交组合：杜鹃红山茶 × 红山茶品种'黄埔之浪'（*C. azalea* × *C. japonica* cultivar 'Huangpu Zhilang'）。

杂交苗编号：DHP-No.4 + 2。

性状：花朵桃红色至红色，单瓣型，喇叭状，中型花，花径9.5～10.0cm，花瓣6～7枚，花瓣阔倒卵形，雄蕊柱状，开花稠密。叶片浓绿色，叶背面灰绿色，革质，椭圆形，长9.5～11.0cm，宽4.0～5.0cm，边缘齿浅。植株立性，枝条稠密，生长旺盛。夏末始花，秋、冬季盛花，春初零星开花。

Abstract: Flowers peach red to red, single form, medium size, 9.5-10.0cm across with 6-7 petals, petals broad obovate, stamens columnar, bloom dense. Leaves dark green, leathery, elliptic, margins shallowly serrate. Plant upright, branches dense and growth vigorous. Starts to bloom from late-summer, fully blooms from autumn to winter and sporadically in early-spring.

HA-53-3. 水墨丹青
Shuimo Danqing (Chinese Ink Painting)

杂交组合： 杜鹃红山茶 × 红山茶品种'黄埔之浪'（ *C. azalea* × *C. japonica* cultivar 'Huangpu Zhilang'）。

杂交苗编号： DHP-No.10。

性状： 花朵深红色至黑红色，具蜡光，中部小花瓣偶有白条纹，半重瓣型至牡丹型，中型至大型花，花径9.0～11.5cm，花朵厚5.5cm，外部大花瓣13～16枚，呈2～3轮排列，倒卵形，外卷，中部小花瓣直立，波浪状，雄蕊散射，花丝淡红色，花药金黄色，开花稠密。叶片浓绿色，叶背面灰绿色，革质，阔椭圆形，长9.0～10.0cm，宽4.0～4.5cm，边缘具浅齿。植株立性，枝条稠密，生长旺盛。夏末始花，秋、冬季盛花，春初零星开花。

> **Abstract:** Flowers deep red to dark red with waxy luster, occasionally with white stripes on the central small petals, semi-double to peony form, medium to large size, 9.0-11.5cm across and 5.5cm deep with 13-16 exterior large petals arranged in 2-3 rows, petals obovate, rolled outward, central small petals erected, wavy, stamens radial, filaments pale red, anthers yellow, bloom dense. Leaves dark green, leathery, broad elliptic, margins shallowly serrate. Plant upright, branches dense and growth vigorous. Starts to bloom from late-summer, fully blooms from autumn to winter and sporadically in early-spring.

杂交组合 HA-54. 杜鹃红山茶 × 红山茶品种'莫顿州长'
HA-54. *C. azalea* × *C. japonica* cultivar 'Governor Mouton'

本杂交组合将介绍1个新品种。

'莫顿州长'品种花朵为艳红色，具有两轮大花瓣，是中型至大型花。

然而，新品种的花朵却是粉红色、单瓣型的，尽管这一结果与我们的预期有出入，但是杂交种的粉红花色非常柔和，而且开花稠密，植株立性，生长旺盛。毫无疑问，它将是园林美化的好品种。

One new hybrid will be introduced in this cross-combination.

'Governor Mouton' cultivar has bright red flowers with two rows of large petals and medium to large size.

The new hybrid, however, is pink and single in its flowers. Although the result is still far away from our original intention, the pink flowers of the new hybrid are very soft, bloom is very dense and plants are upright and strong growth. Without question, it will be a good variety for gardening.

HA-54-1. 粉润心田
Fenrun Xintian (Pink Moistening Hearts)

杂交组合：杜鹃红山茶 × 红山茶品种'莫顿州长'（ *C. azalea* × *C. japonica* cultivar 'Governor Mouton'）。

杂交苗编号：DMDZZ-No.1。

性状：花朵粉红色，瓣面有隐约的小白点，单瓣型，中型至大型花，花径9.5～11.0cm，花瓣7～8枚，阔倒卵形，厚质，外翻，雄蕊基部连生，呈柱状，开花稠密，花朵整朵掉落。叶片浓绿色，叶背面灰绿色，革质，椭圆形，长10.0～11.0cm，宽5.0～5.5cm，基部楔形，边缘叶齿浅。植株立性，枝条开张，生长旺盛。秋初始花，冬季可持续开花，春季亦能零星开花。

Abstract: Flowers pink with many faint white dots on the petal surfaces, single form, medium to large size, 9.5-11.0cm across with 7-8 petals, petals broad obovate, thick texture, and turned outward, stamens united into a column at the base, bloom dense, flowers fall whole. Leaves dark green, leathery, elliptic, cuneate at the base, margins shallowly serrate. Plant upright, branches spread and growth vigorous. Starts to bloom from early-autumn and continuously blooms in winter and sporadically in spring.

⦿ 杂交组合 HA-55. 杜鹃红山茶 × 白花抱茎茶
HA-55. *C. azalea* × *C. lucii* Orel & Curry (2015)

本杂交组合将介绍1个新品种。

为了获得白花色的杜鹃红山茶杂交新品种，本组合以白花抱茎茶为父本，与母本杜鹃红山茶进行杂交。遗憾的是，本组所获得的新品种并没有达到预期目的。这说明，白花抱茎茶的白花色对于杜鹃红山茶的红花色是隐性的，白花色在其 F_1 代是不表达的。

应该提及的是，澳大利亚的两位专家乔治·奥廖尔（George Orel）和安东尼·库里（Anthony S. Curry），通过几年的分子水平上的分类研究，确定白花抱茎茶为一个新种。而我们以前一直把该物种看作是越南抱茎茶的一个白花类型（见第一部书第59页）。

One new hybrid will be introduced in this cross-combination.

In order to get the *C. azalea*'s new hybrids with white flowers, *C. lucii* was as male parent to cross with female parent, *C. azalea* in the combination. Unfortunately, the hybrid we got had not yet achieved the desired aim. It has been shown that white flower color of *C. lucii* is recessive to red flower color of *C. azalea*, therefore, white flower color can not express out in the F_1 hybrid.

It should be mentioned that *C. lucii* is a new camellia species named by the two experts, Dr. George Orel and Anthony S. Curry, Australia, after their researches on the taxonomy in molecular level recent years. However, we had been thinking of the species as a form of *C. amplexicaulis* before they published (See p. 59, our first book).

HA-55-1. 万事顺景
Wanshi Shunjing（Everything Good Fortune）

杂交组合：杜鹃红山茶 × 白花抱茎茶（*C. azalea* × *C. lucii* Orel & Curry）。

杂交苗编号：DBB-No.3。

性状：花朵桃红色，单瓣型至半重瓣型，中型花，花径8.0～9.5cm，花瓣8～13枚，厚质，呈1～2轮排列，花瓣阔倒卵形，雄蕊基部略连生，呈短管状，开花稠密。叶片深绿色，叶背面灰绿色，厚革质，长椭圆形，长12.3～13.5cm，宽4.9～5.5cm，先端钝，上部边缘齿浅。植株立性，枝叶稠密，生长旺盛。夏季始花，秋、冬季盛花，春季零星开花。

> **Abstract:** Flowers peach red, single to semi-double form, medium size, 8.0-9.5cm across with 8-13 petals arranged in 1-2 rows, petals thick texture, broad obovate, stamens slightly united into a short column, bloom dense. Leaves deep green, heavy leathery, long elliptic, apices obtuse, margins shallowly serrate at the upper part. Plant upright, branches dense and growth vigorous. Starts to bloom from summer, fully blooms from autumn to winter and sporadically in spring.

⦿ 杂交组合 HA-56. 杜鹃红山茶 × 非云南山茶杂交种'甜凯特'
HA-56. *C. azalea* × Non-*reticulata* hybrid 'Sweet Emily Kate'

本杂交组合将介绍3个新品种。

'甜凯特'品种具有琉球连蕊茶基因，其花朵淡粉红色，略具芳香，花型为小型的托桂型或牡丹型。

本杂交组合所获得的新品种虽然没有父本芳香性状的表达，但是杂交种花径和叶片变小了。

Three new hybrids will be introduced in this cross-combination.

'Sweet Emily Kate' contains the genes of *C. lutchuensis* and its flowers are light pink with slight fragrance and small size, anemone or peony form.

The new hybrids obtained do not express out any fragrant characteristics, but the sizes of the new hybrids' flowers and leaves both are not large.

HA-56-1. 熊熊火焰
Xiongxiong Huoyan (Leaping Flames)

杂交组合： 杜鹃红山茶 × 非云南山茶杂交种'甜凯特'（ *C. azalea* × Non-*reticulata* hybrid 'Sweet Emily Kate'）。

杂交苗编号： DTKT-No.1。

性状： 花朵黑红色，具蜡质光泽，偶有花瓣具白斑，单瓣型，中型花，花径 8.5～10.0cm，花瓣 6～7 枚，厚质，倒卵形，略皱，雄蕊柱状。叶浓绿色，叶背面灰绿色，厚革质，椭圆形，长 8.5～9.5cm，宽 4.0～4.5cm，边缘齿浅。植株立性，生长旺盛。仲夏始花，秋、冬季盛花，春季零星开花。

Abstract: Flowers dark red with waxy luster, occasionally with white markings, single form, medium size, 8.5-10.0cm across with 6-7 petals, petals thick texture, obovate, petals slightly crinkled, stamens columnar. Leaves dark green, heavy leathery, elliptic, margins shallowly serrate. Plant upright and growth vigorous. Starts to bloom from mid-summer, fully blooms from autumn to winter and sporadically in spring.

HA-56-2. 绛唇映日
Jiangchun Yingri (Crimson Lips Reflected Sunlight)

杂交组合： 杜鹃红山茶 × 非云南山茶杂交种'甜凯特'（*C. azalea* × Non-*reticulata* hybrid 'Sweet Emily Kate'）。

杂交苗编号： DTKT-No.2。

性状： 花朵红色，中部雄蕊瓣上具小白斑，单瓣型至托桂型，小型花，花径6.0～7.0cm，花瓣5～6枚，阔倒卵形，中部雄蕊瓣直立，簇拥，似唇状。叶片浓绿色，叶背面灰绿色，革质，阔椭圆形，长8.0～8.5cm，宽4.0～4.5cm，边缘齿钝。植株立性，生长旺盛。夏初始花，秋、冬季盛花，春季零星开花。

Abstract: Flowers red, with tiny white spots on the central petaloids, single to anemone form, small size, 6.0-7.0cm across with 5-6 petals, petals broad obovate, central petaloids erected and crowded together, which are like many lips in shape. Leaves dark green, leathery, broad elliptic, margins obtusely serrate. Plant upright and growth vigorous. Starts to bloom from early-summer, fully blooms from autumn to winter and sporadically in spring.

HA-56-3. 花季对歌
Huaji Duige (Coupled Singing in Blooming Season)

杂交组合：杜鹃红山茶 × 非云南山茶杂交种'甜凯特'(*C. azalea* × Non-*reticulata* hybrid 'Sweet Emily Kate')。

杂交苗编号：DTKT-No.3。

性状：花朵桃红色，泛紫色调，瓣面具模糊的小白点，中部花瓣淡红色，半重瓣型至牡丹型，小型至中型花，花径 7.0～8.5cm，花朵厚约 3.5cm，外部大花瓣 19 枚，呈 2 轮排列，阔倒卵形，中部小花瓣直立，扭曲，大部分雄蕊瓣化，雄蕊瓣具白斑。叶片浓绿色，叶背面灰绿色，革质，阔椭圆形，长 8.0～8.5cm，宽 5.0～5.8cm，边缘齿钝。植株立性，生长旺盛。夏初始花，秋、冬季盛花，春季零星开花。

Abstract: Flowers peach red with some purple hue and faint tiny white dots on the petal surfaces, central petals light red, semi-double to peony form, small to medium size, 7.0-8.5cm across and 3.5cm deep with 19 exterior large petals arranged in 2 rows, petals broad obovate, central small petals erected, wrinkled, most of stamens changed into petaloids with white markings. Leaves dark green, leathery, broad elliptic, margins obtusely serrate. Plant upright and growth vigorous. Starts to bloom from early-summer, fully blooms from autumn to winter and sporadically in spring.

杂交组合 HA-57. 杜鹃红山茶 × 红山茶品种'波特'
HA-57. *C. azalea* × *C. japonica* cultivar 'Peter Pan'

本杂交组合将介绍 2 个新品种。

第一部书曾介绍过一个'波特'× 杜鹃红山茶的杂交组合，获得了 5 个杂交新品种（见第一部书第 330～335 页）。

本杂交组合是上述杂交组合中的两个亲本的反交组合。可以看到，该反交组合所获得的新品种，其花径要比第一部书中的那 5 个新品种大得多。

Two new hybrids will be introduced in this cross-combination.

A cross-combination of *C. japonica* cultivar 'Peter Pan' × *C. azalea* was introduced in our first book and five new hybrids were obtained (See p.330-335, our first book).

This combination is a reciprocal cross to the above cross-combination. It can be seen that the flower sizes of the new hybrids obtained from the reciprocal cross are larger than which of the five new hybrids in our first book.

HA-57-1. 波特新生
Bote Xinsheng (Peter Pan's Rebirth)

杂交组合：杜鹃红山茶 × 红山茶品种'波特'（ *C. azalea* × *C. japonica* cultivar 'Peter Pan'）。

杂交苗编号：DBT-No.1。

性状：花朵深桃红色，瓣面具小白点，花芯偶有具白斑的雄蕊瓣，半重瓣型，中型至大型花，花径9.5～12.0cm，花朵厚4.5cm，花瓣12～15枚，呈2轮排列，花瓣阔倒卵形，雄蕊基部连生。叶片浓绿色，叶背面灰绿色，革质，阔椭圆形，长10.0～11.0cm，宽4.5～5.0cm，边缘叶齿浅。植株立性，生长旺盛。夏末始花，秋、冬季盛花，春季零星开花。

Abstract: Flowers deep peach red with many tiny white dots on the petal surfaces, occasionally with petaloids which have white markings at the center, semi-double form, medium to large size, 9.5-12.0cm across and 4.5cm deep with 12-15 petals arranged in 2 rows, petals broad obovate, stamens united at the base. Leaves dark green, leathery, broad elliptic, margins shallowly serrate. Plant upright and growth vigorous. Starts to bloom from late-summer, fully blooms from autumn to winter and sporadically in spring.

HA-57-2. 波特新姿
Bote Xinzi (Peter Pan's New Posture)

杂交组合： 杜鹃红山茶 × 红山茶品种'波特'（*C. azalea* × *C. japonica* cultivar 'Peter Pan'）。

杂交苗编号： DBT-No.2。

性状： 花朵淡橘红色，单瓣型，大型至巨型花，花径 11.5～13.4cm，花瓣 6～7 枚，长倒卵形，长 6.5cm，宽 4.5cm，外翻，边缘略波浪状，雄蕊基部连生，呈短柱状。叶片浓绿色，叶背面灰绿色，革质，椭圆形，长 9.0～10.5cm，宽 4.0～5.0cm，边缘叶齿浅。植株立性，生长旺盛。夏末始花，秋、冬季盛花，春季零星开花。

Abstract: Flowers light orange red, single form, large to very large size, 11.5-13.4cm across, petals long obovate, rolled outward, edges slightly wavy, stamens united into a short column at the base. Leaves dark green, leathery, elliptic, margins shallowly serrate. Plant upright and growth vigorous. Starts to bloom from late-summer, fully blooms from autumn to winter and sporadically in spring.

◉ 杂交组合 HA-58. 杜鹃红山茶 × 红山茶品种'道温的曙光'
HA-58. *C. azalea* × *C. japonica* cultivar 'Dawn's Early Light'

　　本杂交组合将介绍5个新品种。

　　'道温的曙光'品种的花朵呈淡紫粉红色,瓣缘具深粉红边,中型至大型花,玫瑰重瓣型或牡丹型,非常漂亮。

　　本组合获得的杂交新品种大部分在性状上很奇特,而且它们多数是半重瓣型至牡丹型、大型至巨型花,给我们带来了新惊喜。

　　Five new hybrids will be introduced in this cross-combination.

　　The flowers of the cultivar 'Dawn's Early Light' are light orchid pink with deep pink borders at petal edges, medium to large size, rose-double to peony form which are very beautiful.

　　Most of the new hybrids obtained in this combination are very special in their characteristics and most of them can open flowers with large to very large size and semi-double to peony form, which give us a new surprise.

HA-58-1. 流光溢彩
Liuguang Yicai (Brilliant Color & Glitzy Light)

杂交组合： 杜鹃红山茶 × 红山茶品种'道温的曙光'（ *C. azalea* × *C. japonica* cultivar 'Dawn's Early Light'）。

杂交苗编号： DDW-No.1。

性状： 花朵桃红色至红色，中部花瓣偶有白条纹或者白斑块，随着花朵开放，中部花瓣渐变为淡红色，半重瓣型至牡丹型，大型至巨型花，花径12.0～14.5cm，花朵厚6.5cm，外部大花瓣25～28枚，呈5～7轮排列，阔倒卵形，中部小花瓣直立，与雄蕊混生。叶片浓绿色，叶背面灰绿色，革质，椭圆形，长9.0～10.0cm，宽4.5～5.0cm，边缘叶齿浅。植株立性，生长旺盛。夏末始花，秋、冬季盛花，春季零星开花。

Abstract: Flowers peach red to red, occasionally with white stripes or markings at the central petals, as the flowers bloom, central petals gradually changing into light red, semi-double to peony form, large to very large size, 12.0-14.5cm across and 6.5cm deep with 25-28 exterior large petals arranged in 5-7 rows, petals broad obovate, central small petals erected and mixed with stamens. Leaves dark green, leathery, elliptic, margins shallowly serrate. Plant upright and growth vigorous. Starts to bloom from late-summer, fully blooms from autumn to winter and sporadically in spring.

HA-58-2. 喜迎曙光
Xiying Shuguang (Happily Welcome Dawn)

杂交组合：杜鹃红山茶 × 红山茶品种'道温的曙光'（*C. azalea* × *C. japonica* cultivar 'Dawn's Early Light'）。

杂交苗编号：DDW-No.2。

性状：花朵深粉红色至淡桃红色，渐向花芯呈粉红色，中部花瓣偶有模糊的白条纹，半重瓣型至牡丹型，中型至大型花，花径9.5～11.5cm，花朵厚5.5cm，外部大花瓣11～13枚，呈2轮有序排列，阔倒卵形，中部小花瓣和雄蕊瓣21～25枚，雄蕊簇生。叶片浓绿色，叶背面灰绿色，革质，椭圆形，长10.5～12.0cm，宽4.5～5.0cm，边缘叶齿浅。植株立性，生长旺盛。夏末始花，秋、冬季盛花，春季零星开花。

Abstract: Flowers deep pink to light peach red, fading to pink at the center, occasionally with fuzzy white stripes on the central petals, semi-double to peony form, medium to large size, 9.5-11.5cm across and 5.5cm deep with 11-13 exterior large petals arranged orderly in 2 rows, petals broad obovate, 21-25 small petals and petaloids at the center, stamens clustered. Leaves dark green, leathery, elliptic, margins shallowly serrate. Plant upright and growth vigorous. Starts to bloom from late-summer, fully blooms from autumn to winter and sporadically in spring.

HA-58-3. 杏粉花雨
Xingfen Huayu（Apricot Pink & Flower Rain）

杂交组合：杜鹃红山茶 × 红山茶品种'道温的曙光'（*C. azalea* × *C. japonica* cultivar 'Dawn's Early Light'）。

杂交苗编号：DDW-No.3。

性状：花朵杏粉红色，单瓣型，中型至大型花，花径9.0～11.0cm，花瓣6枚，阔倒卵形，瓣面可见红色脉纹，雄蕊基部连生，呈柱状，开花稠密，花瓣逐片掉落。叶片浓绿色，叶背面灰绿色，革质，椭圆形，长8.5～10.0cm，宽3.5～4.0cm，边缘叶齿浅。植株立性，生长旺盛。夏末始花，秋、冬季盛花，春季零星开花。

> **Abstract:** Flowers apricot pink, single form, medium to large size, 9.0-11.0cm across with 6 petals, petals broad obovate, red veins visible on the petal surfaces, stamens united into a column at the base, bloom dense, petals fall one by one. Leaves dark green, leathery, elliptic, margins shallowly serrate. Plant upright and growth vigorous. Starts to bloom from late-summer, fully blooms from autumn to winter and sporadically in spring.

HA-58-4. 黎明破晓
Liming Poxiao (Break of Dawn)

杂交组合： 杜鹃红山茶 × 红山茶品种'道温的曙光'（ *C. azalea* × *C. japonica* cultivar 'Dawn's Early Light'）。

杂交苗编号： DDW-No.4。

性状： 花朵淡桃红色至红色，中部小花瓣和雄蕊瓣粉红色，牡丹型，中型至大型花，花径 9.0～12.0cm，花朵厚5.2cm，外部大花瓣18～21枚，呈2～3轮紧实排列，阔倒卵形，中部小花瓣和雄蕊瓣众多，扭曲，偶有雄蕊外露。叶片浓绿色，叶背面灰绿色，厚革质，椭圆形，长 9.5～11.0cm，宽4.0～5.0cm，边缘叶齿浅。植株立性，生长旺盛。夏末始花，秋、冬季盛花，春季零星开花。

Abstract: Flowers light peach red to red, both central small petals and petaloids are pink, peony form, medium to large size, 9.0-12.0cm across and 5.2cm deep with 18-21 exterior large petals arranged tightly in 2-3 rows, petals broad obovate, central small petals and petaloids numerous. Leaves dark green, heavy leathery, elliptic, margins shallowly serrate. Plant upright and growth vigorous. Starts to bloom from late-summer, fully blooms from autumn to winter and sporadically in spring.

HA-58-5. 晨阳闪金
Chenyang Shanjin (Morning Sun Flashing Gold)

杂交组合： 杜鹃红山茶 × 红山茶品种'道温的曙光'（*C. azalea* × *C. japonica* cultivar 'Dawn's Early Light'）。

杂交苗编号： DDW-No.5。

性状： 花朵深红色，略泛紫色调，花瓣中部淡红色，牡丹型，中型至大型花，花径9.5～12.5cm，花朵厚5.0cm，外轮大花瓣7～8枚，阔倒卵形，中部花瓣12～15枚簇拥，与雄蕊混生，波浪状。叶片浓绿色，叶背面灰绿色，厚革质，阔椭圆形，长9.5～11.0cm，宽4.5～5.5cm，边缘叶齿浅。植株立性，生长旺盛。夏末始花，秋、冬季盛花，春季零星开花。

Abstract: Flowers deep red slightly with purple hue, petals are light red in the middle, peony form, medium to large size, 9.5-12.5cm across and 5.0cm deep with 7-8 large petals, petals broad obovate, 12-15 central petals crowded, wavy and mixed with stamens. Leaves dark green, heavy leathery, broad elliptic, margins shallowly serrate. Plant upright and growth vigorous. Starts to bloom from late-summer, fully blooms from autumn to winter and sporadically in spring.

⦿ 杂交组合 HA-59. 杜鹃红山茶 × 红山茶品种'金博士'
HA-59. *C. azalea* × *C. japonica* cultivar 'Dr King'

本杂交组合将介绍 4 个新品种。

红山茶品种'金博士'的花朵呈淡红色，为半重瓣型且花朵硕大。

已经证实，本组获得的杂交新品种的始花期可以提早到夏季，盛花期多在秋、冬季，即便在春季也能零星开花，这是杜鹃红山茶基因表达的结果。另外，新品种的花色艳红，有50%的品种为半重瓣型、大型花，这是父本'金博士'基因表达的结果。

Four new hybrids will be introduced in this cross-combination.

The flowers of *C. japonica* cultivar 'Dr King' are light red, semi-double form and large size.

These new hybrids obtained in the cross-combination have been shown that they start to bloom in summer, peak-bloom in both autumn and winter and sparsely bloom in the spring of the following year, which is the result of *C. azalea*'s gene expression. Moreover, the flowers of the new hybrids are bright red, 50% of them are semi-double form and large size, which is the result of the gene expression of the male parent 'Dr King'.

HA-59-1. 蒸蒸日上
Zhengzheng Rishang (Thriving Upward)

杂交组合： 杜鹃红山茶 × 红山茶品种'金博士'（*C. azalea* × *C. japonica* cultivar 'Dr King'）。

杂交苗编号： DJB-No.1。

性状： 花朵红色，瓣面可见清晰红色脉纹，偶有模糊的白斑，牡丹型，偶有玫瑰重瓣型花朵，大型至巨型花，花径11.0～13.5cm，花朵厚5.5cm，外部大花瓣18枚，呈3轮松散排列，阔倒卵形，外翻，雄蕊瓣与雄蕊混生。叶片浓绿色，叶背面灰绿色，厚革质，阔椭圆形，长9.0～10.5cm，宽4.5～5.0cm，边缘叶齿稀。植株立性，枝条稠密，生长旺盛。夏季始花，秋、冬季盛花，春季零星开花。

> **Abstract:** Flowers red with deep red veins obviously visible on the petal surfaces, occasionally a few hazy white markings, peony form, sometimes rose-double form, large to very large size, 11.0-13.5cm across and 5.5cm deep with 18 exterior large petals arranged loosely in 3 rows, petals broad obovate, turned outward, petaloids mixed with stamens. Leaves dark green, heavy leathery, broad elliptic, margins sparsely serrate. Plant upright, branches dense and growth vigorous. Starts to bloom from summer, fully blooms from autumn to winter and sporadically in spring.

HA-59-2. 粉柔舞裙
Fenrou Wuqun (Soft Pink Dance Dress)

杂交组合： 杜鹃红山茶 × 红山茶品种'金博士'（ *C. azalea* × *C. japonica* cultivar 'Dr King' ）。

杂交苗编号： DJB-No.3。

性状： 花朵柔和的粉红色，单瓣型，中型花，花径8.0～9.0cm，花瓣5～6枚，阔倒卵形，外翻，瓣面略皱，雄蕊基部连生，呈短柱状，开花稠密。叶片浓绿色，叶背面灰绿色，厚革质，阔椭圆形，长10.0～11.0cm，宽5.5～6.8cm，叶脉清晰，基部楔形，边缘叶齿浅。植株立性，枝条稠密，生长旺盛。夏末始花，秋、冬季盛花，春季零星开花。

> **Abstract:** Flowers soft pink, single form, medium size, 8.0-9.0cm across with 5-6 petals, petals broad obovate, rolled outward, petal surface slightly crinkled, stamens united into a short column, bloom dense. Leaves dark green, heavy leathery, broad elliptic, veins visible at the surfaces, cuneate at the base, margins shallowly serrate. Plant upright, branches dense and growth vigorous. Starts to bloom from late-summer, fully blooms from autumn to winter and sporadically in spring.

HA-59-3. 红色主题
Hongse Zhuti (Red Theme)

杂交组合： 杜鹃红山茶 × 红山茶品种'金博士'（ *C. azalea* × *C. japonica* cultivar 'Dr King'）。

杂交苗编号： DJB–No.4。

性状： 花朵红色，单瓣型，中型至大型花，花径 9.5～12.0cm，花瓣 5～6 枚，阔倒卵形，外翻，大波浪状，瓣面略皱，雄蕊基部连生，呈柱状，开花稠密。叶片浓绿色，叶背面灰绿色，厚革质，阔椭圆形，长 11.0～11.5cm，宽 5.0～5.5cm，外翻，基部楔形，边缘叶齿细而浅。植株立性，枝条稠密，生长旺盛。夏末始花，秋、冬季盛花，春季零星开花。

Abstract: Flowers red, single form, medium to large size, 9.5-12.0cm across with 5-6 petals, petals broad obovate, rolled outward, large wavy, petal surfaces slightly crinkled, stamens united into a column at the base, bloom dense. Leaves dark green, heavy leathery, broad elliptic, cuneate at the base, margins thinly and shallowly serrate. Plant upright, branches dense and growth vigorous. Starts to bloom from late-summer, fully blooms from autumn to winter and sporadically in spring.

HA-59-4. 福寿齐天
Fushou Qitian (Happiness & Longevity Comparable to the Universe)

杂交组合： 杜鹃红山茶 × 红山茶品种'金博士'（ *C. azalea* × *C. japonica* cultivar 'Dr King'）。

杂交苗编号： DJB-No.5。

性状： 花朵艳红色，半重瓣型，大型花，花径 11.0～12.0cm，花朵厚 4.5cm，花瓣 12～14 枚，呈 2 轮排列，阔倒卵形，外翻，雄蕊簇生。叶片浓绿色，叶背面灰绿色，革质，阔椭圆形，长 9.5～11.0cm，宽 4.5～5.5cm，边缘叶齿浅。植株立性，枝条稠密，生长旺盛。夏末始花，秋、冬季盛花，春季零星开花。

Abstract: Flowers bright red, semi-double form, large size, 11.0-12.0cm across and 4.5cm deep with 12-14 petals arranged in 2 rows, petals broad obovate, turned outward, stamens clustered. Leaves dark green, leathery, broad elliptic, margins shallowly serrate. Plant upright, branches dense and growth vigorous. Starts to bloom from late-summer, fully blooms from autumn to winter and sporadically in spring.

杂交组合 HA-60. 杜鹃红山茶 × 红山茶类型'耐冬'
HA-60. *C. azalea* × *C. japonica* form 'Naidong'

本杂交组合将介绍1个新品种。

'耐冬'是从分布于我国山东青岛一带的红山茶原生种中发现的一个类型。顾名思义，'耐冬'就是抗寒的意思。正是它的高抗寒性，我们才以它为亲本和杜鹃红山茶杂交，期望能获得既抗寒又能多季开花的新品种。

从获得的新品种上看，尽管它是单瓣型的，但是已经具有了'耐冬'的抗寒特性。山东青岛园林高级工程师吴楠2022年曾在山东科技大学青岛校区对该杂交新品种进行试种，该品种在遭遇到极端低温 –17℃时，仍能正常生长和开花（见'多季耐冬'品种左下角图片）。这为今后培育更多、更好的既抗寒又能四季开花的优良品种奠定了基础。但我们仍需进行更多的试种。

One new hybrid will be introduced in this cross-combination.

'Naidong' is a species form which is found from wild *C. japonica* that distributed in the area of Qingdao City, Shandong Province, China. Just as its name implies, 'Naidong' is the meaning of hardy. It was because of its strong cold resistance that we crossed it with *C. azalea* as a cross parent and expected for getting new hybrids with the characteristics of both cold resistance and multiseason flowering.

Although the new hybrid obtained is single form, it has the cold-resistant trait of 'Naidong'. Wu Nan, a landscape senior engineer of Qingdao, Shandong Province, did a trail planting of the new hybrid at Qingdao Campus, Shandong University of Science and Technology in 2022. The hybrid still grew and blossomed normally under the extreme air temperature at -17°C (See the photo at the lower left corner of 'Duoji Naidong'). The result lays the foundation for breeding more and better cold-resistant hybrids that can bloom all seasons in the future, but it is necessary to do more trial plantings.

HA-60-1. 多季耐冬
Duoji Naidong (Multiseasonal Naidong)

杂交组合：杜鹃红山茶 × 红山茶类型'耐冬'（*C. azalea* × *C. japonica* form 'Naidong'）。

杂交苗编号：DND-No.1。

性状：花朵桃红色，多单朵顶生，单瓣型，呈半开放状态，微型花，花径 5.5～6.0cm，花瓣 6～7枚，阔倒卵形，长 5.5cm，宽 4.5cm，边缘皱褶，雄蕊柱状，基部连生。嫩叶淡红色，成熟叶片浓绿色，叶背面灰绿色，革质，椭圆形，长 10.0～11.0cm，宽 4.5～5.0cm，边缘叶齿浅。植株立性，开张，枝叶稀疏，生长旺盛。夏末始花，秋、冬季盛花，春季零星开花。山东青岛的试种已经证明，该杂交新品种冬季可忍耐 –17℃的极端气温，然而，仍需要进行更多的试种。

Abstract: Flowers peach red, most of flowers are solitary at terminal, single form, semi-opened, miniature size, 5.5-6.0cm across, petals broad obovate, edges crinkled, stamens columnar. Tend leaves pale red, mature leaves dark green, leathery, elliptic, margins shallowly serrate. Plant upright, spread branches sparse and growth vigorous. Starts to bloom from late-summer, fully blooms from autumn to winter and sporadically in spring. The trial planting in Qingdao, Shandong Province has showed that this new hybrid can tolerate -17℃ extreme air temperature in winter, however, it is necessary to do more trial plantings.

杂交组合 HA-61. 杜鹃红山茶 × 红山茶品种'羽衣'
HA-61. *C. azalea* × *C. japonica* cultivar 'Hagoromo'

本杂交组合将介绍1个新品种。

第一部书曾介绍过红山茶品种'羽衣'×杜鹃红山茶的杂交组合，并获得了一个名叫'迷人蔷薇'的杂交新品种（见第一部书第208～209页）。本次的杂交组合是它的反交组合。

本次获得的杂交新品种在花色和花型上与其反交组合所获得的杂交新品种有明显不同。由于这两个杂交组合获得的新品种都含有杜鹃红山茶的基因，所以它们都具有四季开花的性状。

One new hybrid will be introduced in this cross-combination.

A cross-combination of *C. japonica* cultivar 'Hagoromo' × *C. azalea* was introduced and a new hybrid named 'Miren Qiangwei' was obtained (See p.208-209, our first book). This cross-combination is its reciprocal cross-combination.

There are obvious differences between the hybrids obtained this time and its reciprocal cross-combination in flower color and flower form. The new hybrids obtained from these two cross-combinations all contain the genes of *C. azalea*, so they all have ever-blooming characteristics.

HA-61-1. 紫浪吻夏
Zilang Wenxia (Purple Waves Kissing Summer)

杂交组合： 红山茶品种'羽衣'× 杜鹃红山茶（*C. japonica* cultivar 'Hagoromo'× *C. azalea*）。

杂交苗编号： YYD-No.3。

性状： 花朵紫红色，单瓣型，中型花，花径8.0～9.5cm，花瓣5～6枚，阔倒卵形，波浪状，雄蕊基部连生，呈柱状，开花稠密。叶片浓绿色，叶背面灰绿色，厚革质，椭圆形，长8.0cm，宽4.8 cm，上部边缘具浅锯齿，基部光滑。植株紧凑，灌木状，生长旺盛。夏初始花，秋、冬季盛花，春季零星开花。

Abstract: Flowers purple-red, single form, medium size, 8.0-9.5cm across with 5-6 petals, petals broad obovate, wavy, stamens united into a column at the base, bloom dense. Leaves dark green, heavy leathery, margins shallowly serrate at the upper part and smooth at the lower part. Plant compact, bushy and growth vigorous. Starts to bloom from early-summer, fully blooms from autumn to winter and sporadically in spring.

⦿ 杂交组合 HA-62. 杜鹃红山茶 × 红山茶品种'阿兰'
HA-62. *C. azalea* × *C. japonica* cultivar 'Mark Alan'

×

本杂交组合将介绍2个新品种。

'阿兰'品种的花朵呈酒红色，花瓣呈勺状，非常奇特。我们期望本组合获得的杂交新品种能四季开花，且能开出像'阿兰'品种一样花瓣窄长的花朵。

从获得的2个新品种来看，花瓣都是狭长的，但第二个品种花瓣更多，更像其父本'阿兰'品种。由此，也可以证实其杂交新品种的真实性。

Two new hybrid will be introduced in this cross-combination.

The flowers of the cultivar 'Mark Alan' are wine colour with spoon shaped petals, which look very unique. We expect to obtain the new hybrids which not only could ever-blooming, but also could open flowers with narrow long petals as 'Mark Alan' by hybridization.

Seeing the two new hybrids obtained in this combination, the petals of them are all narrow long, however, the second hybrid has more petals, which is more like its male parent, 'Mark Alan'. So, it also can confirm the authenticity of the two new hybrids.

HA-62-1. 星火燎原
Xinghuo Liaoyuan (Catching Fire from a Spark)

杂交组合：杜鹃红山茶 × 红山茶品种'阿兰'（ *C. azalea* × *C. japonica* cultivar 'Mark Alan'）。

杂交苗编号：DAL-No.1。

性状：花朵艳红色至黑红色，单瓣型，小型花，花径 7.0～7.5cm，花瓣 5～6 枚，倒卵形，雄蕊基部连生，呈柱状，开花极稠密。叶片浓绿色，叶背面灰绿色，革质，椭圆形，长 7.5～9.0cm，宽 3.5～4.0cm，上部边缘叶齿细尖，下部边缘光滑。植株立性，生长旺盛。夏末始花，秋、冬季盛花，春季零星开花。

> **Abstract:** Flowers bright red to dark red, single form, small size, 7.0-7.5cm across with 5-6 petals, petals obovate, stamens united into a column at the base, bloom very dense. Leaves dark green, leathery, elliptic, upper margins sharply and finely serrate, lower margins smooth. Plant upright and growth vigorous. Starts to bloom from late-summer, fully blooms from autumn to winter and sporadically in spring.

HA-62-2. 多季阿兰
Duoji Alan（Multiseasonal Mark Alan）

杂交组合：杜鹃红山茶 × 红山茶品种'阿兰'（*C. azalea* × *C. japonica* cultivar 'Mark Alan'）。

杂交苗编号：DAL-No.7。

性状：花朵淡红色至红色，偶有白色条纹，半重瓣型，大型花，花径10.5～12.0cm，花瓣18～25枚，呈3轮松散、有序排列，花朵厚约6.5cm，花瓣长倒卵形，沟槽状，外翻，似其父本'阿兰'，花丝淡粉红色，稀疏，少数瓣化，具白条纹。叶片浓绿色，叶背面灰绿色，革质，椭圆形，长8.5～12.5m，宽4.0～4.5cm，上部叶缘叶齿细尖，下部叶缘光滑。植株立性，生长旺盛。夏末始花，秋、冬季盛花，春季零星开花。

Abstract: Flowers light red to red, occasionally with a few white stripes, semi-double form, large size, 10.5-12.0cm across, 18-25 petals, 6.5cm deep, petals long obovate, groove shaped, turned outward which is like its male parent 'Mark Alan', filaments light pink and with a few white stripes. Leaves dark green, leathery, elliptic, upper margins sharply and finely serrate, lower margins smooth. Plant upright and growth vigorous. Starts to bloom from late-summer, fully blooms from autumn to winter and sporadically in spring.

◉ 杂交组合 HA-63. 杜鹃红山茶 × 红山茶品种'桑迪玛斯'
HA-63. *C. azalea* × *C. japonica* cultivar 'San Dimas'

本杂交组合将介绍 2 个新品种。

红山茶品种'桑迪玛斯'的花朵呈暗红色，半重瓣型，中型至大型花，花芯处雄蕊明显可见，是一个非常有名的品种。我们期望通过杂交获得艳红色或粉红色、大花型、四季开花的新品种。本组合获得的杂交新品种的性状基本符合我们的预期。

Two new hybrids will be introduced in this cross-combination.

The flowers of *C. Japonica* cultivar 'San Dimas' are dark red in color, semi-double in form, medium to large in size and stamens obviously visible at the center, which make 'San Dimas' very famous. We expect to obtain the new hybrids whose flowers are bright red or pink, large and ever-blooming. The characteristics of the new hybrids obtained in this cross-combination basically meet our expectations.

HA-63-1. 美好向往
Meihao Xiangwang (Beautiful Yearning)

杂交组合：杜鹃红山茶 × 红山茶品种'桑迪玛斯'（*C. azalea* × *C. japonica* cultivar 'San Dimas'）。

杂交苗编号：DSDMS-No.2。

性状：花朵红色至黑红色，中部雄蕊瓣偶有模糊的白色条纹，半重瓣型至玫瑰重瓣型，有时会出现牡丹型花朵，大型花，花径10.5～12.5cm，花朵厚5.5～6.0cm，大花瓣30～35枚，呈4～5轮排列，花瓣长倒卵形，刚开放时边缘内卷，全开放后外翻，中部小花瓣半直立，与雄蕊混生，开花稠密。叶片浓绿色，叶背面灰绿色，厚革质，椭圆形，长9.0～9.5cm，宽4.0～5.0cm，基部楔形，上部边缘叶齿钝。植株立性，生长旺盛。夏初始花，秋、冬季盛花，翌年春季偶有开花。

Abstract: Flowers red to dark red, occasionally with misty white stripes on the central petaloids, semi-double to rose-double form, sometimes peony form, large size, 10.5-12.5cm across and 5.5-6.0cm deep with 30-35 large petals arranged in 4-5 rows, when just opening, the petal edges inward rolled and when fully opening, the petal edges outward rolled, central small petals semi-erected and mixed with stamens, bloom dense. Leaves dark green, heavy leathery, elliptic, cuneate at the base, margins obtusely serrate. Plant upright, growth vigorous. Starts to bloom from early-summer, fully blooms from autumn to winter and occasionally in the following spring.

HA-63-2. 大粉丝团
Da Fensituan (Big Group of Fans)

杂交组合：杜鹃红山茶 × 红山茶品种'桑迪玛斯'(*C. azalea* × *C. japonica* cultivar 'San Dimas')。

杂交苗编号：DSDMS-No.3。

性状：花朵桃红色至红色，偶有少量白色条纹，半重瓣型，大型至巨型花，花径 12.5～14.5cm，花朵厚 4.5～5.0cm，大花瓣 18～20 枚，呈 3 轮有序排列，花瓣长倒卵形，平铺，雄蕊簇生，偶有雄蕊瓣，花丝近黄色，花药黄色。叶片浓绿色，叶背面灰绿色，厚革质，阔椭圆形，长 8.0～9.0cm，宽 4.0～4.5cm，基部楔形，上部边缘叶齿尖。植株立性，紧凑，生长旺盛。夏初始花，秋、冬季盛花，翌年春季偶有开花。

Abstract: Flowers peach red to red, occasionally with a few white stripes, semi-double form, large to very large size, 12.5-14.5cm across and 4.5-5.0cm deep with 18-20 large petals arranged orderly in 3 rows, petals flat, long obovate, stamens clustered, occasionally a few of petaloids appearing. Leaves dark green, heavy leathery, broad elliptic, cuneate at the base, margins sharply serrate at the upper part. Plant upright, compact, growth vigorous. Starts to bloom from early-summer, fully blooms from autumn to winter and occasionally in the following spring.

杂交组合 HA-64. 杜鹃红山茶 × 红山茶品种'琼克莱尔'
HA-64. *C. azalea* × *C. japonica* cultivar 'Jean Clere'

本杂交组合将介绍1个新品种。

红山茶品种'琼克莱尔'的花朵呈深粉红色，牡丹型，大型花，花瓣边缘具白色窄边。本组杂交的主要目的是想获得具白色花边、能够四季开花的新品种。

虽然所获得的新品种花瓣并不具白边性状，但是其花朵硕大、花色黑红，且能四季开花。

One new hybrid will be introduced in this cross-combination.

The flowers of *C. japonica* cultivar 'Jean Clere' are deep pink, peony form, large size with a narrow, white border around the petal edges. The combination was expected to obtain the new hybrids whose flowers have white edges on petals and bloom in four seasons.

Although the new hybrid obtained does not have any characteristics of petals with white-edges, it has large, dark-red flowers and can bloom all year round.

HA-64-1. 后起之秀
Houqi Zhixiu (Up-rising Star)

杂交组合：杜鹃红山茶 × 红山茶品种'琼克莱尔'（*C. azalea* × *C. japonica* cultivar 'Jean Clere'）。

杂交苗编号：DQ-No.1。

性状：花朵红色至深红色，中部花瓣偶有白条纹，随着花朵开放，中部花瓣渐渐变为深粉红色，玫瑰重瓣型至牡丹型，大型花，花径 11.5～13.0cm，花朵厚 5.5cm，外部大花瓣 25～30 枚，呈 3～4 轮排列，倒卵形，波浪状，中部小花瓣直立，边缘波浪状，与雄蕊混生。叶片浓绿色，叶背面灰绿色，革质，椭圆形，长 9.5～10.5cm，宽 4.5～5.0cm，边缘叶齿浅。植株立性，生长旺盛。夏末始花，秋、冬季盛花，春季零星开花，最晚可延续到 4 月底。

Abstract: Flowers red to deep red, occasionally with white stripes at the central petals, as the flowers bloom, the central petals gradually changing to deep pink, rose-double to peony form, large size, 11.5-13.0cm across and 5.5cm deep with 25-30 exterior large petals arranged in 3-4 rows, exterior large petals wavy, central small petals erected, edges wavy and mixed with stamens. Leaves dark green, leathery, elliptic, margins shallowly serrate. Plant upright and growth vigorous. Starts to bloom from late-summer, fully blooms from autumn to winter and sporadically in spring, the latest flowering can continue to the end of April.

◉ 杂交组合 HA-65. 杜鹃红山茶 × 红山茶品种'香神'
HA-65. *C. azalea* × *C. japonica* cultivar 'Scentsation'

本杂交组合将介绍3个新品种。

红山茶品种'香神'是一个花朵漂亮、具芳香的品种。本杂交组合期望获得花朵漂亮、芳香，而且能四季开花的新品种。在获得的3个杂交新品种中，虽然只有一个新品种的花朵能表达出极淡的芳香性状，但是它们在其他性状上都符合我们的期盼。

Three new hybrids will be introduced in this cross-combination.

C. Japonica cultivar 'Scentsation' is a beautiful cultivar with fragrance. The combination was expected to obtain the hybrids with beautiful, fragrant flowers, and four seasons open characteristic. In the three new hybrids obtained, although only one hybrid's flowers can express very lightly fragrant characteristic, their other characteristics all can meet our expectations.

HA-65-1. 多季润香
Duoji Runxiang (Multiseasonal Moist Fragrance)

杂交组合：杜鹃红山茶 × 红山茶品种'香神'（ *C. azalea* × *C. japonica* cultivar 'Scentsation' ）。

杂交苗编号：DXS-No.1。

性状：花朵红色，具极淡的微香，中部花瓣偶现白条纹，牡丹型，中型至大型花，花径 9.5～11.0cm，花朵厚 5.7cm，外部大花瓣约 18 枚，呈 3 轮排列，略波浪状，中部小花瓣直立、扭曲，与雄蕊混生。叶片浓绿色，叶背面灰绿色，厚革质，长椭圆形，长 9.0～10.5cm，宽 4.0～5.0cm，基部楔形，上部边缘叶齿尖稀。植株立性，紧凑，生长旺盛。夏末始花，秋、冬季盛花，春季零星开花。

> **Abstract:** Flowers red with very light fragrance, occasionally with white stripes at the central petals, peony form, medium to large size, 9.5-11.0cm across and 5.7cm deep, about 18 large and slightly wavy exterior petals arranged in 3 rows, central petals erected, twisted and mixed with stamens. Leaves dark green, heavy leathery, long elliptic, cuneate at the base, margins thinly and sharply serrate. Plant upright, compact, growth vigorous. Starts to bloom from late-summer, fully blooms from autumn to winter and sporadically in spring.

HA-65-2. 红云闪光
Hongyun Shanguang (Lightening Red Clouds)

杂交组合：杜鹃红山茶 × 红山茶品种'香神'（ *C. azalea* × *C. japonica* cultivar 'Scentsation'）。

杂交苗编号：DXS-No.4。

性状：花朵桃红色，具白色条纹，半重瓣型至玫瑰重瓣型，小型至中型花，花径7.0～8.5cm，花朵厚5.0cm，花瓣28～30枚，呈4轮排列，阔倒卵形，外翻，中部少量花瓣半直立，雄蕊基部连生，呈柱状，花朵整朵掉落。叶片浓绿色，叶背面灰绿色，厚革质，椭圆形，长8.0～9.0cm，宽4.5～5.5cm，基部楔形，边缘叶齿细。植株立性，生长旺盛。夏末始花，秋、冬季盛花，春季零星开花。

Abstract: Flowers peach red with white stripes, semi-double to rose-double form, small to medium size, 7.0-8.5cm across and 5.0cm deep with 28-30 petals arranged in 4 rows, petals broad obovate, rolled outward, a few central small petals erected, stamens united into a column at the base, flowers fall whole. Leaves dark green, heavy leathery, elliptic, cuneate at the base, margins thinly serrate. Plant upright, growth vigorous. Starts to bloom from late-summer, fully blooms from autumn to winter and sporadically in spring.

HA-65-3. 多季绝美
Duoji Juemei (Multiseasonal Absolute Beauty)

杂交组合：杜鹃红山茶 × 红山茶品种'香神'（*C. azalea* × *C. japonica* cultivar 'Scentsation'）。

杂交苗编号：DXS-No.5。

性状：花朵红色，略泛紫色调，具少量白斑或白条纹，牡丹型，中型至大型花，花径 8.5～11.5cm，花朵厚 5.5cm，大花瓣 15～21 枚，呈 3～4 轮排列，花瓣波浪状，外翻，中部雄蕊瓣半直立，雄蕊基部略连生。叶片浓绿色，叶背面灰绿色，革质，椭圆形，长 9.0～9.5cm，宽 3.5～4.5cm，基部楔形，边缘叶齿细。植株立性，生长旺盛。夏末始花，秋、冬季盛花，春季偶有开花。

Abstract: Flowers red slightly with purple hue and with a few white markings or white stripes, peony form, medium to large size, 8.5-11.5cm across and 5.5cm deep with 15-21 large petals arranged in 3-4 rows, petals wavy, rolled outward, central petaloids semi-erected, stamens slightly united at the base. Leaves dark green, leathery, elliptic, cuneate at the base, margins thinly serrate. Plant upright, growth vigorous. Starts to bloom from late-summer, fully blooms from autumn to winter and occasionally in spring.

杂交组合 HA-66. 杜鹃红山茶 × 威廉姆斯杂交种'詹米'
HA-66. *C. azalea* × *C. x williamsii* hybrid 'Jamie'

本杂交组合将介绍1个新品种。

'詹米'品种花朵艳红色、半重瓣型，中型花，花瓣一层套一层，而且抗寒性较强。本组杂交主要是想把'詹米'品种的优良性状转移到杜鹃红山茶杂交新品种上。从获得的这个新品种看，其花色黑红活泼，抗性强，且能重复开花。随着该新品种栽培区域的扩大，它的抗寒性将得到验证。

One new hybrid will be introduced in this cross-combination.

'Jamie' has vivid red, medium size and semi-double of hose-in-hose form flowers and strong cold resistance. The combination was expected to transfer the fine characteristics of 'Jamie' into the hybrids of *C. azalea*. From the obtained hybrid, we can see that it has very vivid dark red flowers, strong resistance and a long flowering period. As the new hybrid's cultivated areas expand, its cold resistance will be verified.

HA-66-1. 玲珑黑妹
Linglong Heimei (Exquisite Black Sister)

杂交组合： 杜鹃红山茶 × 威廉姆斯杂交种'詹米'（*C. azalea* × *C.x williamsii* hybrid 'Jamie'）。

杂交苗编号： DZM-No.1。

性状： 花朵活泼的绯红色至黑红色，具蜡质感，半重瓣型，中型花，花径8.5～10.0cm，花朵厚5.5cm，花瓣15～18枚，呈2～3轮松散排列，花瓣阔倒卵形，雄蕊基部略连生，呈短管状，花丝淡红色，花药金黄色，开花稠密。叶片浓绿色，叶背面灰绿色，厚革质，略波折，长椭圆形，长7.0～8.5cm，宽3.0～3.5cm，上部边缘叶齿尖深。植株紧凑，立性，枝叶稠密，生长旺盛。夏末始花，秋、冬季盛花，春季偶然开花。

Abstract: Flowers vivid red to dark red with waxy feeling, semi-double form, medium size, 8.5-10.0cm across and 5.5cm deep with 15-18 petals arranged loosely in 2-3 rows, petals broad obovate, stamens slightly united in a short column, filaments pale red, anthers golden yellow, bloom dense. Leaves dark green, heavy leathery, slightly twisted, long elliptic, margins sharply and deeply serrate at the upper part. Plant compact, upright, branches dense and growth vigorous. Starts to bloom from late-summer, fully blooms from autumn to winter and occasionally in spring.

杂交组合 HA-67. 杜鹃红山茶 × 非云南山茶杂交种 '龙火珠'
HA-67. *C. azalea* × Non-*reticulata* hybrid 'Dragon Fireball'

本杂交组合将介绍 10 个新品种。

'龙火珠'是由美籍华人王大庄先生从红山茶品种'圣诞快乐'× 琉球连蕊茶这一杂交组合中获得的一个花色黑红、花瓣具白边、牡丹型的品种。

本组杂交试图让所获得的新品种不仅表达出杜鹃红山茶重复开花和叶片浓绿等主要性状，也表达出'龙火珠'品种黑红、具蜡质感的花色。遗憾的是，只有一个新品种能够表达父本花朵白色镶边的性状。另外，新品种具有明显的杂种生长优势。

Ten new hybrids will be introduced in this cross-combination.

'Dragon Fireball' is a cultivar which was obtained through the cross-combination of *C. japonica* cultivar 'Marry Christmas' × *C. lutchuensis* by Mr. John Wang who is a Chinese American. Its flowers are dark red with white borders on petals and peony in form.

This combination tried to obtain new hybrids not only express the main characteristics of *C. azalea* such as ever-blooming and dark green leaves, but also express the traits of 'Dragon Fireball' such as the black red and waxy color. Unfortunately, only one new hybrid can express white border characteristics of male parent. In addition, the hybrids have an obvious heterosis in growth.

HA-67-1. 吉祥如意
Jixiang Ruyi (Good Lucky as Desired)

杂交组合：杜鹃红山茶 × 非云南山茶杂交种'龙火珠'（ *C. azalea* × Non-*reticulata* hybrid 'Dragon Fireball'）。

杂交苗编号：DLHZ-No.1。

性状：花朵深红色，花芯偶有白色斑点，托桂型，中型花，花径 8.0～9.0cm，花朵厚 5.2cm，外部大花瓣 8～10 枚，呈 1～2 轮排列，阔倒卵形，中部小花瓣和雄蕊瓣簇拥成球，雄蕊簇生。叶片浓绿色，叶背面灰绿色，革质，椭圆形，长 8.5～9.5cm，宽 4.2～4.8cm，边缘叶齿钝稀。植株立性，枝叶稠密，生长旺盛。夏末始花，秋、冬季盛花，春季偶然开花。

Abstract: Flowers deep red, occasionally with small white dots at the center, anemone form, medium size, 8.0-9.0cm across and 5.2cm deep, 8-10 exterior large petals arranged in 1-2 rows, broad obovate, central small petals and petaloids crowed into a ball. Leaves dark green, leathery, elliptic, margins obtusely and thinly serrate. Plant upright, branches dense and growth vigorous. Starts to bloom from late-summer, fully blooms from autumn to winter and occasionally in spring.

HA-67-2. 午夜金光
Wuye Jinguang（Midnight Golden Light）

杂交组合： 杜鹃红山茶 × 非云南山茶杂交种'龙火珠'（ C. azalea × Non-reticulata hybrid 'Dragon Fireball'）。

杂交苗编号： DLHZ-No.2。

性状： 花朵深红色至黑红色，具蜡光，偶见白色斑块，半重瓣型至托桂型，偶有牡丹型，颇像品种'午夜魔幻'，中型花，花径 8.5～10.0cm，花朵厚 5.0cm，外部大花瓣 9～12 枚，呈 2 轮排列，阔倒卵形，瓣长 4.0cm，中部少量小花瓣扭曲，雄蕊较短，花丝红色，花药金黄色，开花极稠密。叶片浓绿色，革质，椭圆形，长 8.5～9.0cm，宽 4.0～4.5cm，边缘叶齿钝。植株紧凑，枝叶稠密，生长旺盛。夏初始花，秋、冬季盛花，春季零星开花。

Abstract: Flowers deep red to dark red, with waxy luster, occasionally with some white markings, semi-double to anemone form, occasionally peony form, quite like *C. japonica* cultivar 'Midnight Magic', medium size, 8.5-10.0cm across and 5.0cm deep, 9-12 exterior large petals arranged in 2 rows, broad obovate, small petals and petaloids twisted, stamens short, filaments red, anthers golden yellow, bloom very dense. Leaves dark green, leathery, elliptic, margins obtusely serrate. Plant compact, branches dense and growth vigorous. Starts to bloom from early-summer, fully blooms from autumn to winter and sporadically in spring.

HA-67-3. 知足常乐
Zhizu Changle (Contentment Being Happiness)

杂交组合： 杜鹃红山茶 × 非云南山茶杂交种'龙火珠'(C. azalea × Non-reticulata hybrid 'Dragon Fireball')。

杂交苗编号： DLHZ-No.3。

性状： 花朵红色，单瓣型，中型花，花径8.0～9.5cm，花瓣5～6枚，扭曲，上部外翻，呈玉兰花状，似开口笑，阔倒卵形，雄蕊柱状，基部连生，开花稠密。叶片浓绿色，叶背面灰绿色，革质，阔椭圆形，长8.0～10.0cm，宽4.5～5.0cm，上部边缘叶齿稀。植株立性，枝叶稠密，生长旺盛。夏末始花，秋、冬季盛花，春季零星开花。

> **Abstract:** Flowers red, single form, medium size, 8.0-9.0cm across with 5-6 petals, petals are twisted and roll outward, whose shape is like a magnolia flower and a smile mouth, stamens columnar, united at the base. Leaves dark green, leathery, broad elliptic, margins sparsely serrate at the upper part. Plant upright, branches dense and growth vigorous. Starts to bloom from late-summer, fully blooms from autumn to winter and sporadically in spring.

HA-67-4. 镶边彩扣 Xiangbian Caikou (Frilly Colorful Button)

杂交组合：杜鹃红山茶 × 非云南山茶杂交种'龙火珠'（ *C. azalea* × Non-*reticulata* hybrid 'Dragon Fireball'）。

杂交苗编号：DLHZ-No.4。

性状：花朵红色至黑红色，花瓣边缘具白色窄边，瓣面可见深红色脉纹，单瓣型，微型花，花径5.5～6.0cm，花瓣5～6枚，阔倒卵形，雄蕊短柱状，基部连生。叶片浓绿色，叶背面灰绿色，革质，椭圆形，长9.0～9.5cm，宽4.5～5.0cm，侧脉凸起，叶边缘光滑。植株立性，枝叶稠密，生长旺盛。夏末始花，秋、冬季盛花，春季零星开花。

Abstract: Flowers red to dark red with narrowly white borders at petal edges and deep red veins visible on the petal surfaces, single form, miniature size, 5.5-6.0cm across with 5-6 petals, petals broad obovate, stamens short column-shaped. Leaves dark green, leathery, elliptic, lateral veins raised, margins smooth. Plant upright, branches dense and growth vigorous. Starts to bloom from late-summer, fully blooms from autumn to winter and sporadically in spring.

HA-67-5. 烈焰红唇
Lieyan Hongchun (Flaming Lips)

杂交组合： 杜鹃红山茶 × 非云南山茶杂交种'龙火珠'（ *C. azalea* × Non-*reticulata* hybrid 'Dragon Fireball'）。

杂交苗编号： DLHZ-No.5。

性状： 花朵红色，偶有白斑，托桂型，中型花，花径 8.0～9.5cm，花朵厚 4.8cm，外部大花瓣 8～13 枚，呈 1～2 轮排列，阔倒卵形，外翻，中部小花瓣和雄蕊瓣簇拥成球。叶片浓绿色，叶背面灰绿色，硬革质，椭圆形，长 8.0～9.0cm，宽 3.0～4.0cm，边缘叶齿稀。植株立性，枝叶稠密，生长旺盛。夏季始花，秋、冬季盛花，春季零星开花。

> **Abstract:** Flowers red, occasionally with white markings, anemone form, medium size, 8.0-9.5cm across and 4.8cm deep, 8-13 exterior large petals arranged in 1-2 rows, broad obovate, turned outward, central small petals and petaloids crowed into a ball. Leaves dark gree, heavy leathery, elliptic, margins sparsely serrate. Plant upright, branches dense and growth vigorous. Starts to bloom from summer, fully blooms from autumn to winter and sporadically in spring.

HA-67-6. 龙火贺喜
Longhuo Hexi（Dragon-fire Congratulations）

杂交组合： 杜鹃红山茶 × 非云南山茶杂交种'龙火珠'（*C. azalea* × Non-*reticulata* hybrid 'Dragon Fireball'）。

杂交苗编号： DLHZ-No.7。

性状： 花朵红色至黑红色，单瓣型，中型花，花径8.0～8.5cm，花瓣6枚，阔倒卵形，厚质，雄蕊基部略连生，花丝红色。叶片浓绿色，叶背面灰绿色，革质，椭圆形，长9.0～9.5cm，宽4.0～4.5cm，边缘叶齿浅稀。植株立性，生长旺盛。夏季始花，秋、冬季盛花，春季零星开花。

> **Abstract:** Flowers red to dark red, single form, medium size, 8.0-8.5cm across with 6 petals, petals broad obovate, thick texture, stamens slightly united at the base, filaments red. Leaves dark green, leathery, elliptic, margins shallowly and sparsely serrate. Plant upright and growth vigorous. Starts to bloom from summer, fully blooms from autumn to winter and sporadically in spring.

HA-67-7. 莺歌燕舞
Yingge Yanwu (Orioles Sing & Swallows Dancing)

杂交组合：杜鹃红山茶 × 非云南山茶杂交种'龙火珠'（ *C. azalea* × Non-*reticulata* hybrid 'Dragon Fireball'）。

杂交苗编号：DLHZ-No.10。

性状：花朵粉红色，略泛橘色，瓣面偶有隐约的白斑，单瓣型，中型花，花径8.5～9.5cm，大花瓣6～7枚，外翻或波浪状，阔倒卵形，雄蕊基部连生，呈短柱状，开花稠密。叶片浓绿色，叶背面灰绿色，革质，椭圆形，长8.5～9.0cm，宽3.5～4.0cm，边缘叶齿钝，基部楔形。植株立性，生长旺盛。夏末始花，秋、冬季盛花，春季零星开花。

Abstract: Flowers pink with orange hue, occasionally with hazy white markings, single form, medium size, 8.5-9.5cm across with 6-7 large petals which are broad obovate, rolled outward or wavy, stamens united into a short column at the base, bloom densely. Leaves dark green, leathery, elliptic, margins obtusely serrate, cuneate at the base. Plant upright and growth vigorous. Starts to bloom from late-summer, fully blooms from autumn to winter and sporadically in spring.

HA-67-8. 红装伊人
Hongzhuang Yiren (Beloved Lady Dressed in Red)

杂交组合：杜鹃红山茶 × 非云南山茶杂交种'龙火珠'（*C. azalea* × Non-*reticulata* hybrid 'Dragon Fireball'）。

杂交苗编号：DLHZ-No.11。

性状：花朵艳红色，玫瑰重瓣型至牡丹型，中型至大型花，花径 9.0～11.5cm，花朵厚 5.0cm，花瓣 18～24 枚，呈 2～4 轮排列，阔倒卵形，雄蕊簇生，与中部小花瓣混生。叶片浓绿色，叶背面灰绿色，革质，阔椭圆形，长 8.0～9.0cm，宽 4.0～5.0cm，略扭曲，边缘叶齿浅。植株立性，枝叶稠密，生长旺盛。夏末始花，秋、冬季盛花，春季零星开花。

Abstract: Flowers bright red, rose-double to peony form, medium to large size, 9.0-11.5cm across and 5.0cm deep with 18-24 large petals arranged in 2-4 rows, petals broad obovate, stamens clustered and mixed with central small petals. Leaves dark green, leathery, broad elliptic, slightly twisted, margins shallowly serrate. Plant upright, branches dense and growth vigorous. Starts to bloom from late-summer, fully blooms from autumn to winter and sporadically in spring.

HA-67-9. 夏令趣事
Xialing Qushi（Interesting Stories of Summer Camp）

杂交组合： 杜鹃红山茶 × 非云南山茶杂交种'龙火珠'（*C. azalea* × Non-*reticulata* hybrid 'Dragon Fireball'）。

杂交苗编号： DLHZ-No.13。

性状： 花朵红色，偶有小白斑，半重瓣型至牡丹型，中型至大型花，花径9.5～12.0cm，花朵厚5.3cm，花瓣18～21枚，呈3～4轮排列，阔倒卵形，中部雄蕊瓣扭曲，与少量雄蕊混生。叶片浓绿色，叶背面灰绿色，厚革质，阔椭圆形，长8.5～9.0cm，宽5.0～5.5cm，中脉明显，侧脉模糊，边缘叶齿稀浅。植株立性，枝叶稠密，生长旺盛。夏末始花，秋、冬季盛花，春季零星开花。

Abstract: Flowers red, occasionally with tiny white markings, semi-double to peony form, medium to large size, 9.5-12.0cm across and 5.3cm deep, 18-21 large petals arranged in 3-4 rows, broad obovate, central petaloids twisted and mixed with a few of stamens. Leaves dark green, heavy leathery, broad elliptic, midrib visible, lateral veins blurred, margins sparsely and shallowly serrate. Plant upright, branches dense and growth vigorous. Starts to bloom from late-summer, fully blooms from autumn to winter and sporadically in spring.

HA-67-10. 生日宴会
Shengri Yanhui（Birthday Party）

杂交组合：杜鹃红山茶 × 非云南山茶杂交种'龙火珠'（*C. azalea* × Non-*reticulata* hybrid 'Dragon Fireball'）。

杂交苗编号：DLHZ-No.14 + 15。

性状：花朵黑红色，泛蜡光，雄蕊瓣具白斑，半重瓣型，中型至大型花，花径9.5～12.5cm，花朵厚5.0cm，外部大花瓣17～18枚，呈3轮紧实排列，花瓣阔倒卵形，厚质，外翻，花芯偶有小花瓣和雄蕊瓣，雄蕊辐射状，开花稠密。叶片浓绿色，叶背面灰绿色，革质，椭圆形，长9.5～11.0cm，宽5.0～5.5cm，先端下弯，边缘叶齿细。植株立性，生长旺盛。夏末始花，秋、冬季盛花，春季零星开花。

Abstract: Flowers dark red with waxy luster, petaloids with white markings, semi-double form, medium to large size, 9.5-12.5cm across and 5.0cm deep with 17-18 exterior large petals arranged tightly in 3 rows, petals broad obovate, thick texture, rolled outward, occasionally some small petals and petaloids appear in the center, stamens radial, bloom dense. Leaves dark green, leathery, elliptic, tips bend down, margins thinly serrate. Plant upright and growth vigorous. Starts to bloom from late-summer, fully blooms from autumn to winter and sporadically in spring.

杂交组合 HA-68. 非云南山茶杂交种'龙火珠'× 杜鹃红山茶
HA-68. Non-*reticulata* hybrid 'Dragon Fireball' × *C. azalea*

本杂交组合将介绍4个新品种。

这是HA-67杂交组合的反交组合。'龙火珠'品种容易坐果，因此，以它为母本与杜鹃红山茶杂交，试图获得大花朵、花瓣具白边、能重复开花的新品种。可惜的是，所获得的4个杂交植株，除了具有多季开花和叶片浓绿的性状外，大部分开中小型花朵，而且有50%是单瓣类型的。这究竟是为什么？有待进一步研究。

Four new hybrids will be introduced in this cross-combination.

This is a reciprocal cross-combination of HA-67. 'Dragon Fireball' can set up fruits easily, so it was crossed with *C. azalea* as female parent for getting the new hybrids with large flowers, white bordered petals and ever-blooming characteristics. Unfortunately, beside their multi-season flowering and dark green leaves, most of the four hybrid plants obtained open small and medium size flowers, and 50% of them are single form. Why is that? Further research is needed.

HA-68-1. 五福临门
Wufu Linmen（Five Blessings Arriving at Home）

杂交组合： 非云南山茶杂交种'龙火珠'×杜鹃红山茶（Non-*reticulata* hybrid 'Dragon Fireball' × *C. azalea*）。

杂交苗编号： LHZD-No.1。

性状： 花朵艳红色，小花瓣偶有白斑，托桂型至牡丹型，小型至中型花，花径7.0～8.0cm，花朵厚4.0cm，外轮大花瓣9～12枚，呈1～2轮排列，阔倒卵形，先端近圆，中部小花瓣和雄蕊瓣与雄蕊混生，扭曲或波浪状，簇拥成团。叶片浓绿色，叶背面灰绿色，革质，阔椭圆形，长9.5～11.0cm，宽4.0～4.5cm，叶缘齿钝。植株立性，生长旺盛。夏末始花，秋、冬季盛花，春季零星开花。

Abstract: Flowers bright red, occasionally with white markings on the central small petals, anemone to peony form, small to medium size, 7.0-8.0cm across and 4.0cm deep with 9-12 exterior large petals arranged in 1-2 rows, petals broad obovate, tips nearly round, central small petals and petaloids wrinkled or wavy, mixed with stamens into a ball. Leaves dark green, leathery, broad elliptic, margins obtusely serrate. Plant upright and growth vigorous. Starts to bloom in late-summer, fully blooms from autumn to winter and sporadically in spring.

HA-68-2. 瑞气祥云
Ruiqi Xiangyun（Happy Atmosphere & Auspicious Cloud）

杂交组合： 非云南山茶杂交种'龙火珠'×杜鹃红山茶（Non-*reticulata* hybrid 'Dragon Fireball' × *C. azalea*）。

杂交苗编号： LHZD-No.2。

性状： 花朵黑红色，具蜡质感，中部小花瓣上有白色或粉红色条纹，半重瓣型至托桂型，中型至大型花，花径9.8～11.5cm，花朵厚5.0cm，大花瓣6枚，瓣面可见黑红色脉纹，倒卵形，中部雄蕊瓣呈辐射状，花丝红色。叶片浓绿色，叶背面灰绿色，革质，阔椭圆形，长10.5～11.5cm，宽5.0～5.1cm，叶缘齿钝。植株立性，生长旺盛。夏末始花，秋、冬季盛花，春季零星开花。

Abstract: Flowers dark red with waxy luster, some white or pink stripes on the central small petals, semi-double to anemone form, medium to large size, 9.8-11.5cm across and 5.0cm deep with 6 large petals, petals obovate, central petaloids radial, filaments red. Leaves dark green, leathery, broad elliptic, margins obtusely serrate. Plant upright and growth vigorous. Starts to bloom in late-summer, fully blooms from autumn to winter and sporadically in spring.

HA-68-3. 夏日粉妹
Xiari Fenmei（Summer's Pink Sister）

杂交组合：非云南山茶杂交种'龙火珠'×杜鹃红山茶（Non-*reticulata* hybrid 'Dragon Fireball' × *C. azalea*）。

杂交苗编号：LHZD-No.3。

性状：花朵深粉红色，单瓣型，中型花，花径9.5～10.0cm，花瓣5～6枚，近圆形，厚质，雄蕊基部连生呈柱状，雌蕊长3.0cm，开花稠密。叶片浓绿色，叶背面灰绿色，革质，椭圆形，长7.5～8.5cm，宽3.0～4.0cm，基部楔形，叶缘齿浅。植株立性，生长旺盛。夏末始花，秋、冬季盛花，春季零星开花。

Abstract: Flowers deep pink, single form, medium size, 9.5-10.0cm across with 5-6 petals, petals nearly round, thick texture, stamens united into a column at the base, pistils 3.0cm long, bloom dense. Leaves dark green, leathery, elliptic, cuneate at the base, margins shallowly serrate. Plant upright and growth vigorous. Starts to bloom in late-summer, fully blooms from autumn to winter and sporadically in spring.

HA-68-4. 夏日红妹
Xiari Hongmei (Summer's Red Sister)

杂交组合： 非云南山茶杂交种'龙火珠'×杜鹃红山茶（Non-*reticulata* hybrid 'Dragon Fireball' × *C. azalea*）。

杂交苗编号： LHZD-No.4。

性状： 花朵艳红色，略显蜡光，单瓣型，中型花，花径8.5～9.5cm，花瓣6枚，阔倒卵形，略波浪状，雄蕊基部连生呈柱状，开花稠密。叶片浓绿色，叶背面灰绿色，革质，阔椭圆形，长9.5～10.5cm，宽4.0～4.5cm，叶缘齿浅。植株立性，生长旺盛。夏末始花，秋、冬季盛花，春季零星开花。

Abstract: Flowers bright red, slightly with waxy luster, single form, medium size, 8.5-9.5cm across with 6 petals, petals broad obovate, slightly wavy, stamens united into a column at the base, bloom dense. Leaves dark green, leathery, broad elliptic, margins shallowly serrate. Plant upright and growth vigorous. Starts to bloom in late-summer, fully blooms from autumn to winter and sporadically in spring.

⦿ 杂交组合 HA-69. 杜鹃红山茶 × 红山茶品种'火把'
HA-69. *C. azalea* × *C. japonica* cultivar 'Firebrand'

×

本杂交组合将介绍 1 个新品种。

红山茶品种'火把',具有艳红色、半重瓣型、大型花的特点。我们试图通过杂交把该品种改造为不仅具有'火把'品种的性状,而且能够四季开花的新品种。

获得的杂交新品种具有桃红色、半重瓣型、中型至大型花的性状,而且也能四季开花,基本符合我们的育种目标。

One new hybrid will be introduced in this cross-combination.

C. japonica cultivar 'Firebrand' has the traits of scarlet red, semi-double form and large sized flowers. We tried to reinvent it into a new hybrid which not only has the characteristics of 'Firebrand' cultivar, but also can bloom year-round by crossing.

The flowers of the new hybrid obtained are peach red, semi-double to rose-double form, medium to large size and open in four seasons, so our breeding target has been reached.

HA-69-1. 西子晚霞
Xizi Wanxia (West Lake's Sunset Glow)

杂交组合： 杜鹃红山茶 × 红山茶品种'火把'(*C. azalea* × *C. japonica* cultivar 'Firebrand')。

杂交苗编号： DHB-No.1。

性状： 花朵桃红色至深桃红色，偶有白斑，半重瓣型，中型至大型花，花径8.5～11.5cm，花朵厚4.5cm，花瓣15～18枚，呈3轮有序排列，花瓣阔倒卵形，厚质，外翻，雄蕊少，花朵整朵掉落。叶片浓绿色，叶背面灰绿色，厚革质，长椭圆形，长8.5～9.5cm，宽4.5～5.0cm，基部楔形，上部边缘叶齿浅而稀。植株立性，生长旺盛。夏末始花，秋、冬季盛花，春季零星开花。

Abstract: Flowers peach red to deep peach red, occasionally with white markings, semi-double form, medium to large size, 8.5-11.5cm across and 4.5cm deep with 15-18 petals arranged orderly in 3 rows, petals broad obovate, thick texture, rolled outward, stamens a few, flowers fall whole. Leaves dark green, heavy leathery, long elliptic, cuneate at the base, margins shallowly and thinly serrate at the upper part. Plant upright and growth vigorous. Starts to bloom from late-summer, fully blooms from autumn to winter and sporadically in spring.

杂交组合 HA-70. 杜鹃红山茶 × 红山茶品种'艳口红'
HA-70. *C. azalea* × *C. japonica* cultivar 'Lipstick'

本杂交组合将介绍 2 个新品种。

红山茶品种'艳口红'的花朵浓红色、托桂型、微型花。我们试图通过杂交获得更漂亮、四季开花的好品种。

获得的两个杂交种，其花色、花型和花径都是两个亲本的中间型，而且花期从夏季一直延续到冬末，这是杜鹃红山茶性状表达的结果。

Two new hybrids will be introduced in this cross-combination.

The flowers of *C. japonica* cultivar 'Lipstick' are dark red, anemone form and miniature size. We tried to get better hybrids which are more beautiful and bloom year-round.

The two hybrids obtained are intermediates between the two parents on flower colour, flower form and flower size, and their flowering can last from summer to the end of winter, which is the result of characteristics expression of *C. azalea*.

HA-70-1. 盈盈笑口
Yingying Xiaokou (Smiling Mouths)

杂交组合： 杜鹃红山茶 × 红山茶品种'艳口红'（ *C. azalea* × *C. japonica* cultivar 'Lipstick'）。

杂交苗编号： DYKH-No.1。

性状： 花朵亮红色，单瓣型，微型花，花径 3.0～4.0cm，花瓣5枚，多为半开放状态，呈扁嘴状，雄蕊基部连生呈圆柱状，雄蕊瓣很小，黄白色。叶片绿色，叶背面灰绿色，革质，椭圆形，长 8.0～8.5cm，宽 3.0～3.5cm，边缘上部齿浅而密。植株立性，生长旺盛。夏末始花，秋、冬季盛花，春季零星开花。

Abstract: Flowers bright red, single form, miniature size, 3.0-4.0cm across with 5 petals that are semi-opened as flat mouths, stamens united into a column at the base, petaloids yellow and white. Leaves green, leathery, elliptic, margins shallowly and densely serrate at the top part. Plant upright and growth vigorous. Starts to bloom from late-summer, fully blooms from autumn to winter and sporadically in spring.

HA-70-2. 旭日东升
Xuri Dongsheng (The Sun Rising from East)

杂交组合： 杜鹃红山茶 × 红山茶品种'艳口红'（ *C. azalea* × *C. japonica* cultivar 'Lipstick'）。

杂交苗编号： DYKH-No.2。

性状： 花朵大红色，雄蕊瓣偶现白斑，半重瓣型，大型至巨型花，花径12.0～14.0cm，花朵厚5.0cm，花瓣18～21枚，呈3～4轮排列，阔倒卵形，厚质，外翻，雄蕊似肥后茶的辐射状，花丝略泛红色，花药金黄色，花朵整朵掉落。叶片浓绿色，叶背面灰绿色，厚革质，阔椭圆形，长9.5～10.0cm，宽4.5～5.0cm，边缘上部叶齿深尖。植株立性，生长旺盛。夏末始花，秋、冬季盛花，春季零星开花。

Abstract: Flowers true red, petaloids occasionally with a few white patches, semi-double form, large to very large size, 12.0-14.0cm across and 5.0cm deep with 18-21 petals arranged in 3-4 rows, petals broad obovate, thick texture, rolled outward, stamens radial which is like Higo camellias, filaments slightly reddish, anthers golden yellow, flowers fall whole. Leaves dark green, heavy leathery, broad elliptic, margins deeply and sharply serrate at the upper part. Plant upright and growth vigorous. Starts to bloom from late-summer, fully blooms from autumn to winter and sporadically in spring.

⦿ 杂交组合 HA-71. 红山茶品种'艳口红'× 杜鹃红山茶
HA-71. *C. japonica* cultivar 'Lipstick' × *C. azalea*

本杂交组合将介绍 1 个新品种。

本杂交组合是杂交组合 HA-70 的一个反交组合。红山茶品种'艳口红'的雄蕊虽然大部分瓣化了，但是它仍可以坐果。我们试图通过这组杂交把'艳口红'品种的优良性状添加到杂交种上。

获得的新品种在花型和花径大小方面，遗传了'艳口红'品种的基本性状，而在开花期和叶色方面，则遗传了杜鹃红山茶的性状。

One new hybrid will be introduced in this cross-combination.

This is a reciprocal cross-combination of HA-70. Although most of the stamens are petaloids in *C. japonica* cultivar 'Lipstick', it still can fruit. We tried to add the good characteristics of 'Lipstick' into its hybrid by crossing.

The obtained hybrid has inherited the basic characteristics of 'Lipstick' in flower form and flower size, while the hybrid has inherited the characteristics of *C. azalea* in the flowering period and leaf color.

HA-71-1. 红林瓦寨
Honglin Wazhai (Tile Village in Red Forest)

杂交组合：红山茶品种'艳口红' × 杜鹃红山茶（ *C. japonica* cultivar 'Lipstick' × *C. azalea* ）。

杂交苗编号：YKHD-No.2。

性状：花朵红色至深红色，中部瓣化雄蕊具白色和淡红色斑点，托桂型，小型至中型花，花径6.5～8.0cm，花朵厚3.5cm，大花瓣5～6枚，长卵形，中部雄蕊全部瓣化，花丝红色，开花稠密。叶片浓绿色，叶背面灰绿色，革质，长椭圆形，长9.0～10.0cm，宽4.0～5.0cm，基部急楔形，边缘叶齿浅。植株立性，生长旺盛。夏末始花，秋、冬季盛花，春季零星开花。

Abstract: Flowers red to deep red, central petaloids blotched white and pale red, anemone form, small to medium size, 6.5-8.0cm across and 3.5cm deep with 5-6 large petals, petals long obovate, stamens are all petaloids, bloom dense. Leaves dark green, leathery, long elliptic, urgently cuneate at the base, margins shallowly serrate. Plant upright and growth vigorous. Starts to bloom from late-summer, fully blooms from autumn to winter and sporadically in spring.

杂交组合 HA-72. 杜鹃红山茶 × 非云南山茶杂交种'克拉丽'
HA-72. *C. azalea* × Non-*reticulata* hybrid 'Clarrie Fawcett'

本杂交组合将介绍 1 个新品种。

'克拉丽'的花朵为淡粉红色、半重瓣型、中型花，开花非常稠密。

本组杂交想获得花朵像'克拉丽'一样漂亮，花期像杜鹃红山茶那样能够多季开花的新品种。

本杂交组合获得的该新品种，具有杜鹃红山茶和怒江红山茶的基因，因此，它开花期更长，花朵更漂亮，而且抗性也大幅度提高了。

One new hybrid will be introduced in this cross-combination.

The flowers of 'Clarrie Fawcett' are light pink, semi-double form, medium size and very dense.

The aim of the cross-combination is to get new hybrids that flowers are as beautiful as 'Clarrie Fawcett' and the flowering period is ever-blooming like *C. azalea*.

This new hybrid obtained from the cross-combination has the genes of both *C. azalea* and *C. saluenensis*, therefore, it has a longer flowering period, more beautiful flowers and greatly improved resistance.

HA-72-1. 多季玲珑
Duoji Linglong（Multiseasonal Exquisiteness）

杂交组合： 杜鹃红山茶 × 非云南山茶杂交种'克拉丽'（*C. azalea* × Non-*reticulata* hybrid 'Clarrie Fawcett'）。

杂交苗编号： DKLL-No.1。

性状： 花朵淡红色至红色，瓣面可见少量白条纹，花朵接近凋萎前，变为粉红色，玫瑰重瓣型，小型至中型花，花径7.0～9.5cm，大花瓣18～20枚，呈3～4轮排列，花芯雄蕊少，偶有雄蕊瓣。叶片浓绿色，叶背面灰绿色，革质，椭圆形，长10.0～11.0cm，宽4.5～5.0cm，边缘上半部叶齿明显，下半部光滑。植株半开张，枝叶稠密，圆头形，生长旺盛。夏末始花，秋、冬季盛花，春季零星开花。

Abstract: Flowers light red to red, white stripes visible on the surfaces, before the flowers wither, their color changing into pink, rose-double form, small to medium size, 7.0-9.5cm across with 18-20 large petals arranged in 3-4 rows, a few stamens, occasionally some petaloids appear in the center. Leaves dark green, leathery, elliptic, margins obviously serrate at the upper part and the lower margins smooth. Plant semi-spread, branches dense and growth vigorous. Starts to bloom from late-summer, fully blooms from autumn to winter and sporadically in spring.

◉ 杂交组合 HA-73. 杜鹃红山茶 × 红山茶品种'红绒贝蒂'
HA-73. *C. azalea* × *C. japonica* cultivar 'Betty Foy Sanders'

本杂交组合将介绍1个新品种。

红山茶品种'红绒贝蒂'是引自美国的一个花朵白色，具放射状、细长红色条纹和斑点，半重瓣型，开花稠密的品种。本组杂交期望获得既有父本花朵性状，又有母本花期的新品种。

可惜的是，我们所获得的这个杂交种花朵为深红色，花瓣上也没有出现放射状的条纹，但是能多季开花。

One new hybrid will be introduced in this cross-combination.

C. Japonica cultivar, 'Betty Foy Sanders' is an American cultivar that flowers are white with radial and elongated red splotches, semi-double form and dense blooming. The new hybrids obtained in this cross-combination were expected to have the flower characteristics of the male parent and the flowering period of the female parent.

Unfortunately, the flowers of the hybrid obtained are deep red and do not show any radial stripes on its petals, but it blooms in multi-seasons.

HA-73-1. 加冕红毯
Jiamian Hongtan (The Red Carpet for Crowning)

杂交组合： 杜鹃红山茶 × 红山茶品种'红绒贝蒂'（*C. azalea* × *C. japonica* cultivar 'Betty Foy Sanders'）。

杂交苗编号： DHRBD-No.2。

性状： 花朵暗红色，略有绒光，偶具少量白条纹，半重瓣型，中型花，花径 8.5～10.0cm，花朵厚 5.5cm，大花瓣 15～18 枚，呈 3 轮排列，花瓣阔倒卵形，雄蕊基部连生，花朵整朵掉落。叶片浓绿色，叶背面灰绿色，厚革质，椭圆形，长 9.5～10.5cm，宽 4.0～4.5cm，基部急楔形，边缘叶齿钝。植株立性，生长旺盛。夏末始花，秋、冬季盛花，春季偶尔开花。

> **Abstract:** Flowers dark red, slightly with velvet luster, occasionally with a few white stripes, semi-double form, medium size, 8.5-10.0cm across and 5.5cm deep with 15-18 large petals arranged in 3 rows, petals broad obovate, stamens united at the base, flowers fall whole. Leaves dark green, heavy leathery, elliptic, urgently cuneate at the base, margins obtusely serrate. Plant upright, growth vigorous. Starts to bloom from late-summer, fully blooms from autumn to winter and occasionally in spring.

⦿ 杂交组合 HA-74. 杜鹃红山茶 × 红山茶品种'锦鱼叶椿'
HA-74. *C. azalea* × *C. japonica* cultivar 'Kingyo-tsubaki'

本杂交组合将介绍1个新品种。

红山茶品种'锦鱼叶椿'源自日本，其为花朵红色、单瓣型，叶片先端鱼尾状的奇特茶花品种。本组杂交是为了获得能四季开花，而且叶片呈鱼尾状的新品种。

获得的这个新品种，其花朵为红色至深红色，半重瓣型，中型花，绝大部分叶片是鱼尾状的，而且能多季开花。

One new hybrid will be introduced in this cross-combination.

C. japonica cultivar 'Kingyo-tsubaki' is originated from Japan. It is an unique cultivar that flowers are red, single form and leaves are fishtail at the apices. The cross-combination was designed to obtain new hybrids that can bloom every season and have fishtail shaped leaves.

The flowers of the obtained new hybrid are red to deep red, semi-double form, and medium size. Most of its leaves are fishtail, and it can bloom in multi-seasons.

HA-74-1. 年年有余
Niannian Youyu (Having Surplus Every Year)

杂交组合： 杜鹃红山茶 × 红山茶品种'锦鱼叶椿'（ *C. azalea* × *C. japonica* cultivar 'Kingyo-tsubaki'）。

杂交苗编号： DHJY-No.1。

性状： 花朵红色至深红色，偶有少量白点，半重瓣型，中型花，花径 7.5～10.0cm，花朵厚 4.5cm，花瓣 20～23 枚，呈 4 轮排列，花瓣阔倒卵形，厚质，中部花瓣瓣面略皱褶，雄蕊少量，花丝淡红色，花药黄色，开花稠密，花瓣逐片掉落。叶片浓绿色，叶背面灰绿色，厚革质，长椭圆形，长 8.5～9.5cm，宽 3.5～4.0cm，上部边缘叶齿浅，先端鱼尾状，基部楔形。植株立性，生长旺盛。夏末始花，秋、冬季盛花，春季零星开花。

Abstract: Flowers red to deep red, occasionally with little white spots, semi-double form, medium size, 7.5-10.0cm across and 4.5cm deep with 20-23 petals arranged in 4 rows, petals broad obovate, thick texture, central petals slightly wrinkled at the surfaces, a few stamens, filaments pale red, anthers yellow, bloom dense, petals fall one by one. Leaves dark green, heavy leathery, long elliptic, margins shallowly serrate at the upper part, apices fishtail in shape. Plant upright and growth vigorous. Starts to bloom from late-summer, fully blooms from autumn to winter and sporadically in spring.

◉ 杂交组合 HA-75. 杜鹃红山茶 × 红山茶品种'迪士尼乐园'
HA-75. *C. azalea* × *C. japonica* cultivar 'Disneyland'

本杂交组合将介绍 1 个新品种。

红山茶品种'迪士尼乐园'系美国品种，其花朵为红色至黑红色，半重瓣型。

本组杂交期望获得花朵黑红色、开花稠密、能够四季开花的新品种。我们的育种目标基本达到了。

One new hybrid will be introduced in this cross-combination.

C. Japonica cultivar, 'Disneyland' was imported from USA and its flowers are red to dark red and semi-double form.

The cross-combination was expected to obtain the new hybrids which can open dark red flowers, bloom densely and year round. Our breeding goal is basically achieved.

HA-75-1. 赤红花海
Chihong Huahai (Crimson Flower Sea)

杂交组合： 杜鹃红山茶 × 红山茶品种'迪士尼乐园'（*C. azalea* × *C. japonica* cultivar 'Disneyland'）。

杂交苗编号： DDL-No.6。

性状： 花朵红色至黑红色，泛紫色调，具蜡光，单瓣型，大型花，花径12.0～12.8cm，花瓣6～7枚，阔倒卵形，上部外翻，大波浪状，瓣面可见清晰的纵向脉纹，边缘略皱褶，雄蕊呈柱状，开花极稠密，花朵整朵掉落。叶片浓绿色，叶背面灰绿色，厚革质，倒卵形，长8.5～9.5cm，宽4.5～5.0cm，上部边缘具稀疏锯齿。植株立性，枝叶稠密，生长旺盛。夏末始花，秋、冬季盛花，春季零星开花。

Abstract: Flowers red to dark red, with purple hue and with some waxy luster, single form, large size, 12.0-12.8cm across with 6-7 petals, petals broad obovate, rolled outward at the upper part, big wavy, the surfaces clearly longitudinally veined, slightly ruffled at the margins, stamens columnar, bloom very dense, flowers fall whole. Leaves dark green, heavy leathery, obovate, margins sparsely serrate at the upper part. Plant upright, branches dense and growth vigorous. Starts to bloom from late-summer, fully blooms from autumn to winter and sporadically in spring.

⦿ 杂交组合 HA-76. 杜鹃红山茶 × 红山茶品种'夸特'
HA-76. *C. azalea* × *C. japonica* cultivar 'Bill Quattlebaum'

本杂交组合将介绍1个新品种。

红山茶品种'夸特'是一个花朵酒红色至深红色、托桂型、大花型、花瓣偶现白斑的漂亮品种。此次杂交试图把该品种的优良性状添加到新品种上。

本组仅获得了一个花朵深粉红色至红色、偶具模糊白斑、单瓣型的新品种，可喜的是，该杂交新品种开花稠密，而且夏秋始花，叶片浓绿，生长旺盛，遗传了其母本杜鹃红山茶的性状。

One new hybrid will be introduced in this cross-combination.

C. japonica cultivar 'Bill Quattlebaum' is a beautiful cultivar that flowers are wine red to deep red, anemone form, large size, and occasionally with white patches. The cross-combination was expected to add its good characteristics into the new hybrids.

The only one new hybrid obtained from this cross-combination is deep pink to red in color, occasionally with faint white patches and single in form. Fortunately, the new hybrid blooms densely and begins to bloom in summer and autumn with dark green leaves and vigorous growth, which inherit the characteristics of its female parent *C. azalea*.

HA-76-1. 美秀映照
Meixiu Yingzhao (The Beauty from Reflecting)

杂交组合：杜鹃红山茶×红山茶品种'夸特'（*C. azalea* × *C. japonica* cultivar 'Bill Quattlebaum'）。

杂交苗编号：DKT-No.1。

性状：花朵深粉红色至红色，偶有模糊的白斑，单瓣型，小型至中型花，花径7.0～8.0cm，花瓣6～7枚，倒卵形，外翻，雄蕊基部连生，呈柱状，花丝淡红色，花药黄色，开花稠密。叶片浓绿色，叶背面灰绿色，厚革质，椭圆形，长8.0～9.0cm，宽3.0～3.5cm，边缘上部叶齿尖，基部楔形。植株立性，生长旺盛。夏末始花，秋、冬季盛花，春季零星开花。

Abstract: Flowers deep pink to red, occasionally with faint white patches, single form, small to medium size, 7.0-8.0cm across with 6-7 petals, petals obovate, rolled outward, stamens united into a column at the base, filaments pale red, anthers yellow, bloom dense. Leaves dark green, heavy leathery, elliptic, margins sharply serrate at the upper part. Plant upright and growth vigorous. Starts to bloom from late-summer, fully blooms from autumn to winter and sporadically in spring.

◉ 杂交组合 HA-77. 杜鹃红山茶 × 非云南山茶杂交种'香四射'
HA-77. *C. azalea* × Non-*reticulata* hybrid 'Xiangsishe'

本杂交组合将介绍3个新品种。

'香四射'是我们在2016年公布的一个具芳香的非云南山茶杂交种。它是通过红山茶品种'白斑康乃馨'×非云南山茶杂交种'香漩涡'这一杂交组合获得的（见第一部书第513～514页）。

我们试图通过这次杂交获得花朵具芳香，且能四季开花的好品种。本杂交组合基本达到了我们的育种目标。

Three new hybrids will be introduced in this cross-combination.

The variety 'Xiangsishe' which is a Non-*reticulata* hybrid with fragrant characteristics was published by us in 2016. It was obtained from the cross-combination *C. japonica* cultivar 'Ville de Nantes' × Non-*reticulata* hybrid 'Scented Swirl' (See p.513-514, our first book).

We tried to obtain super new hybrids with fragrant and could bloom year-round. The cross-combination basically achieved our breeding goals.

HA-77-1. 良口红秋
Liangkou Hongqiu (Liangkou Town's Red Autumn)

杂交组合： 杜鹃红山茶 × 非云南山茶杂交种'香四射'（ *C. azalea* × Non-*reticulata* hybrid 'Xiangsishe'）。

杂交苗编号： DXSS-No.1。

性状： 花朵红色至黑红色，具蜡光和淡香，半重瓣型，中型至大型花，花径 8.5～11.5cm，花朵厚 4.5～5.0cm，花瓣 16～20 枚，呈 3～4 轮排列，阔倒卵形，厚质，雄蕊基部略连生，花丝淡红色，花药金黄色，开花稠密。叶片浓绿色，叶背面灰绿色，革质，阔椭圆形，长 9.5～10.5cm，宽 4.0～4.5cm，上部边缘叶齿浅，基部楔形。植株紧凑，枝叶稠密，生长旺盛。夏末始花，秋、冬季盛花，春季零星开花。

Abstract: Flowers red to dark red with waxy luster and slight-fragrance, semi-double form, medium to large size, 8.5-11.5cm across and 4.5-5.0cm deep with 16-20 petals arranged in 3-4 rows, petals broad obovate, thick texture, stamens slightly united at the base, filaments light red, anthers golden yellow, bloom dense. Leaves dark green, leathery, broad elliptic, margins shallowly serrate at the upper part, cuneate at the base. Plant compact, branches dense and growth vigorous. Starts to bloom from late-summer, fully blooms from autumn to winter and sporadically in spring.

HA-77-2. 红瓣香心
Hongban Xiangxin (Red Petals with Fragrant Heart)

杂交组合： 杜鹃红山茶 × 非云南山茶杂交种'香四射'（ *C. azalea* × Non-*reticulata* hybrid 'Xiangsishe'）。

杂交苗编号： DXSS-No.2。

性状： 花朵红色，微香，单瓣型，中型花，花径8.5～9.5cm，花瓣5～6枚，阔倒卵形，厚质，瓣面略皱，先端外翻，雄蕊基部连生呈筒状，花丝淡红色，花药黄色，开花稠密。叶片浓绿色，叶背面灰绿色，厚革质，椭圆形，长8.5～10.0cm，宽4.0～5.0cm，上部边缘具钝齿，下部边缘光滑。植株立性，枝叶稠密，生长旺盛。夏末始花，秋、冬季盛花，春季零星开花。

Abstract: Flowers red, slightly fragrant, single form, medium size, 8.5-9.5cm across with 5-6 petals, petals broad obovate, thick texture, slightly wrinkled at the surfaces, stamens united into a column at the base, filaments pale red, anthers yellow, bloom dense. Leaves dark green, heavy leathery, elliptic, margins shallowly serrate at the upper part and smooth at the lower part. Plant upright, branches dense and growth vigorous. Starts to bloom from late-summer, fully blooms from autumn to winter and sporadically in spring.

HA-77-3. 赤焰炉火
Chiyan Luhuo (Red Flame Stove Fire)

杂交组合： 杜鹃红山茶 × 非云南山茶杂交种'香四射'（ *C. azalea* × Non-*reticulata* hybrid 'Xiangsishe'）。

杂交苗编号： DXSS-No.3 + 4 + 6。

性状： 花朵火红色，略泛蜡光，微香，单瓣型，中型花，花径 8.0～9.0cm，花瓣 5～6 枚，长倒卵形，厚质，两侧内扣呈沟槽状，先端外翻，雄蕊基部连生呈筒状，花丝淡红色，花药黄色，开花稠密。叶片浓绿色，叶背面灰绿色，厚革质，中部略扭曲，椭圆形，长 10.0～10.5cm，宽 4.5～5.0cm，上部边缘具钝齿，下部边缘光滑。植株立性，枝叶稠密，生长旺盛。夏末始花，秋、冬季盛花，春季零星开花。

Abstract: Flowers flame-red, faintly fragrant, single form, medium size, 8.0-9.0cm across with 5-6 petals, petals long obovate, thick texture, both sides of petal edges curved into a groove shape inward and the top part curved outward, stamens united into a column at the base, filaments pale red, anthers yellow, bloom dense. Leaves dark green, heavy leathery, slightly twisted in the middle, elliptic, margins shallowly serrate at the upper part and smooth at the lower part. Plant upright, branches dense and growth vigorous. Starts to bloom from late-summer, fully blooms from autumn to winter and sporadically in spring.

⊙ 杂交组合 HA-78. 杜鹃红山茶 × 云南山茶品种'山茶之都'
HA-78. *C. azalea* × *C. reticulata* cultivar 'Massee Lane'

本杂交组合将介绍 1 个新品种。

云南山茶品种'山茶之都'是一个非常著名的品种，它的花朵深粉红色、牡丹型、巨大。

本组期望获得花朵巨大、花色艳丽、开花稠密，而且能够四季开花的杂交新品种。从获得的这个新品种看，它完全实现了当初的育种目标。

One new hybrid will be introduced in this cross-combination.

C. reticulata 'Massee Lane' is an outstanding cultivar that flowers are deep pink, peony form, and very large.

We expected to get the new hybrid which should be large flowers, bright color, dense bloom and could bloom year-round. From the new hybrid obtained, it achieved the original breeding objectives perfectly.

HA-78-1. 佛植盈瑞
Fozhi Yingrui (Foshan Botanical Garden's Full Auspiciousness)

杂交组合： 杜鹃红山茶 × 云南山茶品种'山茶之都'（*C. azalea* × *C. reticulata* cultivar 'Massee Lane'）。

杂交苗编号： Fozhi –No.2。

性状： 花朵西瓜红色，向上部花色渐淡，松散的托桂型至牡丹型，中部雄蕊瓣具白条纹，大型至巨型花，花径12.0～14.0cm，花朵厚5.0cm，大花瓣13～18枚，呈2轮排列，雄蕊瓣簇拥呈松散的球状，花丝淡黄色，花药金黄色。叶片浓绿色，叶背面灰绿色，厚革质，倒披针形，基部楔形，长9.5～12.0cm，宽3.4～5.2cm，上部叶缘具稀疏尖齿。植株半开张，生长中等。夏季始花，秋季盛花，冬季零星开花。

注：本品种是由广东省佛山市佛山植物园培育的。

Abstract: Flowers watermelon red, fading to light color at the top part, loose-anemone to peony form, central petaloids with white stripes, large to very large size, 12.0-14.0cm across and 5.0cm deep, 13-18 large petals arranged in 2 rows, the petaloids crowed into a loose ball at the center. Leaves dark green, heavy leathery, oblanceolate, cuneate at the base, margins thinly and sparsely serrate at the upper part. Plant semi-spread and growth normal. Starts to bloom from summer, fully blooms in autumn and sporadically in winter.

Mark: The hybrid is bred by Foshan Botanical Garden, Foshan City, Guangdong Province.

⦿ 杂交组合 HA-79. 红山茶品种'牛西奥转马'× 杜鹃红山茶
HA-79. *C. japonica* cultivar 'Nuccio's Carousel' × *C. azalea*

×

本杂交组合将介绍4个新品种。

'牛西奥转马'是一个非常驰名的红山茶品种，其花朵呈柔和的粉红色，花瓣边缘深粉红色，半重瓣型。

本组杂交试图获得既能开出漂亮的花，也能四季开花的好品种。

获得的这4个新品种，它们的花朵为红色，花瓣宽厚，半重瓣型至牡丹型，叶片浓绿，能够多季开花，证明杂交新品种已经具备两个亲本的主要性状，也说明本杂交组合是十分完美的。

Four new hybrids will be introduced in this cross-combination.

The 'Nuccio's Carousel' is a *C. japonica* cultivar which is very famous and its flowers are soft pink, deeper pink at edges and semi-double form.

The cross attempted to obtain nice new hybrids which could open beautiful flowers and bloom year-round.

The flowers of the four new hybrids obtained are red, petals are broad and thick and flower forms are semi-double to peony, they have dark green leaves and ever-blooming characteristics, these all show that the hybrids have the major characteristics from their two cross parents. It is also shown that this cross-combination is very perfect.

HA-79-1. 和谐家园
Hexie Jiayuan (Harmonious Homeland)

杂交组合： 红山茶品种'牛西奥转马' × 杜鹃红山茶（ *C. japonica* cultivar 'Nuccio's Carousel' × *C. azalea*)。

杂交苗编号： NZMD-No.1。

性状： 花朵桃红色至红色，具隐约的白条纹，中部雄蕊瓣具小白斑，半重瓣型，中型至大型花，花径9.5~12.0cm，花朵厚4.0~4.5cm，花瓣15~18枚，呈2~3轮排列，阔倒卵形，外翻，雄蕊偶然出现少量雄蕊瓣。叶片浓绿色，叶背面灰绿色，革质，阔椭圆形，长8.5~9.5cm，宽4.5~5.5cm，上部叶齿尖而浅。植株紧凑，枝叶稠密，生长旺盛。夏末始花，秋、冬季盛花，春季零星开花。

Abstract: Flowers peach red to red with hazy white stripes, central petaloids with small white dots, semi-double form, medium to large size, 9.5-12.0cm across and 4.0-4.5cm deep with 15-18 petals arranged in 2-3 rows, petals broad obovate, rolled outward, stamens occasionally with some petaloids. Leaves dark green, leathery, broad elliptic, margins sparsely and shallowly serrate at the upper part. Plant compact, branches dense and growth vigorous. Starts to bloom from late-summer, fully blooms from autumn to winter and sporadically in spring.

HA-79-2. 金艳靓影
Jinyan Liangying (Jinyan's Pretty Shadow)

杂交组合：红山茶品种'牛西奥转马'×杜鹃红山茶（*C. japonica* cultivar 'Nuccio's Carousel'× *C. azalea*)。

杂交苗编号：NZMD-No.2。

性状：花朵深桃红色，玫瑰重瓣型，大型花，花径10.0～12.5cm，花朵厚5.0cm，花瓣45枚以上，呈4～5轮有序排列，完全开放后，少量短雄蕊显露，花丝淡红色。叶片浓绿色，叶背面灰绿色，厚革质，阔椭圆形，龟背状，长9.0～10.0cm，宽5.0～5.5cm，上部边缘叶齿尖。植株立性，枝叶稠密，生长旺盛。夏末始花，秋、冬季盛花，春季零星开花。

Abstract: Flowers deep peach red, rose-double form, large size, 10.0-12.5cm across and 5.0cm deep with over 45 petals arranged orderly in 4-5 rows, when a flower fully opens, a few of short stamens appears, filaments pale red. Leaves dark green, heavy leathery, broad elliptic which are shaped as tortoiseshell, margins sharply serrate at the upper part. Plant upright, branches dense and growth vigorous. Starts to bloom from late-summer, fully blooms from autumn to winter and sporadically in spring.

HA-79-3. 红色勋章
Hongse Xunzhang（Red Medal）

杂交组合： 红山茶品种'牛西奥转马'× 杜鹃红山茶（*C. japonica* cultivar 'Nuccio's Carousel' × *C. azalea*）。

杂交苗编号： NZMD-No.5。

性状： 花朵艳红色至暗红色，玫瑰重瓣型，大型花，花径10.0～12.5cm，花朵厚5.5cm，大花瓣18～21枚，呈3～4轮星状排列，阔倒卵形，沟槽状，厚肉质，花芯大花瓣半直立，雄蕊瓣簇拥，偶具白斑，花瓣逐片掉落。叶片浓绿色，叶背面灰绿色，厚革质，阔椭圆形，长9.3～10.0cm，宽5.0～5.8cm，边缘叶齿钝。植株立性，枝叶稠密，生长旺盛。夏末始花，秋、冬季盛花，春季零星开花。

Abstract: Flowers bright red to dark red, rose-double form, large size, 10.0-12.5cm across and 5.5cm deep with 18-21 large petals arranged in 3-4 rows as star shaped, petals broad obovate, grooved, thickly fleshy, central large petals semi-erected, petaloids clustered, occasionally with small white patches, petals fall one by one. Leaves dark green, heavy leathery, broad elliptic, margins obtusely serrate. Plant upright, branches dense and growth vigorous. Starts to bloom from late-summer, fully blooms from autumn to winter and sporadically in spring.

HA-79-4. 迷你雅秀
Mini Yaxiu (Miniature Elegant Posture)

杂交组合：红山茶品种'牛西奥转马'×杜鹃红山茶（*C. japonica* cultivar 'Nuccio's Carousel' × *C. azalea*）。

杂交苗编号：NZMD-No.6。

性状：花朵粉红色至深粉红色，具模糊的白条纹，半重瓣型，微型至小型花，花径5.5～7.0cm，花朵厚3.5～4.0cm，花瓣25枚以上，呈3～4轮有序排列，倒卵形，雄蕊少量。叶片浓绿色，叶背面灰绿色，革质，椭圆形，长8.0～8.5cm，宽3.5～4.5cm，边缘上部具尖齿。植株立性，生长旺盛。秋初始花，秋末至冬季盛花，春季零星开花。

> **Abstract:** Flowers pink to deep pink with fuzzy white stripes, semi-double form, miniature to small size, 5.5-7.0cm across and 3.5-4.0cm deep with over 25 petals arranged orderly in 3-4 rows, petals obovate, a few stamens. Leaves dark green, leathery, elliptic, margins sharply serrate at the upper part. Plant upright and growth vigorous. Starts to bloom from early-autumn, fully blooms from later-autumn to winter and sporadically in spring.

⦿ 杂交组合 HA-80. 红山茶品种'忠实'× 杜鹃红山茶
HA-80. *C. japonica* cultivar 'Yours Truly' × *C. azalea*

　　本杂交组合将介绍2个新品种。

　　红山茶品种'忠实',其花朵具深粉红色条纹和白边。

　　本组获得的2个新品种,其开花期与杜鹃红山茶非常相似。这已经是一个改进了。

Two new hybrids will be introduced in this cross-combination.

The flowers of the cultivar 'Yours Truly' have deep red stripes and white borders.

The blooming periods of the two new hybrids obtained are similar to *C. azalea*'s. This is an improvement.

HA-80-1. 古庙红钟
Gumiao Hongzhong (Red Bell of Ancient Temple)

杂交组合： 红山茶品种'忠实'× 杜鹃红山茶（*C. japonica* cultivar 'Yours Truly'× *C. azalea*）。

杂交苗编号： ZSD-No.5。

性状： 花朵浓红色至黑红色，泛蜡光，单瓣型至托桂型，小型至中型花，花径7.0～8.5cm，花瓣6枚，阔倒卵形，厚质，雄蕊柱状，有时瓣化，花丝红色，雌蕊高出雄蕊1～2cm，花朵整朵掉落。叶片浓绿色，但容易出现具黄斑的复色叶片，革质，椭圆形，长8.5～10.5cm，宽3.5～4.5cm，边缘叶齿钝。植株立性，生长旺盛。夏末始花，秋、冬季盛花，春季零星开花。

Abstract: Flowers deep red to dark red with some waxy luster, single to anemone form, small to medium size, 7.0-8.5cm across with 6 petals, petals broad obovate, thick texture, stamens united into a column, some stamens changes into petaloids, filaments red, pistil is 1-2cm higher than the stamens, flowers fall whole. Leaves dark green, but easily to be variegated with yellow stripes, leathery, elliptic, margins obtusely serrate. Plant upright and growth vigorous. Starts to bloom from late-summer, fully blooms from autumn to winter and sporadically in spring.

HA-80-2. 不夜红城
Buye Hongcheng (Ever-bright Red City)

杂交组合： 红山茶品种'忠实'× 杜鹃红山茶（*C. japonica* cultivar 'Yours Truly' × *C. azalea* ）。

杂交苗编号： ZSD-No.6。

性状： 花朵深红色至黑红色，泛蜡光，中部花瓣偶有白条纹，花开几天后，中部花瓣花色变为粉红色，半重瓣型，中型花，花径8.5～10.0cm，花朵厚4.8cm，花瓣22～25枚，呈3轮排列，阔倒卵形，略波浪状，雄蕊短，基部连生，偶有雄蕊瓣，花瓣逐片掉落。叶片浓绿色，容易出现黄色复色化白条纹，叶背面灰绿色，革质，椭圆形，长8.5～9.0cm，宽4.0～4.5cm，略波浪状，边缘叶齿稀而浅。植株立性，枝叶稠密，生长旺盛。夏末始花，秋、冬季盛花，春季零星开花。

Abstract: Flowers deep red to dark red with waxy luster, occasionally with white stripes at the center, central petals will become pink after several days of blooming, semi-double form, medium size, 8.5-10.0cm across and 4.8cm deep with 22-25 petals arranged in 3 rows, petals broad obovate, slightly wavy, stamens short with some petaloids, petals fall one by one. Leaves dark green, some of them easily to be variegated with white stripes, leathery, elliptic, slightly wavy, margins sparsely and shallowly serrate. Plant upright, branches dense and growth vigorous. Starts to bloom from late-summer, fully blooms from autumn to winter and sporadically in spring.

◉ 杂交组合 HA-81. 红山茶品种 '孔雀玉浦' × 杜鹃红山茶
HA-81. *C. japonica* cultivar 'Tama Peacock' × *C. azalea*

本杂交组合将介绍8个新品种。

'孔雀玉浦'品种的花瓣边缘具白边，枝条下垂，灌木状。本组杂交试图让杂交新品种保持'孔雀玉浦'品种的花色、株形等性状，同时也保持杜鹃红山茶四季开花的性状。

在这8个新品种中，多数品种的花朵具明显的白边或者花瓣具白斑，花径为小型至中型，这些性状与母本'孔雀玉浦'相似。而开花期从秋初开始，一直到冬季，即便在翌年春季，仍有零星开花，这些性状与其父本杜鹃红山茶相似。很明显，杂交种已经具有了两个杂交亲本的主要性状。

Eight new hybrids will be introduced in this cross-combination.

The cultivar 'Tama Peacock' has the characteristics of bordered white petals, hanging and shrubby branches. The cross was to let the hybrids keep the flower color and plant shape of 'Tama Peacock' and the ever-blooming characteristics of *C. azalea*.

Most of the eight new hybrids have white bordered petals or white patches on the central petaloids, small to medium size, which are similar to the female parent 'Tama Peacock'. However, a long blooming period that starts to bloom in early-autumn, continuously to winter and even sporadically in the spring of the following year, which are similar to the male parent *C. azalea*. It is obvious that the new hybrids have the major characteristics of their two cross parents.

HA-81-1. 如画梦境
Ruhua Mengjing (Picturesque Dream)

杂交组合： 红山茶品种'孔雀玉浦'×杜鹃红山茶（*C. japonica* cultivar 'Tama Peacock'× *C. azalea*）。

杂交苗编号： KQYPD-No.3。

性状： 花朵红色，花瓣边缘具整齐的宽白边，玫瑰重瓣型，小型花，花径6.0～7.0cm，花朵厚3.5cm，花瓣19～22枚，呈4轮有序排列，阔倒卵形，雄蕊少量，基部略连生。叶片浓绿色，叶背面灰绿色，革质，阔椭圆形，长7.5～8.0cm，宽4.5～5.0cm，基部楔形，边缘叶齿浅。植株紧凑，枝叶稠密，生长旺盛。秋初始花，冬季盛花，春季零星开花。

Abstract: Flowers red with a wide white border on each petal, rose-double form, small size, 6.0-7.0cm across and 3.5cm deep with 19-22 petals arranged orderly in 4 rows, petals broad obovate, a few stamens, slightly united at the base. Leaves dark green, leathery, broad elliptic, cuneate at the base, margins shallowly serrate. Plant compact, branches dense and growth vigorous. Starts to bloom from early-autumn, fully blooms in winter and sporadically in spring.

HA-81-2. 微雕作品
Weidiao Zuopin (Miniature-sculpture Work)

杂交组合： 红山茶品种'孔雀玉浦'×杜鹃红山茶（*C. japonica* cultivar 'Tama Peacock'× *C. azalea*）。

杂交苗编号： KQYPD-No.2 + 4。

性状： 花朵红色至深红色，中部花瓣偶显隐约的白斑，玫瑰重瓣型，小型至中型花，花径6.5～7.8cm，花朵厚4.0cm，大花瓣15～16枚，呈3轮排列，花瓣外翻，阔倒卵形，雄蕊少数。叶片浓绿色，叶背面灰绿色，革质，阔椭圆形，长8.5～9.5cm，宽4.0～4.5cm，基部楔形，边缘叶齿浅。植株立性，生长旺盛。秋初始花，冬季盛花，春季零星开花。

Abstract: Flowers red to deep red, occasionally with white patches on the central petals, rose-double form, small to medium size, 6.5-7.8cm across and 4.0cm deep with 15-16 large petals arranged in 3 rows, petals broad obovate, curved outward, a few stamens. Leaves dark green, leathery, elliptic, cuneate at the base, margins shallowly serrate. Plant upright and growth vigorous. Starts to bloom in early-autumn, fully blooms in winter and sporadically in spring.

HA-81-3. 乖巧女孩
Guaiqiao Nühai (Well-behaved & Clever Girl)

杂交组合：红山茶品种'孔雀玉浦'×杜鹃红山茶（*C. japonica* cultivar 'Tama Peacock' × *C. azalea*）。

杂交苗编号：KQYPD-No.5。

性状：花朵淡红色，瓣面有扩散性白斑，边缘有宽窄不等的白边，单瓣型，小型花，花径6.0～6.5cm，花瓣5～6枚，阔倒卵形，雄蕊基部连生。叶片浓绿色，叶背面灰绿色，革质，椭圆形，长8.5～10.0cm，宽4.0～4.5cm，基部楔形，边缘叶齿浅。植株立性，生长旺盛。秋初始花，冬季盛花，春季零星开花。

> **Abstract:** Flowers light red with white diffuse markings, and differently wide white borders on edges, single form, small size, 6.0-6.5cm across with 5-6 petals, petals broad obovate, stamens united at the base. Leaves dark green, leathery, elliptic, cuneate at the base, margins shallowly serrate. Plant upright and growth vigorous. Starts to bloom from early-autumn, fully blooms in winter and sporadically in spring.

HA-81-4. 银边绣衣
Yinbian Xiuyi (Embroidered Clothes with Silver Borders)

杂交组合：红山茶品种'孔雀玉浦'×杜鹃红山茶（*C. japonica* cultivar 'Tama Peacock' × *C. azalea*）。

杂交苗编号：KQYPD-No.7。

性状：花朵淡红色，花瓣边缘具宽度为0.3～0.5cm的白边，单瓣型，中型至大型花，花径9.5～11.0cm，花瓣7枚，倒卵形，雄蕊基部连生，呈柱状。叶片浓绿色，叶背面灰绿色，革质，椭圆形，长9.5～11.0cm，宽4.5～5.0cm，基部急楔形，边缘叶齿稀疏。植株开张，枝条略下垂，生长旺盛。秋初始花，冬季盛花，春季零星开花。

Abstract: Flowers light red with 0.3-0.5cm wide white borders, single form, medium to large size, 9.5-11.0cm across with 7 petals, petals obovate, stamens united into a column at the base. Leaves dark green, leathery, elliptic, urgently cuneate at the base, margins sparsely serrate. Plant spread, branches slightly hanging down and growth vigorous. Starts to bloom from early-autumn, fully blooms in winter and sporadically in spring.

HA-81-5. 五谷丰登
Wugu Fengdeng (Abundant Harvest of All Crops)

杂交组合： 红山茶品种'孔雀玉浦'×杜鹃红山茶（*C. japonica* cultivar 'Tama Peacock' × *C. azalea*）。

杂交苗编号： KQYPD-No.11。

性状： 花朵红色至黑红色，具蜡光，中部雄蕊瓣具白斑，托桂型，小型至中型花，花径6.0～8.0cm，花朵厚4.5cm，外轮大花瓣5～6枚，呈1轮排列，花瓣阔倒卵形，厚质，边缘略外翻，雄蕊完全瓣化，簇拥成球，颇似谷物红高粱，偶有较大的雄蕊瓣。叶片浓绿色，叶背面灰绿色，革质，椭圆形，长7.5～8.5cm，宽3.5～4.0cm，基部楔形，边缘叶齿浅疏。植株开张，枝条软而稠密，生长旺盛。秋初始花，冬季盛花，春季零星开花。

Abstract: Flowers red to dark red with waxy luster, central petaloids with white patches, anemone form, small to medium size, 6.0-8.0cm across and 4.5cm deep with 5-6 exterior large petals arranged in 1 row, petals broad obovate, thick texture, edges curved outward, stamens fully petalody which clustered into a ball and is similar to grain sorghum, occasionally some larger petaloids appeared at the center. Leaves dark green, leathery, elliptic, cuneate at the base, margins shallowly and sparsely serrate. Plant spread, branches soft and dense and growth vigorous. Starts to bloom from early-autumn, fully blooms in winter and sporadically in spring.

HA-81-6. 亭亭玉立 Tingting Yuli (Girls with Beautiful Standing Postures)

杂交组合：红山茶品种'孔雀玉浦'×杜鹃红山茶（*C. japonica* cultivar 'Tama Peacock' × *C. azalea*）。

杂交苗编号：KQYPD-No.12。

性状：花朵深红色至酱红色，有绒质感，中部花瓣偶有白窄边，半重瓣型至玫瑰重瓣型，小型花，花径6.0～7.5cm，花朵厚4.5cm，大花瓣15～16枚，呈3轮排列，厚质，外翻，阔倒卵形，花芯处几枚雄蕊，小花瓣有白斑，花瓣逐片掉落。叶片浓绿色，叶背面灰绿色，厚革质，狭长椭圆形，长8.5～9.5cm，宽3.5～4.9cm，基部楔形，边缘具浅齿。植株立性，生长旺盛。秋初始花，冬季盛花，春季零星开花。

Abstract: Flowers deep red to sauce red with velvet texture, occasionally with narrow white edges on central petals, semi-double to rose-double form, small size, 6.0-7.5cm across and 4.5cm deep with 15-16 large petals arranged in 3 rows, petals thick texture, broad obovate, several central petaloids with white markings, petals fall one by one. Leaves dark green, heavy leathery, narrow-long elliptic, cuneate at the base, margins shallowly serrate. Plant upright and growth vigorous. Starts to bloom from early-autumn, fully blooms in winter and sporadically in spring.

HA-81-7. 玉浦新秀
Yupu Xinxiu（Tama-no-ura's New Rookie）

杂交组合：红山茶品种'孔雀玉浦'× 杜鹃红山茶（*C. japonica* cultivar 'Tama Peacock' × *C. azalea*）。

杂交苗编号：KQYPD-No.14。

性状：花朵淡红色，花瓣边缘具较宽的白边，或者花瓣中部具明显的细红条纹，单瓣型，中型至大型花，花径9.0～10.5cm，花瓣6～7枚，阔倒卵形，雄蕊基部连生，呈柱状，花朵整朵掉落。叶片浓绿色，叶背面灰绿色，革质，椭圆形，长9.0～9.5cm，宽4.0～4.5cm，基部楔形，边缘叶齿钝。植株开张，生长旺盛。秋初始花，冬季盛花，春季零星开花。

> **Abstract:** Flowers light red with wider white borders, or with obvious thin red stripes at the middle of petal surfaces, single form, medium to large size, 9.0-10.5cm across with 6-7 petals, petals broad obovate, stamens united into a column at the base, flowers fall whole. Leaves dark green, leathery, elliptic, cuneate at the base, margins obtusely serrate. Plants spread and growth vigorous. Starts to bloom from early-autumn, fully blooms in winter and sporadically in spring.

HA-81-8. 民族服饰
Minzu Fushi (National Dress)

杂交组合：红山茶品种'孔雀玉浦'×杜鹃红山茶（*C. japonica* cultivar 'Tama Peacock' × *C. azalea* ）。

杂交苗编号：KQYPD-No.18。

性状：花朵红色至深红色，花瓣边缘偶具较窄的白边，中部花瓣可见模糊的白斑和清晰的白条纹，玫瑰重瓣型，小型至中型花，花径6.0～8.0cm，花朵厚4.0cm，花瓣21枚，呈3轮有序排列，阔倒卵形，边缘略内卷，雄蕊少量。叶片浓绿色，叶背面灰绿色，革质，椭圆形，长8.5～9.5cm，宽3.0～3.5cm，基部急楔形，边缘叶齿浅疏。植株立性，枝条软而稠密，生长旺盛。秋初始花，冬季盛花，春季零星开花。

Abstract: Flowers red to deep red, occasionally with narrow white borders, faint white blotches and clear white stripes are visible on the central petals, rose-double form, small to medium size, 6.0-8.0cm across and 4.0cm deep with 21 petals arranged orderly in 3 rows, petals broad obovate, edges slightly rolled inward, a few stamens. Leaves dark green, leathery, elliptic, urgently cuneate at the base, margins shallowly and sparsely serrate. Plant upright, branches soft and dense and growth vigorous. Starts to bloom from early-autumn, fully blooms in winter and sporadically in spring.

⦿ 杂交组合 HA-82. 红山茶品种'玉之浦'× 杜鹃红山茶
HA-82. *C. japonica* cultivar 'Tama-no-ura' × *C. azalea*

本杂交组合将介绍 1 个新品种。

母本'玉之浦'最具吸引力的性状是花朵红色，花瓣边缘具清晰白边。本组杂交的目的是想获得具有白色花边且能四季开花的新品种。

所获得的这个新品种，花色黑红，花瓣边缘具窄白边，也能四季开花，这说明，新品种不仅已表达了双亲的性状，而且超越了它的亲本。

One new hybrid will be introduced in this cross-combination.

The most attractive characteristics of the female parent, 'Tama-no-ura' are red flowers edged with clear white borders. The aim of the cross is to obtain the new hybrid that flowers have white borders and bloom in four seasons.

The new hybrid obtained is dark red in flower color and petals have narrow and white edges, and can also bloom year-round, which indicates that the new hybrid not only expresses the characteristics of both parents, but also is beyond its parents.

HA-82-1. 红星奖章
Hongxing Jiangzhang (Red Star Medal)

杂交组合：红山茶品种'玉之浦' × 杜鹃红山茶（*C. japonica* cultivar 'Tama-no-ura' × *C. azalea*）。

杂交苗编号：YZPD-No.4。

性状：花朵黑红色，泛蜡光，花瓣边缘具宽窄不等的白边，单瓣型，小型至中型花，花径 6.5～8.0cm，花瓣 5～6 枚，阔倒卵形，雄蕊基部连生，呈短柱状。叶片浓绿色，叶背面灰绿色，革质，长椭圆形，长 8.0～9.5cm，宽 3.5～4.5cm，基部楔形，边缘具稀齿。植株立性，生长旺盛。夏末始花，秋、冬季盛花，春季偶有开花。

> **Abstract:** Flowers dark red with waxy luster, petals edges with white borders that are unequal width, single form, small to medium size, 6.5-8.0cm with 5-6 petals, petals broad obovate, stamens united into a short column at the base. Leaves dark green, leathery, long elliptic, cuneate at the base, margins sparsely serrate. Plant upright and growth vigorous. Starts to bloom from late-summer, fully blooms from autumn to winter and occasionally in spring.

杂交组合 HA-83. 红山茶品种'黑魔法'× 杜鹃红山茶
HA-83. *C. japonica* cultivar 'Black Magic'× *C. azalea*

本杂交组合将介绍2个新品种。

'黑魔法'的花朵黑红色，花瓣泛蜡光，叶片边缘具深锯齿。

本杂交组合获得的2个新品种中，'星源晚秋'是主要的，其花朵呈黑红色，叶缘叶齿尖深，颇像母本'黑魔法'品种。而多季开花，叶片浓绿，则很像其父本杜鹃红山茶。因为它极易发生复色突变，所以从中分离出了'晚秋白云'品种。

Two new hybrids will be introduced in this cross-combination.

'Black Magic' has dark red color flowers with waxy luster, deep and sharp serrations on its leaf margins. Of the two new hybrids obtained in this cross-combination, 'Xingyuan Wanqiu' is the main one. Its flowers are dark red and its leaves have deep serrations, which is quite like its female parent 'Black Magic'. However, its blooming period is multi-seasons and its leaves are dark green, which is similar to its male parent *C. azalea*. Because it is easy to mutate in flower color, a variety named 'Wanqiu Baiyun' had been isolated from it.

HA-83-1. 星源晚秋
Xingyuan Wanqiu (Xingyuan's Late Autumn)

杂交组合：红山茶品种'黑魔法' × 杜鹃红山茶（ *C. japonica* cultivar 'Black Magic' × *C. azalea* ）。

杂交苗编号：HMFD-No.1。

性状：花朵黑红色，具蜡质感，偶有少量隐约的白斑，半重瓣型，中型至大型花，花径 9.5～11.5cm，花朵厚 5.5cm，大花瓣约 18 枚，呈 2～3 轮排列，雄蕊基部连生，花瓣逐片掉落，开花稠密。叶片浓绿色，叶背面灰绿色，厚革质，椭圆形，长 9.0～10.0cm，宽 3.5～4.0cm，基部心形，上部叶缘齿深尖。植株紧凑，枝条稠密，生长旺盛。夏末始花，秋、冬季盛花，春季零星开花。

注：本品种由上海茶花园培育，并已获得国家林业和草原局的专利授权。

Abstract: Flowers dark red with some waxy feelings, occasionally with a few faint white markings, semi-double form, medium to large size, 9.5-11.5cm across and 5.5cm deep with 18 large petals arranged in 2-3 rows, stamens united at the base, petals fall one by one, bloom dense. Leaves dark green, heavy leathery, elliptic, base heart-shaped, margins deeply and sharply serrate at the upper part. Plant compact, branches dense and growth vigorous. Starts to bloom from late-summer, fully blooms from autumn to winter and sporadically in spring.

Mark: The hybrid is bred by Shanghai Camellia Garden and its patent right has been obtained from National Forestry and Grassland Administration.

HA-83-2. 晚秋白云
Wanqiu Baiyun (Late-autumn's White Clouds)

杂交组合： 红山茶品种'黑魔法' × 杜鹃红山茶（*C. japonica* cultivar 'Black Magic' × *C. azalea*）。

杂交苗编号： HMFD-No.1Mu。

性状： 本品种是'星源晚秋'的一个复色突变。其大部分性状与'星源晚秋'品种相似，但是花瓣上具大量大理石云状白斑，叶片偶然出现少量黄斑。

注：本品种于2018年由上海茶花园培育。

Abstract: It is a variegated mutation obtained from the hybrid 'Xingyuan Wanqiu'. Most of its characteristics are similar to 'Xingyuan Wanqiu', but there are a lot of marbled white markings on the petal surfaces and occasionally a few of yellow patches on the leaves.

Mark: The hybrid was bred by Shanghai Camellia Garden in 2018.

◉ 杂交组合 HA-84. 红山茶品种'咖啡杯' × 杜鹃红山茶
HA-84. *C. japonica* cultivar 'Demi-Tasse' × *C. azalea*

本杂交组合将介绍1个新品种。

'咖啡杯'品种，其花朵呈桃粉红色，花瓣边缘粉色加深，半重瓣型，小型至中型花，植株紧凑、立性，生长旺盛。本组杂交主要是想获得能四季开花，而且植株立性、生长旺盛、开花稠密的新品种。

获得的新品种的花朵呈嫩粉红色至淡红色，尽管是单瓣型的，但是开花极为稠密，而且植株立性，是未来园林美化的宝贵品种。

One new hybrid will be introduced in this cross-combination.

The flowers of 'Demi-Tasse' are peach pink and fading to deep pink at petal edges, semi-double form, small to medium size, its plants are both compact and upright and grow vigorously. The cross-combination is to get the hybrids that not only bloom year-round and densely, but also grow uprightly and vigorously.

The new hybrid obtained has tend pink to light-red flowers. Its flowers are single form, but blooms very densely and grows uprightly, which will be a valuable variety for beautifying gardens in the future.

HA-84-1. 星源花歌
Xingyuan Huage (Xingyuan's Flowers Song)

杂交组合： 红山茶品种'咖啡杯'×杜鹃红山茶（*C. japonica* cultivar 'Demi-Tasse' × *C. azalea*）。

性状： 花朵嫩红色至淡红色，单瓣型，微型至小型花，花径5.5～6.5cm，花瓣6枚，倒卵形，雄蕊基部连生，呈圆柱状，花朵整朵掉落，开花稠密。叶片浓绿色，叶背面灰绿色，厚革质，长椭圆形，长10.0～11.0cm，宽4.0～4.5cm，基部楔形，上部边缘叶齿钝。植株立性，高大，生长旺盛。夏末始花，秋、冬季盛花，春季零星开花。

注：本品种由上海茶花园培育，并已获得国家林业和草原局专利授权。

Abstract: Flowers tender red to light red, single form, miniature to small size, 5.5-6.5cm with 6 petals, petals obovate, stamens united into a column at the base, flowers fall whole, bloom dense. Leaves dark green, heavy leathery, long elliptic, cuneate at the base, margins obtusely serrate at the upper part. Plant upright, tall and growth vigorous. Starts to bloom from late-summer, fully blooms from autumn to winter and sporadically in spring.

Mark: The hybrid is bred by Shanghai Camellia Garden and its patent right has been obtained from National Forestry and Grassland Administration.

⦿ 杂交组合 HA-85. 杜鹃红山茶 × 雪山茶品种'万代'
HA-85. *C. azalea* × *C. rusticana* cultivar 'Bandai'

本杂交组合将介绍1个新品种。

'万代'是我国引自日本的一个抗寒性很强的雪山茶品种。该品种花朵呈朱砂红色，单瓣型，容易坐果。本组杂交是想获得四季开花和冬季抗寒性较强的杂交新品种。

从获得的杂交植株看，其花瓣厚质，与父本'万代'品种相似。而叶片浓绿、稠密、基部楔形、重复开花等性状，则很像它的母本杜鹃红山茶。这些性状可以证明它确实是真实的杜鹃红山茶杂交种。我们期待能在寒冷地区成功试种，从而在抗寒育种方面有所突破。

One new hybrid will be introduced in this cross-combination.

'Bandai' was imported from Japan and it is a *C. rusticana* cultivar with strong cold hardy. The cultivar's flowers are cinnabar red, single form and easily setting fruits. The combination was to obtain the new hybrids which could ever-bloom and strongly resist to cold temperature in winter.

From the hybrid plant, it can be seen that the flower petals are thick texture which are similar to the male parent 'Bandai'. The characteristics such as leaves are dark green, dense, base cuneate and the plants are ever-blooming, are similar to its female parent *C. azalea*. It improves that the hybrid is, indeed, from the hybridization between *C. azalea* and 'Bandai'. We are expecting the result of its trail in cold area and hope the hybrid could have a breakthrough on the breeding of cold hardy.

HA-85-1. 桃红凝夏
Taohong Ningxia (Peach Red Coagulating Summer)

杂交组合： 杜鹃红山茶 × 雪山茶品种'万代'（ *C. azalea* × *C. rusticana* cultivar 'Bandai' ）。

杂交苗编号： DWD-No.3。

性状： 花朵深桃红色，单瓣型，中型花，花径 8.5～9.0cm，花瓣 7 枚，厚质，阔倒卵形，上半部外翻，似喇叭状，先端微凹，雄蕊基部连生，花丝淡红色，花药黄色，开花稠密。叶片浓绿色，叶背面灰绿色，革质，椭圆形，长 8.0～8.5cm，宽 3.0～3.5.cm，基部楔形，上部边缘叶齿浅。植株立性，生长旺盛。夏末始花，秋、冬季盛花，春季零星开花。

Abstract: Flowers deep peach red, single form, medium size, 8.5-9.0cm with 7 petals, petals broad obovate, thick texture, curved outward at the petal's upper part, stamens united at the base, filaments pale red, anthers yellow, bloom dense. Leaves dark green, leathery, elliptic, cuneate at the base, margins shallowly serrate at the upper part. Plant upright and growth vigorous. Starts to bloom from late-summer, fully blooms from autumn to winter and sporadically in spring.

◉ 杂交组合 HA-86. 雪山茶品种'万代'× 杜鹃红山茶
HA-86. *C. rusticana* cultivar 'Bandai' × *C. azalea*

本杂交组合将介绍 1 个新品种。

本杂交组合是本书 HA-85 的一个反交组合。同样，本组杂交也是为了获得抗寒性较强的杂交新品种。

从获得的杂交植株看，花朵呈朱砂红色，略泛蜡光，花瓣厚质，与母本'万代'品种相似。而叶片浓绿、稠密、基部楔形，且重复开花，则很像它的父本杜鹃红山茶。我们期待能在寒冷地区试种成功，从而在抗寒育种方面有所突破。

One new hybrid will be introduced in this cross-combination.

The cross-combination is reciprocal crossed to the HA-85 in this book. Likewise, the combination is for getting the strong cold hardy new hybrids.

From the hybrid plant, it can be seen that its flowers are cinnabar red with slight waxy luster and petals are thick texture which are similar to the female parent 'Bandai' and its leaves are dark green, dense, and base cuneate as well as its ever-blooming trait, which are simliar to its male parent *C. azalea*. We are expecting the result of its trail in cold area and hope the hybrid could have a breakthrough on the breeding of cold hardy.

HA-86-1. 半卷红帘
Banjuan Honglian (Semi-rolled Red Curtain)

杂交组合： 雪山茶品种'万代' × 杜鹃红山茶（*C. rusticana* cultivar 'Bandai' × *C. azalea*）。

杂交苗编号： WDD-No.1 + 4。

性状： 花朵朱砂红色，略泛蜡光，瓣面可见深红脉纹，单瓣型，中型花，花径8.5～10.0cm，花瓣7枚，厚质，阔倒卵形，上半部外卷，雄蕊基部连生，呈短柱状，柱头3深裂。叶片浓绿色，叶背面灰绿色，革质，狭长椭圆形，长9.0～10.2cm，宽3.5～4.0cm，基部楔形，上半部边缘叶齿疏。植株开张，矮性，生长旺盛。夏末始花，秋、冬季盛花，春季零星开花。

Abstract: Flowers cinnabar red slightly with waxy luster, deep red veins visible on the petal surfaces, single form, medium size, 8.5-10.0cm with 7 petals, petals broad obovate, roll outward at the petal's upper part, stamens united into a short column at the base, stigma 3 deep splits. Leaves dark green, leathery, long narrow elliptic, cuneate at the base, margins sparsely serrate at the upper part. Plant spread, dwarf and growth vigorous. Starts to bloom from late-summer, fully blooms from autumn to winter and sporadically in spring.

● 杂交组合 HA-87. 云南山茶品种'豪斯' × 杜鹃红山茶
HA-87. *C. reticulata* cultivar 'Frank Houser' × *C. azalea*

本杂交组合将介绍1个新品种。

'豪斯'品种是由云南山茶品种'和尚'×红山茶品种'史蒂夫'得到的,其花朵呈深粉红色,巨型,半重瓣型至牡丹型。

本组杂交试图获得能够多季开花,且花朵巨大、花色鲜艳的新品种。我们的目标基本达到了。

One hybrid will be introduced in this cross-combination.

The *C. reticulata* cultivar 'Frank Houser' was obtained through *C. reticulata* cultivar 'Buddah' × *C. japonica* cultivar 'Steve Blount'. Its flowers are deep pink color, very large size and semi-double to peony form.

The cross-combination attempted to obtain the new hybrids that could ever-bloom, and open very large and bright colour flowers. Our goal is basically achieved.

HA-87-1. 张灯结彩
Zhangdeng Jiecai (Decorated Scenes with Lanterns & Streamers)

杂交组合： 云南山茶品种'豪斯' × 杜鹃红山茶（*C. reticulata* cultivar 'Frank Houser' × *C. azalea*）。

杂交苗编号： HSD-No.1。

性状： 花朵红色，偶有白斑，半重瓣型至牡丹型，大型至巨型花，花径11.5～13.5cm，花朵厚6.0cm，大花瓣60～70枚，呈7轮排列，花瓣阔倒卵形，先端近圆或波浪状，中部小花瓣直立、扭曲，与雄蕊混生。叶片浓绿色，叶背面灰绿色，厚革质，阔椭圆形，长9.0～10.0cm，宽5.0～5.5cm，上部边缘叶齿稀而钝，下部边缘光滑。植株紧凑，立性，生长旺盛。夏末始花，秋、冬季盛花，春季零星开花。

> **Abstract:** Flowers red, occasionally with white markings, semi-double to peony form, large to very large size, 11.5-13.5cm across and 6.0cm deep with 60-70 large petals arranged in 7 rows, petals broad obovate, apices round or wavy, central small petals erected, twisted and mixed with stamens. Leaves dark green, heavy leathery, broad elliptic, margins sparsely and obtusely serrate at the upper part, smooth at the lower part. Plant compact, upright and growth vigorous. Starts to bloom from late-summer, fully blooms from autumn to winter and sporadically in spring.

杜鹃红山茶 F_1 代回交新品种育种概况

The Breeding Outline on the Backcross New Hybrids of *C. azalea* F_1

杜鹃红山茶 F_1 代回交组合及获得的杜鹃红山茶 F_1 代回交新品种的数量

应该指出，这里所说的回交，是把杜鹃红山茶和常规茶花品种作为亲本，杂交后获得的新品种，即杜鹃红山茶 F_1 代新品种（见第一部书 HA-01 至 HA-52 和本书 HA-01 至 HA-82），再分别与杜鹃红山茶或者常规茶花品种进行回交。

本书将介绍 183 个杜鹃红山茶 F_1 代回交新品种。这些新品种共涉及 38 个回交组合。

The Quantities of Backcross Combinations and Obtained Backcross New Hybrids of *C. azalea* F_1

It should be pointed out that the backcross mentioned here is to use *C. azalea* and normal camellia cultivars as cross parents to cross, the new hybrids obtained from the crosses, that is, *C. azalea* F_1 new hybrids (See HA-01 to HA-52, our first book and HA-01 to HA-82, this book), and then using the F_1 to cross with *C. azalea* or normal camellia cultivar respectively.

One hundred and eighty-three backcross new hybrids of *C. azalea* F_1 will be introduced in this book. These new hybrids are involved in 38 backcross combinations.

杜鹃红山茶 F_1 代回交的 3 种方式

在这些杂交组合中，有 3 种回交方式。应该指出的是，无论哪种回交方式，作为母本的亲本都应该能够坐果，作为父本的亲本都应该有花粉，而且母本与父本应该具有一定的杂交亲和性。

第一种方式是杜鹃红山茶与杜鹃红山茶 F_1 代之间的回交（见本书 HAR-1 至 HAR-27 和 HAR-30 至 HAR-33），共涉及 31 个回交组合。在理论上，这种回交方式所获得的回交新品种的杜鹃红山茶基因频率可增加到 75%，常规茶花品种基因频率则降至 25%。第二种方式是常规茶花品种与杜鹃红山茶 F_1 代之间的回交（见本书 HAR-34 至 HAR-38），共有 5 个回交组合。在理论上，这种回交方式所获得的回交新品种的杜鹃红山茶基因频率降低到 25%，而常规茶花品种基因频率则增加到 75%。第三种方式是杜鹃红山茶 F_1 代之间的杂交，即姊妹间杂交（见本书 HAR-28 至 HAR-29），共 2 个回交组合。在理论上，这种杂交方式所获得的新品种，其杜鹃红山茶的基因频率和常规茶花品种的基因频率各为 50%。

Three Backcross Patterns of *C. azalea* F_1

There are three backcross patterns in these backcross combinations. It should be pointed out that no matter which backcross pattern, the female parent should be able to bear fruits, the male parent should have pollens, and the female parent and the male parent should have a certain compatibility.

The first pattern is the backcross between *C. azalea* and *C. azalea*'s F_1 (See HAR-1 to HAR-27 and HAR-30 to HAR-33, this book) which have involved in 31 backcross combinations. In theory the new backcross hybrids obtained from this pattern can increase to 75% in *C. azalea*'s gene frequency and decrease to 25% in normal camellia cultivar's gene frequency. The second pattern is the backcross between normal camellia cultivar and *C. azalea*'s F_1 (See HAR-34 to HAR-38, this book) which have involved in five backcross combinations. In theory the new backcross hybrids obtained from the second pattern can decrease to 25% in *C. azalea*'s genes frequency and increase to 75% in normal camellia cultivar's gene frequency. The third pattern is the cross between two *C. azalea* F_1 new hybrids, that is, sister-cross (See HAR-28 to HAR-29, this book), which have two cross combinations. The new hybrids obtained from this kind of pattern, in theory, their *C. azalea*'s genes frequency and normal camellia cultivar's gene frequency will be 50% respectively.

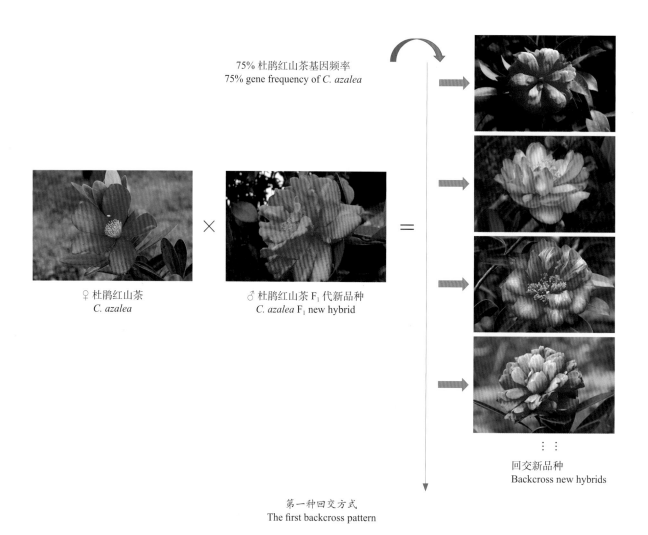

第一种回交方式
The first backcross pattern

第二种回交方式
The second backcross pattern

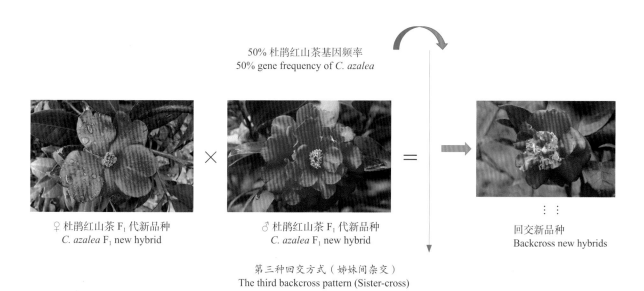

第三种回交方式（姊妹间杂交）
The third backcross pattern (Sister-cross)

杜鹃红山茶 F_1 代回交新品种的性状表达

我们的实践已经证明，杜鹃红山茶的基因频率直接影响着新品种的性状表达，如回交新品种的开花期、叶片、植株和生长态势等性状的表达。杜鹃红山茶基因频率占75%的回交新品种，其叶片、植株和开花期性状的表达，更倾向于杜鹃红山茶；而其花色、花型、花径等性状，则倾向于杜鹃红山茶 F_1 代的性状。杜鹃红山茶基因频率仅占25%的回交新品种，其开花期、叶色等性状仍倾向于杜鹃红山茶；而在花色、花型、花径、株形等方面，则明显倾向于常规茶花品种。因此，用杜鹃红山茶 F_1 代与其两个亲本进行回交，可以进一步改良回交新品种的性状。应该指出，杜鹃红山茶 F_1 代之间的杂交（第三种方式），其杜鹃红山茶的基因频率仍为50%，因此，它们的性状表达大体与它们的亲本相似。

Characteristic Expressions of the Backcross New Hybrids of *C. azalea* F_1

Our practice has proved that the gene frequencies of *C. azalea* are directly affecting the characteristic expressions of the new hybrids, such as the blooming period, leaves, plants, growth potential and so on. The characteristic expressions of leaves, plant and the blooming period are more inclined to *C. azalea* in the new hybrids with 75% *C. azalea* gene, but flower color, flower form and plant shape obviously tend to normal camellia cultivars in the new hybrids only with 25% *C. azalea* gene. Therefore, *C. azalea* F_1 backcross with its two parents can further improve the characteristics of the new hybrids. It should be pointed out that the gene frequency of *C. azalea* in the crosses between two *C. azalea* F_1 hybrids (The third pattern) is still 50%, so their characteristic expressions are roughly similar to those of their parents.

现以开花期为例，让我们看看这三类回交新品种究竟是怎样的。请看表2：

Now, take the blooming period as an example, let's see the differences among these three cross patterns. Please see the table below:

表2　3种回交方式获得的回交新品种与常规茶花品种、杜鹃红山茶的开花期的比较

Table 2　The Blooming Periods Comparison of the Backcross New Hybrids Obtained by the Three Cross Patterns, Normal Camellia Cultivars and *C. azalea*

杂交亲本或杂交新品种 Cross-parents or New hybrids	杜鹃红山茶基因频率/% Gene frequency of *C. azalea* /%	春季 3—5月份 Spring Mar.–May	夏季 6—8月份 Summer Jun.–Aug.	秋季 9—11月份 Autumn Sept.–Nov.	冬季 12—2月份 Winter Dec.–Feb.	全花期/月数 Entire blooming /Mon.	盛花期/月数 Fully blooming /Mon.	无花期/月数 No blooming /Mon.
常规茶花品种 Normal Camellia cultivar	0					7	3	5
杜鹃红山茶 *C. azalea*	100					10	8	2

续表

杂交亲本或杂交新品种 Cross-parents or New hybrids	杜鹃红山茶基因频率 /% Gene frequency of C. azalea /%	春季 3—5月份 Spring Mar.–May	夏季 6—8月份 Summer Jun.–Aug.	秋季 9—11月份 Autumn Sept.–Nov.	冬季 12—2月份 Winter Dec.–Feb.	全花期/月数 Entire blooming /Mon.	盛花期/月数 Fully blooming /Mon.	无花期/月数 No blooming /Mon.
由杜鹃红山茶 × 杜鹃红山茶 F₁ 代获得的回交新品种 The backcross new hybrids obtained from C. azalea × C. azalea F_1 hybrid	75					10	7	2
由常规茶花品种 × 杜鹃红山茶 F₁ 代获得的新品种 The backcross new hybrids obtained from normal camellia cultivar × C. azalea F_1 hybrid	25					8	5	4
由 2 个杜鹃红山茶 F₁ 代之间杂交获得的新品种 The new hybrids obtained from the cross between two C. azalea F_1 hybrids	50					9	6	3

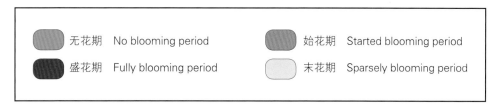

无花期 No blooming period　　始花期 Started blooming period
盛花期 Fully blooming period　　末花期 Sparsely blooming period

从上表可以清楚地看出，第一种回交方式所获得的回交新品种的全花期与杜鹃红山茶一样可达 10 个月，盛花期也仅比杜鹃红山茶短 1 个月。因此，可以说，第一种方式获得的回交新品种基本上也能够全年开花。

第二种回交方式所获得的回交新品种，全花期可达 8 个月，盛花期可达 5 个月，比常规茶花品种分别长 1 个月和 2 个月。尽管该回交方式获得的回交新品种比杜鹃红山茶开花期短一些，但是，它们在秋、冬、春 3 个季节仍然可开花，也表达出了多季开花的性状。

再比较这两种回交方式获得的回交新品种，我们可以看到，第一种回交方式的全花期和盛花期，分别比第二种回交方式长 2 个月。可是，第二种方式的全花期和盛花期也占据了 3 个季节，这说明即便回交新品种的杜鹃红山茶基因频率降低到 25%，回交新品种仍然能保持重复开花的性状。

最后应该提及的是，第三种回交方式属于姊妹间的杂交。此类新品种的杜鹃红山茶基因频率在理论上是 50%。经观测，其开花期大体与杜鹃红山茶 F₁ 代相似（见本书第 11 页），而其他性状则与其两个杂交亲本的性状有密切关系。

From the table above, it can be clearly seen that the entire blooming period of the backcross new hybrids obtained from the first backcross pattern can reach to ten months which is as long as *C. azalea*'s, and the fully blooming period of the backcross new hybrids is only one month shorter than *C. azalea*. Therefore, It can be concluded that the backcross new hybrids obtained from the first backcross pattern can also bloom year-round.

In the second backcross pattern, the entire blooming period of the backcross new hybrids obtained can reach to eight months and the fully blooming period of them can reach to five months, which are one month and two months longer than those of normal camellia cultivars respectively. Although the backcross new hybrids obtained by this backcross pattern have a shorter blooming period than *C. azalea*'s, they can still bloom in autumn, winter and spring, which have expressed the characteristics of multi-season blooming.

Let us compare the backcross new hybrids obtained by the two backcross patterns again, we can see that the entire blooming period and the fully blooming period of the first backcross pattern are two months longer than that of the second backcross pattern respectively. However, the entire blooming period and the fully blooming period of the second pattern have also lasted for three seasons, which indicates that even if the *C. azalea*'s gene frequency is reduced to 25%, the backcross new hybrids can also keep the ever-blooming characteristics.

Finally, it should be mentioned that the third cross pattern belongs to the sister-cross. The *C. azalea*'s gene frequency in this cross pattern, in theory, is 50%. It can be seen that the blooming period of these hybrids obtained is similar to what of *C. azalea* F_1 generation (See p.11, this book), while their other characteristics are closely related to the characteristics of their two hybrid parents.

将军风度
(Jiangjun Fengdu) HAR-09-1

大福大贵
(Dafu Dagui) HAR-02-20

暑期红艳
(Shuqi Hongyan) HAR-14-7

秋风送霞
(Qiufeng Songxia) HAR-02-3

杜鹃红山茶 F_1 代回交新品种详解

The Detailed Descriptions for the Backcross New Hybrids of *C. azalea* F_1

⦿ 回交组合 HAR-01. 杜鹃红山茶 × 杜鹃红山茶 F_1 代新品种'吉利牡丹'
HAR-01. *C. azalea* × *C. azalea* F_1 new hybrid 'Jili Mudan'

本回交组合将介绍1个新品种。

'吉利牡丹'是从杜鹃红山茶 × 云南山茶品种'帕克斯先生'杂交组合中得到的（见第一部书第252～253页）。

可以看到，本回交组合所获得的新品种的花性状主要倾向于父本'吉利牡丹'，而其叶片和植株性状则倾向于母本杜鹃红山茶。

One new hybrid will be introduced in this backcross combination.

'Jili Mudan' was obtained from the cross-combination between *C. azalea* and *C. reticulata* cultivar 'Dr Clifford Parks' (See p.252-253, our first book).

It can be seen that the flowers' characteristics of the new hybrid obtained in this backcross combination mainly tend to its male parent 'Jili Mudan' and the leaves and plant characteristics tend to its female parent *C. azalea*.

HAR-01-1. 回龙晨曦
Huilong Chenxi (Huilong Town's Morning Light)

回交组合：杜鹃红山茶 × 杜鹃红山茶 F_1 代新品种'吉利牡丹'(C. azalea × C. azalea F_1 new hybrid 'Jili Mudan')。

回交苗编号：DJLMD-No.9。

性状：花朵淡红色至橘红色，中部小花瓣偶有白条纹，随着花朵开放，会渐渐变为淡粉红色，半重瓣型至牡丹型，中型至大型花，花径 9.0～12.0cm，花朵厚 5.0cm，外部大花瓣 20 枚左右，呈 3 轮排列，花瓣倒卵形，中部小花瓣扭曲，雄蕊簇状。叶片浓绿色，叶背面灰绿色，厚革质，长椭圆形，长 9.5～11.5cm，宽 3.5～4.0cm，基部急楔形，先端钝尖，边缘光滑。植株紧凑，矮性，生长旺盛。春末至夏初始花，秋、冬季盛花，春季零星开花。

Abstract: Flowers light red to orange red, occasionally with white stripes at the central small petals, as flowers opening, the flower color gradually changes into pale pink, semi-double to peony form, medium to large size, 9.0-12.0cm across and 5.0cm deep with about 20 exterior large petals arranged in 3 rows, petals obovate, central small petals wrinkled, stamens clustered. Leaves dark green, heavy leathery, urgently cuneate at the base, margins smooth. Plant compact, dwarf and growth vigorous. Starts to bloom from late-spring to early-summer, fully blooms from autumn to winter and sporadically in spring.

⦿ 回交组合 HAR-02. 杜鹃红山茶 × 杜鹃红山茶 F₁ 代新品种 '夏风热浪'

HAR-02. *C. azalea* × *C. azalea* F₁ new hybrid 'Xiafeng Relang'

本回交组合将介绍27个新品种。

'夏风热浪'是从杜鹃红山茶 × 红山茶品种'花牡丹'杂交组合中得到的（见第一部书第193～194页）。

本回交组合是我们获得新品种数量最多的回交组合之一，这些新品种几乎包括了茶花的全部花型，其中，只有4个品种是单瓣花型的，不到总数的15%。这说明，父本'夏风热浪'品种在红山茶品种'花牡丹'基因的介入下，而杂合性较高，从而导致回交新品种性状呈现多样性。

×

Twenty seven new hybrids will be introduced in this backcross combination.

'Xiafeng Relang' was obtained from the cross-combination between *C. azalea* and *C. japonica* cultivar 'Daikagura' (See p.193-194, our first book).

This backcross combination is one of the backcross combinations in which we have obtained the largest number of new hybrids. These new hybrids almost include all the flower forms that camellias have, only four of them are single form, which are less than 15% of the total. It indicates that, owing to the intervention of *C. japonica* cultivar 'Daikagura' gene to 'Xiafeng Relang', it has higher heterozygosity, which leads to the diversity of the characteristics of the new backcross hybrids.

HAR-02-1. 窈窕淑女
Yaotiao Shunü (Fair Maiden)

回交组合：杜鹃红山茶 × 杜鹃红山茶 F_1 代新品种'夏风热浪'（ *C. azalea* × *C. azalea* F_1 new hybrid 'Xiafeng Relang'）。

回交苗编号：DXFRL-No.2。

性状：花朵橘红色，偶有白斑，单瓣型，大型花，花径 10.5～12.5cm，花瓣 6 枚，长倒卵形，先端"V"形凹口，顶部勺状或外翻，雄蕊基部略连生，花丝淡粉红色，开花稠密。叶片浓绿色，叶背面灰绿色，稠密，半直立，革质，椭圆形，长 10.0～11.5cm，宽 4.0～5.0cm，基部急楔形，上部叶缘具浅齿。植株立性，灌木状，生长旺盛。夏初始花，秋、冬季盛花，春季零星开花。

> **Abstract:** Flowers orange red, occasionally with a white marking, single form, large size, 10.5-12.5cm across with 6 petals, petals long obovate, apices notched as V-shaped, top part spoon-like or rolled outward, stamens slightly united at the base, filaments pale red, bloom dense. Leaves dark green, dense, semi-erected, leathery, urgently cuneate at the base, margins shallowly serrate at the upper part. Plant upright, bushy and growth vigorous. Starts to bloom from early-summer, fully blooms from autumn to winter and sporadically in spring.

HAR-02-2. 大喜临门
Daxi Linmen (Great Happy Event is Coming)

回交组合：杜鹃红山茶 × 杜鹃红山茶 F_1 代新品种'夏风热浪'（ C. azalea × C. azalea F_1 new hybrid 'Xiafeng Relang'）。

回交苗编号：DXFRL-No.3 + 22。

性状：花朵桃红色至红色，偶有白斑，单瓣型，中型花，花径 8.5～10.0cm，花瓣 5～6 枚，阔倒卵形，雄蕊基部略连生，花丝淡粉红色，开花稠密。叶片浓绿色，叶背面灰绿色，厚革质，阔椭圆形，长 9.0～10.5cm，宽 5.0～5.5cm，基部略楔形，叶缘光滑。植株立性，灌状，生长旺盛。夏初始花，秋、冬季盛花，春初零星开花。

Abstract: Flowers peach red to red, occasionally with a white marking, single form, medium size, 8.5-10.0cm across with 5-6 broad obovate petals, stamens slightly united at the base, bloom dense. Leaves dark green, back surfaces gray-green, heavy leathery, slightly cuneate at the base, margins smooth. Plant upright, bushy and growth vigorous. Starts to bloom from early-summer, fully blooms from autumn to winter and sporadically in early-spring.

HAR-02-3. 秋风送霞
Qiufeng Songxia (Autumn Breeze Blowing Rosy Clouds)

回交组合：杜鹃红山茶 × 杜鹃红山茶 F_1 代新品种'夏风热浪'(*C. azalea* × *C. azalea* F_1 new hybrid 'Xiafeng Relang')。

回交苗编号：DXFRL-No.4。

性状：花朵红色，偶有大理石白斑，随着花朵开放，中部花瓣花色逐渐变为淡红色，完全重瓣型，中型花，花径8.0～10.0cm，花朵厚5.0cm，花瓣70枚以上，呈10～15轮覆瓦状排列，常见多花芯现象，在冷凉季节，有时呈六角形排列，花瓣逐片掉落。叶片浓绿色，叶背面灰绿色，革质，椭圆形，长7.5～8.5cm，宽2.5～4.0cm，基部楔形，叶缘具细齿，叶尖渐尖。植株紧凑，矮灌状，枝叶稠密，生长旺盛。夏初始花，秋、冬季盛花，春季零星开花。

Abstract: Flowers red, occasionally with marbled white markings, as flowers opening, central petals gradually change into light red, formal double form, medium size, 8.0-10.0cm across and 5.0cm deep with over 70 imbricate petals arranged in 10-15 rows. The phenomenon of multiple flower cores is common, sometimes petals arranged in a hexagon in cool season, petals fall one by one. Leaves dark green, leathery, elliptic, cuneate at the base, margins thinly serrate. Plant compact, bushy, branches dense and growth vigorous. Starts to bloom from early-summer, fully blooms from autumn to winter and sporadically in spring.

HAR-02-4. 紫瓣融雪
Ziban Rongxue (Purple Petals Melting Snow)

回交组合：杜鹃红山茶 × 杜鹃红山茶 F_1 代新品种'夏风热浪'(*C. azalea* × *C. azalea* F_1 new hybrid 'Xiafeng Relang')。

回交苗编号：DXFRL-No.7。

性 状：花朵粉红色，泛紫色调，花瓣的中部渐呈淡粉白色，半重瓣型，大型花，花径 11.0～12.5cm，花朵厚4.5cm，大花瓣18～20枚，呈3轮松散有序排列，边缘内卷，中部小花瓣直立，雄蕊近散射。叶片浓绿色，叶背面灰绿色，厚革质，椭圆形，长9.5～11.0cm，宽3.5～4.5cm，基部急楔形，边缘具浅齿。植株紧凑，矮灌状，生长旺盛。夏初始花，秋、冬季盛花，春季零星开花。

Abstract: Flowers pink with purple hue, gradually fading to pink-white at mid-part of the petals, semi-double form, large size, 11.0-12.5cm across and 4.5cm deep with 18-20 large petals arranged loosely and orderly in 3 rows, petal edges rolled inward, central small petals erected, stamens slightly scattered. Leaves dark green, heavy leathery, elliptic, urgently cuneate at the base, margins shallowly serrate. Plant compact, bushy and growth vigorous. Starts to bloom from early-summer, fully blooms from autumn to winter and sporadically in spring.

HAR-02-5. 橘粉醉夏
Jufen Zuixia（Orange Pink Drunken Summer）

回交组合： 杜鹃红山茶 × 杜鹃红山茶 F_1 代新品种'夏风热浪'（*C. azalea* × *C. azalea* F_1 new hybrid 'Xiafeng Relang'）。

回交苗编号： DXFRL-No.8。

性状： 花朵橘粉红色，花瓣中部粉色较淡，有时会有白条纹，半重瓣型至牡丹型，中型花，花径8.5～10.0cm，花朵厚4.3cm，大花瓣15～19枚，呈2～3轮松散排列，阔倒卵形，中部小花瓣半直立，与雄蕊混生。叶片浓绿色，叶背面灰绿色，厚革质，椭圆形，长9.5～10.5cm，宽3.5～4.5cm，基部楔形，上部叶缘锯齿浅。植株紧凑，生长旺盛。夏初始花，秋、冬季盛花，春季零星开花。

Abstract: Flowers orange pink and light pink at the mid-part of the petals, sometimes, some white stripes visible on the petals, semi-double to peony form, medium size, 8.5-10.0cm across and 4.3cm deep with 15-19 large petals arranged loosely in 2-3 rows, petals broad obovate, central small petals semi-erected, mixed with stamens. Leaves dark green, heavy leathery, elliptic, cuneate at the base, margins shallowly serrate at the upper part. Plants compact and growth vigorous. Starts to bloom from early-summer, fully blooms from autumn to winter and sporadically in spring.

HAR-02-6. 灿烂阳光
Canlan Yangguang（Brilliant Sunshine）

回交组合： 杜鹃红山茶 × 杜鹃红山茶 F_1 代新品种'夏风热浪'（*C. azalea* × *C. azalea* F_1 new hybrid 'Xiafeng Relang'）。

回交苗编号： DXFRL-No.13。

性状： 花朵深桃红色至红色，半重瓣型，中型至大型花，花径 8.5～12.5cm，花朵厚 5.0cm，大花瓣 13～15 枚，呈 2 轮排列，花瓣阔倒卵形，厚质，先端微凹，雄蕊基部略连生，花丝辐射状，花药黄色。叶片浓绿色，叶背面灰绿色，革质，阔椭圆形，长 10.5～11.5cm，宽 3.5～4.0cm，基部楔形，上部边缘锯齿浅。植株立性，枝叶稠密，生长旺盛。夏初始花，秋、冬季盛花，春季零星开花。

Abstract: Flowers deep peach red to red, semi-double form, medium to large size, 8.5-12.5cm across and 5.0cm deep with 13-15 large petals arranged in 2 rows, petals broad obovate, thick texture, apices emarginate, stamens slightly united at the base, filaments radial, anthers yellow. Leaves dark green, leathery, broad elliptic, cuneate at the base, margins shallowly serrate at the upper part. Plants upright, branches dense and growth vigorous. Starts to bloom from early-summer, fully blooms from autumn to winter and sporadically in spring.

HAR-02-7. 锦绣河山
Jinxiu Heshan (Land of Splendours)

回交组合：杜鹃红山茶 × 杜鹃红山茶 F₁ 代新品种'夏风热浪'（C. azalea × C. azalea F₁ new hybrid 'Xiafeng Relang'）。

回交苗编号：DXFRL-No.14。

性状：夏季花朵深粉红色，秋冬季节花朵红色，泛紫色调，瓣面具白色条纹、斑块和深红色脉纹，花朵半开放时，似完全重瓣型，完全开放后为松散牡丹型，大型花，花径 10.0～11.5cm，花朵厚 5.7cm，花瓣 28～33 枚，呈 5～6 轮松散排列，狭倒卵形，边缘内卷，雄蕊簇生。叶片浓绿色，叶背面灰绿色，厚革质，椭圆形，长 9.0～10.0cm，宽 3.5～4.0cm，基部楔形，边缘近光滑。植株紧凑，矮灌状，生长旺盛。夏初始花，秋、冬季盛花，春季零星开花。

Abstract: Flowers deep pink in summer and red with purple hue in autumn and winter, petal surfaces with white stripes or markings, deep red veins visible, when semi-opened, the flowers are looked as formal double form, when fully opened, the flowers are loose peony form, large size, 10.0-11.5cm across and 5.7cm deep with 28-33 petals arranged loosely in 5-6 rows, petals narrow obovate, edges rolled inward, stamens clustered. Leaves dark green, heavy leathery, elliptic, cuneate at the base, margins nearly smooth. Plant compact, bushy and growth vigorous. Starts to bloom from early-summer, fully blooms from autumn to winter and sporadically in spring.

HAR-02-8. 幸福玛黛琳
Xingfu Madailin（Happy Madeline）

回交组合：杜鹃红山茶 × 杜鹃红山茶 F_1 代新品种'夏风热浪'（*C. azalea* × *C. azalea* F_1 new hybrid 'Xiafeng Relang'）。

回交苗编号：DXFRL-No.15。

性状：花朵粉红色，渐渐变为柔嫩的淡粉红色，雄蕊瓣偶有少量白斑，半重瓣型至玫瑰重瓣型，大型花，花径 10.0～12.0cm，花朵厚 4.8cm，外部大花瓣 18～20 枚，呈 3 轮有序排列，花瓣阔倒卵形，外翻，开花稠密。叶片浓绿色，叶背面灰绿色，厚革质，阔椭圆形，长 8.5～9.0cm，宽 4.0～4.5cm，上部边缘具浅齿，下部边缘光滑，基部楔形。植株紧凑，立性，生长旺盛。夏初始花，秋、冬季盛花，春季零星开花。

注：为祝贺美国茶花协会前执行理事塞莱斯特·理查德（Celeste M. Richard）女士的女儿诞生，经她同意，本品种特用其女儿的名字命名。

Abstract: Flowers pink and gradually changing into lightly soft pink, occasionally with a few of white markings on the petaloids, semi-double to rose-double form, large size, 10.0-12.0cm across and 4.8cm deep with 18-20 exterior large petals arranged orderly in 3 rows, petals broad obovate, turned outward, bloom dense. Leaves dark green, heavy leathery, broad elliptic, margins shallowly serrate at the upper part and cuneate at the base. Plant compact, upright and growth vigorous. Starts to bloom from early-summer, fully blooms from autumn to winter and sporadically in spring.

Mark: To congratulate Mrs Celeste M. Richard, the former executive director of ACS, on the birth of her daughter, the hybrid is specially named after her daughter with her consent.

HAR-02-9. 双面佳人
Shuangmian Jiaren (Beautiful Maiden with Double Faces)

回交组合：杜鹃红山茶 × 杜鹃红山茶 F_1 代新品种'夏风热浪'（*C. azalea* × *C. azalea* F_1 new hybrid 'Xiafeng Relang'）。

回交苗编号：DXFRL-No.17。

性状：花朵红色，渐渐变为深粉红色，中部花瓣偶为淡粉红色，完全重瓣型，大型至巨型花，花径 11.5～13.5cm，花朵厚 5.0cm，花瓣 60 枚以上，呈 6～7 轮覆瓦状紧实排列，花瓣倒卵形。叶片浓绿色，叶背面灰绿色，厚革质，椭圆形，长 9.0～9.8cm，宽 3.5～4.0cm，基部楔形，叶缘光滑。植株立性，紧凑，灌木状，生长旺盛。夏初始花，秋、冬季盛花，春季零星开花。

Abstract: Flowers red, gradually changed into deep pink, the central petals occasionally light pink, formal double form, large to very large size, 11.5-13.5cm across and 5.0cm deep with over 60 imbricate petals arranged tightly in 6-7 rows. Leaves dark green, heavy leathery, cuneate at the base, margins smooth. Plant upright, compact, bushy and growth vigorous. Starts to bloom from early-summer, fully blooms from autumn to winter and sporadically in spring.

HAR-02-10. 多季盛情
Duoji Shengqing (Multiseasonal Great Kindness)

回交组合： 杜鹃红山茶 × 杜鹃红山茶 F_1 代新品种'夏风热浪'(*C. azalea* × *C. azalea* F_1 new hybrid 'Xiafeng Relang')。

回交苗编号： DXFRL-No.19。

性状： 冷凉季节花朵艳红似火，炎热季节花瓣上部具大量白斑，半重瓣型，大型花，花径11.5～13.0cm，花朵厚5.5cm，花瓣15～18枚，呈3轮排列，花瓣长倒卵形，瓣长5.5cm，宽3.5cm，厚质，坚挺，雄蕊基部略连生，辐射状，开花稠密。叶片浓绿色，叶背面灰绿色，厚革质，阔椭圆形，长9.7～10.5cm，宽4.5～5.5cm，基部楔形，叶缘光滑。植株紧凑，灌木状，生长旺盛。夏初始花，秋、冬季盛花，春季零星开花。

> **Abstract:** Flowers bright red that is like fire in cold season, but with a lot of white markings on petals in hot season, semi-double form, large size, 11.5-13.0cm across and 5.5cm deep with 15-18 petals arranged in 3 rows, petals long obovate, thick texture, firm, stamens slightly united at the base, bloom dense. Leaves dark green, heavy leathery, broad elliptic, cuneate at the base, margins smooth. Plants compact, bushy and growth vigorous. Starts to bloom from early-summer, fully blooms from autumn to winter and sporadically in spring.

HAR-02-11. 粉颜仙姿
Fenyan Xianzi（Pink Face & Fairy Posture）

回交组合：杜鹃红山茶 × 杜鹃红山茶 F_1 代新品种'夏风热浪'（ C. azalea × C. azalea F_1 new hybrid 'Xiafeng Relang'）。

回交苗编号：DXFRL-No.20。

性状：花朵淡桃红色，瓣面可见隐约的红脉纹，半重瓣型，中型至大型花，花径8.5～11.5cm，花朵厚5.3cm，大花瓣20～25枚，呈4～5轮排列，雄蕊基部连生。叶片浓绿色，叶背面灰绿色，革质，狭长椭圆形，长11.5～12.5cm，宽3.0～3.5cm，基部急楔形，上部边缘锯齿浅。植株立性，枝叶稠密，生长旺盛。夏初始花，秋、冬季盛花，春季零星开花。

> **Abstract:** Flowers light peach red with faint red veins at petal surfaces, semi-double form, medium to large size, 8.5-11.5cm across and 5.3cm deep with 20-25 large petals arranged in 4-5 rows, stamens united at the base. Leaves dark green, leathery, narrow-long elliptic, urgently cuneate at the base, margins shallowly serrate at the upper part. Plants upright, branches dense and growth vigorous. Starts to bloom from early-summer, fully blooms from autumn to winter and sporadically in spring.

HAR-02-12. 凤冠霞帔
Fengguan Xiapei (Phoenix Coronet & Robes)

回交组合： 杜鹃红山茶 × 杜鹃红山茶 F_1 代新品种'夏风热浪'（ *C. azalea* × *C. azalea* F_1 new hybrid 'Xiafeng Relang'）。

回交苗编号： DXFRL-No.23。

性状： 花朵红色，渐变为淡紫粉红色，瓣面可见深红色脉纹，偶具模糊的白条纹，单瓣型至半重瓣型，中型至大型花，花径 9.0～12.0cm，花朵厚 5.5cm，大花瓣 21～24 枚，呈 4 轮排列，花瓣长倒卵形，雄蕊散射。叶片浓绿色，叶背面灰绿色，革质，长椭圆形，长 9.5～10.5cm，宽 3.5～4.5cm，基部急楔形，上部边缘锯齿浅。植株立性，枝叶稠密，生长旺盛。春末始花，夏、秋、冬季盛花，春季零星开花。

Abstract: Flowers red, gradually changed into lightly purple-pink, with deep red veins visible at petal surfaces, occasionally with hazy white stripes, single to semi-double form, medium to large size, 9.0-12.0cm across and 5.5cm deep with 21-24 large petals arranged in 4 rows, petals long obovate, stamens radial. Leaves dark green, leathery, long elliptic, urgently cuneate at the base, margins shallowly serrate at the upper part. Plants upright, branches dense and growth vigorous. Starts to bloom from late-spring, fully blooms in summer, autumn and winter and sporadically in spring.

HAR-02-13. 当代俏丽
Dangdai Qiaoli (Contemporary Pretty Girl)

回交组合：杜鹃红山茶 × 杜鹃红山茶 F_1 代新品种'夏风热浪'（ C. azalea × C. azalea F_1 new hybrid 'Xiafeng Relang'）。

回交苗编号：DXFRL-No.26。

性状：花朵艳红色，偶具白斑，随着花朵开放，渐渐变为淡红色，通常炎夏花朵红色较淡，秋冬红色较浓，半重瓣型，中型花，花径 8.5～10.0cm，花朵厚 4.5cm，花瓣 18～22 枚，呈 3～4 轮有序排列，花瓣阔倒卵形，雄蕊基部略连生，偶有少量雄蕊瓣。叶片浓绿色，叶背面灰绿色，厚革质，椭圆形，长 8.5～9.5cm，宽 4.0～5.0cm，基部急楔形，叶缘光滑。植株紧凑，灌木状，生长旺盛。夏初始花，秋、冬季盛花，春季零星开花。

Abstract: Flowers bright red, occasionally with white markings, as flowers opening, the color gradually changes into light red, usually flowers are light red in hot summer and deep red in autumn and winter, semi-double form, medium size, 8.5-10.0cm across and 4.5cm deep with 18-22 petals arranged orderly in 3-4 rows, petals broad obovate, stamens slightly united at the base, occasionally a few of petaloids appeared at the center. Leaves dark green, heavy leathery, urgently cuneate at the base, margins smooth. Plants compact, bushy and growth vigorous. Starts to bloom from early-summer, fully blooms from autumn to winter and sporadically in spring.

HAR-02-14. 清丽少女
Qingli Shaonü (Charming & Young Girl)

回交组合： 杜鹃红山茶 × 杜鹃红山茶 F_1 代新品种'夏风热浪'(C. azalea × C. azalea F_1 new hybrid 'Xiafeng Relang')。

回交苗编号： DXFRL-No.27。

性状： 花朵粉红色，花芯偶有少量白斑，半重瓣型至牡丹型，中型至大型花，花径9.5～11.0cm，花朵厚4.8cm，外部大花瓣13～15枚，呈2轮松散有序排列，平铺或外翻，倒卵形，中部小花瓣3～5枚直立、扭曲或波浪状，雄蕊基部略连生，花丝散射。叶片浓绿色，叶背面灰绿色，厚革质，阔椭圆形，长9.0～10.5cm，宽4.5～5.0cm，基部楔形，边缘具浅齿，侧脉黄绿色、清晰。植株紧凑，矮灌状，生长旺盛。夏初始花，秋、冬季盛花，春季零星开花。

Abstract: Flowers pink, occasionally with a few white markings on the central petals, semi-double to peony form, medium to large size, 9.5-11.0cm across and 4.8cm deep with 13-15 exterior large petals arranged loosely and orderly in 2 rows, obovate, 3-5 small petals at the center erected, wrinkled or wavy, stamens slightly united at the base, filaments scattered. Leaves dark green, heavy leathery, broad elliptic, cuneate at the base, margins shallowly serrate, lateral veins yellow green and visible. Plant compact, bushy and growth vigorous. Starts to bloom from early-summer, fully blooms from autumn to winter and sporadically in spring.

HAR-02-15. 双红荔城
Shuanghong Licheng（Double Red Litchi City）

回交组合： 杜鹃红山茶 × 杜鹃红山茶 F_1 代新品种'夏风热浪'（ C. azalea × C. azalea F_1 new hybrid 'Xiafeng Relang'）。

回交苗编号： DXFRL-No.28。

性状： 花朵艳红色至橘红色，渐渐变为深红色，具白斑或白条纹，半重瓣型至松散的牡丹型，中型至大型花，花径9.0～12.0cm，花朵厚6.0cm，大花瓣16～18枚，呈3轮松散排列，花瓣倒卵形，雄蕊簇生。叶片浓绿色，叶背面灰绿色，厚革质，椭圆形，长9.5～10.5cm，宽4.5～5.0cm，基部楔形，边缘近光滑，中脉黄绿色，侧脉不明显。植株立性，生长旺盛。夏初始花，秋、冬季盛花，春季零星开花。

Abstract: Flowers bright red to orange red and gradually changing into deep red, with white markings or white stripes, semi-double to loose peony form, medium to large size, 9.0-12.0cm across and 6.0cm deep with 16-18 large petals arranged loosely in 3 rows, petals obovate, stamens clustered. Leaves dark green, heavy leathery, elliptic, cuneate at the base, margins smooth, midrib yellow-green, lateral veins unconspicuous. Plant upright and growth vigorous. Starts to bloom from early-summer, fully blooms from autumn to winter and sporadically in spring.

HAR-02-16. 橘色晓霞
Juse Xiaoxia (Orange Color Foredawn)

回交组合：杜鹃红山茶 × 杜鹃红山茶 F₁ 代新品种'夏风热浪'(*C. azalea* × *C. azalea* F₁ new hybrid 'Xiafeng Relang')。

回交苗编号：DXFRL-No.30。

性状：花朵淡橘粉红色，花瓣上部淡粉红色至白色，半重瓣型，中型花，花径 8.0～10.0cm，花朵厚 5.0cm，大花瓣约 15 枚，呈 2 轮排列，阔倒卵形，边缘内扣，雄蕊辐射状，基部略连生，开花稠密。叶片浓绿色，叶背面灰绿色，厚革质，窄椭圆形，长 9.5～10.5cm，宽 3.0～4.0cm，基部急楔形，叶缘近光滑。植株立性，生长旺盛。夏初始花，秋、冬季盛花，春季零星开花。

Abstract: Flowers light orange pink but light pink or nearly white at the upper part of petals, semi-double form, medium size, 8.0-10.0cm across and 5.0cm deep with about 15 large petals arranged in 2 rows, petals broad obovate, edges rolled inward, stamens radial-shaped, bloom dense. Leaves dark-green, heavy leathery, narrow elliptic, urgently cuneate at the base, margins nearly smooth. Plant upright and growth vigorous. Starts to bloom from early-summer, fully blooms from autumn to winter and sporadically in spring.

HAR-02-17. 飞溅火花
Feijian Huohua (Flying Sparks)

回交组合： 杜鹃红山茶 × 杜鹃红山茶 F_1 代新品种'夏风热浪'（ C. azalea × C. azalea F_1 new hybrid 'Xiafeng Relang'）。

回交苗编号： DXFRL-No.33。

性状： 花朵深红色，渐向花瓣基部泛白色，半重瓣型，花芯偶有雄蕊瓣，中型至大型花，花径 8.5～10.5cm，花朵厚 4.5cm，大花瓣约 18 枚，呈 3 轮排列，狭长倒卵形，两侧边缘外扣，全开放后，多数花瓣整体下弯，雄蕊辐射状，基部略连生，似四溅的火花，开花稠密。叶片浓绿色，叶背面灰绿色，厚革质，窄椭圆形，长 9.0～10.0cm，宽 3.0～4.0cm，基部急楔形，叶缘近光滑。植株立性，生长旺盛。夏初始花，秋、冬季盛花，春季零星开花。

Abstract: Flowers deep red, fading to white at petal base, semi-double form, occasionally with a few of petaloids at the center, medium to large size, 8.5-10.5cm across and 4.5cm deep with about 18 large petals arranged in 3 rows, petals narrow-long obovate, edges rolled outward, most of petals downward-curved after fully opened, stamens radial-shaped, which are like sparks flying, bloom dense. Leaves dark-green, heavy leathery, narrow elliptic, urgent cuneate at the base, margins nearly smooth. Plant upright and growth vigorous. Starts to bloom from early-summer, fully blooms from autumn to winter and sporadically in spring.

HAR-02-18. 花容月貌
Huarong Yuemao（Beautiful Feature as Flower & Moon）

回交组合：杜鹃红山茶 × 杜鹃红山茶 F_1 代新品种'夏风热浪'（*C. azalea* × *C. azalea* F_1 new hybrid 'Xiafeng Relang'）。

回交苗编号：DXFRL-No. 37。

性状：花朵桃红色至红色，略泛橘红色调，具模糊的脉纹，半重瓣型，中型至大型花，花径 8.0～11.0cm，花朵厚 5.2cm，大花瓣 16～18 枚，呈 3 轮排列，花瓣阔倒卵形，瓣面略皱褶，雄蕊基部略连生，花丝淡红色，花药金黄色。叶片浓绿色，叶背面灰绿色，革质，窄椭圆形，长 8.5～9.5cm，宽 3.0～3.5cm，基部急楔形，上部叶缘具浅齿。植株立性，生长旺盛。夏初始花，秋、冬季盛花，春季零星开花。

Abstract: Flowers peach red to red, slightly with orange hue and faint veins, semi-double form, medium to large size, 8.0-11.0cm across and 5.2cm deep with 16-18 large petals arranged in 3 rows, petals broad obovate, petal surfaces slightly wrinkled, stamens slightly united at the base, filaments pale red, anthers golden yellow. Leaves dark-green, leathery, narrow elliptic, urgently cuneate at the base, margins shallowly serrate at the upper part. Plant upright and growth vigorous. Starts to bloom from early-summer, fully blooms from autumn to winter and sporadically in spring.

HAR-02-19. 英雄本色
Yingxiong Bense（Hero's True Quality）

回交组合：杜鹃红山茶 × 杜鹃红山茶 F_1 代新品种'夏风热浪'（ C. azalea × C. azalea F_1 new hybrid 'Xiafeng Relang'）。

回交苗编号：DXFRL-No.38。

性状：花朵红色，瓣面偶然出现小白点，单瓣型，大型花，花径 11.5～12.5cm，花瓣 6～7 枚，阔倒卵形，厚质，雄蕊基部连生，呈短柱状。叶片浓绿色，叶背面灰绿色，厚革质，阔椭圆形，长 9.0～11.0cm，宽 4.0～4.5cm，基部楔形，叶缘齿浅。植株开张，生长旺盛。夏初始花，秋、冬季盛花，春季零星开花。

Abstract: Flowers red, occasionally with small white dots on petal surfaces, single form, large size, 11.5-12.5cm across with 6-7 broad obovate petals, petals thick texture, stamens united into short column at the base. Leaves dark green, heavy leathery, broad elliptic, cuneate at the base, margins shallowly serrate. Plant spread and growth vigorous. Starts to bloom from early-summer, fully blooms from autumn to winter and sporadically in spring.

HAR-02-20. 大福大贵
Dafu Dagui (Big Happiness & Big Millionaire)

回交组合： 杜鹃红山茶 × 杜鹃红山茶 F_1 代新品种'夏风热浪'（C. azalea × C. azalea F_1 new hybrid 'Xiafeng Relang'）。

回交苗编号： DXFRL-No.41。

性状： 刚开放时，花朵桃红色，3～4天后逐渐变为深粉红色或淡红色，渐向花瓣先端呈粉红色，牡丹型，大型至巨型花，花径12.0～13.5cm，花朵厚6.5～7.0cm，花瓣37～41枚，呈6轮以上松散排列，花瓣多为半直立状态，长倒卵形，边缘两侧外卷，呈辐射状，花朵整朵掉落。叶片浓绿色，叶背面灰绿色，厚革质，狭长椭圆形，长9.5～10.5cm，宽3.5～4.0cm，基部急楔形，边缘近光滑，侧脉绿色，略突起。植株立性，枝叶稠密，生长旺盛。夏初始花，秋、冬季盛花，春季零星开花。

Abstract: When just opening, the flowers are peach red, then gradually change into deep pink or light red after 3-4 days, fading to pink at the top of the petals, peony form, large to very large size, 12.0-13.5cm across and 6.5-7.0cm deep with 37-41 petals arranged loosely in over 6 rows, petals semi-erected, long obovate, petal edges rolled outward, which are radial, flowers fall whole. Leaves dark green, heavy leathery, narrow-long elliptic, urgently cuneate at the base, margins nearly smooth, lateral veins green and raised. Plant upright, branches dense and growth vigorous. Starts to bloom from early-summer, fully blooms from autumn to winter and sporadically in spring.

HAR-02-21. 回眸之丽
Huimou Zhili (Beautiful Glancing Back)

回交组合： 杜鹃红山茶 × 杜鹃红山茶 F₁ 代新品种 '夏风热浪'（ C. azalea × C. azalea F₁ new hybrid 'Xiafeng Relang)。

回交苗编号： DXFRL-No. 50。

性状： 花朵艳红色，具模糊的白条纹，半重瓣型，中型至大型花，花径 9.3～11.5cm，花朵厚 5.0cm，大花瓣 17～19 枚，呈 3 轮松散排列，花瓣长倒卵形，呈勺状，雄蕊簇生，花丝粉白色，花药金黄色。叶片浓绿色，叶背面灰绿色，革质，椭圆形，长 8.0～9.0cm，宽 3.5～4.5cm，基部楔形，上部叶缘具浅齿。植株立性，生长旺盛。夏初始花，秋、冬季盛花，春季零星开花。

> **Abstract:** Flowers bright red, with faint white stripes, semi-double form, medium to large size, 9.3-11.5cm across and 5.0cm deep with 17-19 large petals arranged loosely in 3 rows, petals long obovate which are spoon-like, stamens clustered, filaments pink-white, anthers golden yellow. Leaves dark-green, leathery, elliptic, cuneate at the base, margins shallowly serrate at the upper part. Plant upright and growth vigorous. Starts to bloom from early-summer, fully blooms from autumn to winter and sporadically in spring.

HAR-02-22. 秋月闻莺
Qiuyue Wenying（Orioles Singing from the Autumn Moon）

回交组合： 杜鹃红山茶 × 杜鹃红山茶 F_1 代新品种'夏风热浪'（ C. azalea × C. azalea F_1 new hybrid 'Xiafeng Relang'）。

回交苗编号： DXFRL-No.52。

性状： 花朵刚开放时深粉红色至橘红色，花芯多为珠球状，而后逐渐变为淡粉红色，偶具白斑，半重瓣型，大型花，花径 11.0～12.5cm，花朵厚 5.0cm，大花瓣 30 枚以上，呈 3～4 轮排列，偶有雄蕊瓣，花瓣长倒卵形，雄蕊基部近离生。叶片浓绿色，叶背面灰绿色，厚革质，长椭圆形，长 11.0～11.5cm，宽 3.5～4.0cm，基部急楔形，上部边缘锯齿浅。植株立性，枝叶稠密，生长旺盛。春末始花，夏、秋、冬季盛花，春季零星开花。

> **Abstract:** When just opening, the flowers are deep pink to orange red with a globular flower core, and then gradually change into light pink, occasionally with a few of white markings, semi-double form, large size, 11.0-12.5cm across and 5.0cm deep with over 30 large petals arranged in 3-4 rows, occasionally with some petaloids, petals long obovate, stamens nearly free at the base. Leaves dark-green, heavy leathery, long elliptic, urgently cuneate at the base, margins shallowly serrate at the upper part. Plants upright, branches dense and growth vigorous. Starts to bloom from later-spring, fully blooms in summer, autumn and winter and sporadically in spring.

HAR-02-23. 夏日海滩
Xiari Haitan（Summer's Beach）

回交组合： 杜鹃红山茶 × 杜鹃红山茶 F$_1$ 代新品种'夏风热浪'（*C. azalea* × *C. azalea* F$_1$ new hybrid 'Xiafeng Relang'）。

回交苗编号： DXFRL-No.53。

性状： 花朵淡桃红色，外部花瓣具白色宽边，牡丹型，大型花，花径10.0～11.0cm，花朵厚5.7cm，外部大花瓣12～15枚，呈2～3轮排列，花瓣倒卵形，平展，中部小花瓣边缘皱褶，雄蕊簇生。叶片浓绿色，叶背面灰绿色，革质，椭圆形，长9.0～9.5cm，宽4.0～4.5cm，基部楔形，边缘近光滑。植株紧凑，矮性，生长旺盛。夏初始花，秋、冬季盛花，春季零星开花。

Abstract: Flowers light peach red, exterior petals with broad white edges, peony form, large size, 10.0-11.0cm across and 5.7cm deep with 12-15 exterior large petals arranged in 2-3 rows, petals obovate, explanate, central small petals wrinkled, stamens clustered. Leaves dark green, leathery, elliptic, cuneate at the base, margins nearly smooth. Plant compact, dwarf and growth vigorous. Starts to bloom from early-summer, fully blooms from autumn to winter and sporadically in spring.

HAR-02-24. 闲花映池
Xianhua Yingchi（Wild Flowers Reflected from Pool）

回交组合：杜鹃红山茶 × 杜鹃红山茶 F_1 代新品种'夏风热浪'（ C. azalea × C. azalea F_1 new hybrid 'Xiafeng Relang'）。

回交苗编号：DXFRL-No.56 + 51。

性状：花朵淡粉红色，渐向瓣面中部呈白色，单瓣型，中型至大型花，花径9.5～12.0cm，花瓣5～6枚，阔倒卵形，雄蕊基部略连生，花丝淡粉红色，花药黄色，开花稠密。叶片浓绿色，叶背面灰绿色，厚革质，长椭圆形，长8.0～9.5cm，宽3.5～4.5cm，基部楔形，叶缘光滑。植株略立性，生长旺盛。夏初始花，秋、冬季盛花，春季零星开花。

Abstract: Flowers pale pink, fading to white at the mid-surfaces of petals, single form, medium to large size, 9.5-12.0cm across with 5-6 broad obovate petals, stamens slightly united at the base, filaments pale pink, anthers yellow, bloom dense. Leaves dark green, heavy leathery, cuneate at the base, margins smooth. Plant slightly upright and growth vigorous. Starts to bloom from early-summer, fully blooms from autumn to winter and sporadically in the spring.

HAR-02-25. 绚丽夏秋
Xuanli Xiaqiu (Gorgeous Summer & Autumn)

回交组合：杜鹃红山茶 × 杜鹃红山茶 F_1 代新品种'夏风热浪'（ C. azalea × C. azalea F_1 new hybrid 'Xiafeng Relang' ）。

回交苗编号：DXFRL-No.59。

性状：花朵鲜艳的桃红色，单瓣型，中型花，花径 8.0～9.5cm，花瓣 5～6 枚，阔倒卵形，上部外翻，瓣面皱褶，雄蕊基部连生，呈短柱状，开花稠密。叶片浓绿色，叶背面灰绿色，革质，椭圆形，长 8.0～9.5cm，宽 4.0～4.5cm，基部楔形，叶缘光滑。植株立性，灌木状，生长旺盛。夏初始花，秋、冬季盛花，春季零星开花。

Abstract: Flowers bright peach red, single form, medium size, 8.0-9.5cm across with 5-6 broad obovate petals, petals rolled outward and wrinkled at the upper part, stamens united into a column at the base, bloom dense. Leaves dark green, leathery, cuneate at the base, margins smooth. Plant upright, bushy and growth vigorous. Starts to bloom from early-summer, fully blooms from autumn to winter and sporadically in spring.

HAR-02-26. 多姿丽影
Duozi Liying (Multi-Poses & Pretty Shadow)

回交组合：杜鹃红山茶 × 杜鹃红山茶 F_1 代新品种'夏风热浪'（ C. azalea × C. azalea F_1 new hybrid 'Xiafeng Relang'）。

回交苗编号：DXFRL-No.65。

性状：花朵刚开放时，呈淡桃红色，略泛橘红色调，渐渐变为粉红色，中部花瓣具隐约的白条纹，松散的牡丹型，中型花，花径 8.5～10.0cm，花朵厚 5.5cm，外部大花瓣 15～18 枚，呈 3 轮松散排列，花瓣倒卵形，中部小花瓣半直立，与簇生雄蕊混生。叶片浓绿色，叶背面灰绿色，厚革质，椭圆形，长 9.5～10.0cm，宽 3.8～4.0cm，基部楔形，边缘锯齿浅而钝。植株立性，生长旺盛。夏初始花，秋、冬季盛花，春季零星开花。

> **Abstract:** Flowers light peach red slightly with orange hue when just opening and gradually change into pink, central petals with faint white stripes, loose peony form, medium size, 8.5-10.0cm across and 5.5cm deep with 15-18 exterior large petals arranged loosely in 3 rows, petals obovate, central small petals semi-erected and mixed with the clustered stamens. Leaves dark green, heavy leathery, elliptic, cuneate at the base, margins shallowly and obtusely serrate. Plant upright and growth vigorous. Starts to bloom from early-summer, fully blooms from autumn to winter and sporadically in spring.

HAR-02-27. 羞涩桃腮
Xiuse Taosai (Peach Cheeks with Shyness)

回交组合：杜鹃红山茶 × 杜鹃红山茶 F_1 代新品种'夏风热浪'（*C. azalea* × *C. azalea* F_1 new hybrid 'Xiafeng Relang'）。

回交苗编号：DXFRL-No.79。

性状：花朵深粉红色至淡红色，半重瓣型，大型花，花径 11.0～12.0cm，花朵厚 4.7cm，花瓣 14～16 枚，呈 2～3 轮松散排列，倒卵形，厚质，外翻或波浪状，雄蕊基部略连生。叶片浓绿色，叶背面灰绿色，厚革质，椭圆形，长 9.5～10.0cm，宽 4.2～5.0cm，基部楔形，叶缘光滑。植株紧凑，灌木状，生长旺盛。夏初始花，秋、冬季盛花，春季零星开花。

Abstract: Flowers deep pink to light red, semi-double form, large size, 11.0-12.0cm across and 4.7cm deep with 14-16 petals arranged loosely in 2-3 rows, petals obovate, rolled outward or wavy, stamens slightly united at the base. Leaves dark green, heavy leathery, cuneate at the base, margins smooth. Plants compact, bushy and growth vigorous. Starts to bloom from early-summer, fully blooms from autumn to winter and sporadically in spring.

◉ 回交组合 HAR-03. 杜鹃红山茶 × 杜鹃红山茶 F₁ 代新品种 '夏日粉黛'

HAR-03. *C. azalea* × *C. azalea* F₁ new hybrid 'Xiari Fendai'

本回交组合将介绍 8 个新品种。

'夏日粉黛'是从杜鹃红山茶 × 非云南山茶杂交种'玉盘金华'杂交组合中得到的（见第一部书第 232～233 页）。在第一部书中，本组曾介绍过 5 个回交新品种（见第一部书第 367～372 页和第 374～375 页）。

因为本回交组合中的亲本杜鹃红山茶和'夏日粉黛'品种的花朵都是单瓣型的，所以，这 8 个回交新品种的花型表达全部是单瓣型的。毫无疑问，它们将受到人们的青睐，也会广泛地应用于园林造景，如应用于花篱、花坛、色块等。

Eight new hybrids will be introduced in this backcross combination.

'Xiari Fendai' was obtained from the cross-combination between *C. azalea* and Non-*reticulata* hybrid 'Yupan Jinhua' (See p.232-233, our first book). Five backcross new hybrids were introduced in the combination of our first book (See p.367-372 and p.374-375, our first book).

Because the flower form of the two cross-parents, *C. azalea* and the cultivar 'Xiari Fendai' is single form, the flowers of eight new hybrids obtained are all single form. It is without doubt that they will be favored by people, and will be widely used for landscaping, such as for hedgerows, flower beds, color blocks and so on.

HAR-03-1. 波吉特爱心
Bojite Aixin (Mrs Birgit Linthe's Loving Heart)

回交组合：杜鹃红山茶 × 杜鹃红山茶 F_1 代新品种'夏日粉黛'（ *C. azalea* × *C. azalea* F_1 new hybrid 'Xiari Fendai'）。

回交苗编号：DFD-No.24。

性状：花朵淡粉红色，泛淡橘红色调，偶有白斑，略有茶香，单瓣型至半重瓣型，小型花，花径6.0～7.5cm，花瓣8～13枚，长倒卵形，长4.1cm，宽1.8cm，近平铺，雄蕊基部连生，呈短柱状，开花稠密。叶片浓绿色，叶背面灰绿色，革质，长倒卵形，长7.5～8.0cm，宽2.5～3.0cm，基部急楔形，叶缘光滑，近似于杜鹃红山茶。植株紧凑，矮性，生长旺盛。春末始花，夏、秋、冬季盛花，翌年春初仍可持续开花。

注：为纪念德国茶花爱好者休伯特·林特（Hubert Linthe）先生与我们之间多年的友谊，特将此品种用他夫人的姓名命名。

Abstract: Flowers light-pink with orange hue, occasionally with white markings, slightly with tea fragrance, single to semi-double form, small size, 6.0-7.5cm across with 8-13 petals, petals long obovate, stamens united into a short column at the base, bloom dense. Leaves dark green, leathery, long obovate, urgently cuneate at the base, margins smooth. Plant compact, dwarf and growth vigorous. Starts to bloom from late-spring, fully blooms in summer, autumn and winter and can still bloom continuously in early-spring of the following year.

Mark: For commemorating years of friendship with Dr. Hubert Linthe, a camellia enthusiast of Germany, we specially named this camellia hybrid after his wife's name.

HAR-03-2. 端庄秀丽
Duanzhuang Xiuli (Dignified Pretty)

回交组合：杜鹃红山茶 × 杜鹃红山茶 F_1 代新品种'夏日粉黛'（ *C. azalea* × *C. azalea* F_1 new hybrid 'Xiari Fendai'）。

回交苗编号：DFD-No. 26 + 34 + 48 + 89。

性状：花朵淡粉红色至粉红色，单瓣型，中型花，花径 8.7～9.8cm，花瓣 6～7 枚，长倒卵形，瓣面近平铺，深粉色，脉纹清晰可见，雄蕊基部略连生，开花稠密。叶片浓绿色，叶背面灰绿色，革质，椭圆形，长 7.6～8.5cm，宽 3.2～3.8cm，基部楔形，叶缘近光滑。植株开张，矮性，生长旺盛。夏初始花，秋、冬季盛花，春季零星开花。

> **Abstract:** Flowers light-pink to pink, single form, medium size, 8.7-9.8cm across with 6-7 long obovate petals, petal surfaces nearly flat, stamens slightly united at the base, bloom dense. Leaves dark green, leathery, elliptic, cuneate at the base, margins nearly smooth. Plant spread, dwarf and growth vigorous. Starts to bloom from early-summer, fully blooms from autumn to winter and sporadically in spring.

HAR-03-3. 火红年代
Huohong Niandai (Fire Red Era)

回交组合：杜鹃红山茶 × 杜鹃红山茶 F_1 代新品种'夏日粉黛'（*C. azalea* × *C. azalea* F_1 new hybrid 'Xiari Fendai'）。

回交苗编号：DFD-No.35。

性状：花朵艳红色至深红色，瓣面可见深红色脉纹，单瓣型，中型花，花径 7.5～8.5cm，花瓣 7～8 枚，中部外翻和扭曲，呈角状，雄蕊基部连生，呈管状，开花稠密。叶片浓绿色，革质，椭圆形，长 7.5～9.0cm，宽 2.5～3.0cm，略扭曲，叶缘光滑，中脉黄绿色。植株紧凑，矮性，生长旺盛。夏初始花，秋、冬季盛花，春季零星开花。

Abstract: Flowers bright red to deep red with some visible deep red veins, single form, medium size, 7.5-8.5cm across with 7-8 petals, mid-parts rolled outward and twisted which is like angular shape, stamens united at the base, bloom dense. Leaves dark green, leathery, elliptic, slightly twisted, margins smooth, midrib yellow green. Plant compact, dwarf and growth vigorous. Starts to bloom from early-summer, fully blooms from autumn to winter and sporadically in spring.

HAR-03-4. 夏云秋色
Xiayun Qiuse (Summer's Cloud & Autumn's Color)

回交组合：杜鹃红山茶 × 杜鹃红山茶 F₁ 代新品种'夏日粉黛'（ $C.$ azalea × $C.$ azalea F₁ new hybrid 'Xiari Fendai'）。

回交苗编号：DFD-No.37。

性状：花朵深粉红色，瓣面具无数小白点，随着花朵的开放，这些小白点越来越明显，单瓣型，大型至巨型花，花径 12.5～13.5cm，花瓣 7～8 枚，阔倒卵形，瓣面上部向外翻卷，雄蕊基部略连生，开花稠密。叶片中等绿色，叶背面灰绿色，硬革质，长椭圆形，长 9.5～10.5cm，宽 3.5～4.0cm，上部叶缘近光滑。植株开张，矮性，生长旺盛。夏初始花，秋、冬季盛花，春季零星开花。

Abstract: Flowers deep pink with countless tiny white spots on the petal surfaces, as flowers opening, the white spots become more and more obvious, single form, large to very large size, 12.5-13.5cm across with 7-8 broad obovate petals, the petal surfaces rolled outward at the upper part, stamens slightly united at the base, bloom dense. Leaves normal green, hard leathery, long elliptic, margins nearly smooth at the upper part. Plant spread, dwarf and growth vigorous. Starts to bloom from early-summer, fully blooms from autumn to winter and sporadically in spring.

HAR-03-5. 一等颜值
Yideng Yanzhi (First Class Pretty)

回交组合： 杜鹃红山茶 × 杜鹃红山茶 F_1 代新品种'夏日粉黛'（*C. azalea* × *C. azalea* F_1 new hybrid 'Xiari Fendai'）。

回交苗编号： DFD-No.61 + 59 + 60。

性状： 花朵深粉红色至淡红色，单瓣型，中型花，花径 8.5～9.7cm，花瓣 6～7 枚，阔倒卵形，雄蕊基部略连生，开花稠密。叶片浓绿色，叶背面灰绿色，革质，椭圆形，长 7.5～8.0cm，宽 3.5～4.0cm，基部急楔形，叶缘光滑。植株开张，枝叶稠密，生长旺盛。夏初始花，秋、冬季盛花，春季零星开花。

> **Abstract:** Flowers deep pink to light red, single form, medium size, 8.5-9.7cm across with 6-7 broad obovate petals, stamens slightly united at the base, bloom dense. Leaves dark green, leathery, elliptic, urgently cuneate at the base, margins smooth. Plant spread, branches dense and growth vigorous. Starts to bloom from early-summer, fully blooms from autumn to winter and sporadically in spring.

HAR-03-6. 六朝脂粉
Liuchao Zhifen (Six Dynasties' Face Powder)

回交组合：杜鹃红山茶 × 杜鹃红山茶 F_1 代新品种'夏日粉黛'（C. azalea × C. azalea F_1 new hybrid 'Xiari Fendai'）。

回交苗编号：DFD-No.64 + 13 + 43。

性状：花蕾细纺锤状，萼片淡绿色。花朵嫩粉红色，单瓣型，中型至大型花，花径 8.5～11.0cm，花瓣 6 枚，阔倒卵形，长 5.5cm，宽 4.0cm，外翻，先端近圆，雄蕊基部略连生，散射状，花丝淡粉红色，花药米黄色，开花稠密。叶片浓绿色，叶背面灰绿色，革质，椭圆形，长 7.5～8.5cm，宽 3.5～4.5cm，基部楔形，叶缘光滑。植株紧凑，枝叶稠密，生长旺盛。夏初始花，秋、冬季盛花，春季零星开花。

Abstract: Flowers tender pink, single form, medium to large size, 8.5-11.0cm across with 6 broad obovate petals, petals turned outward, apices nearly round, stamens slightly united at the base, radial shaped, filaments pale pink, bloom dense. Leaves dark green, back surfaces pale green, leathery, elliptic, cuneate at the base, margins smooth. Plant compact, branches dense and growth vigorous. Starts to bloom from early-summer, fully blooms from autumn to winter and sporadically in spring.

HAR-03-7. 日月重光
Riyue Chongguang (The Sun & The Moon Appear Together)

回交组合： 杜鹃红山茶 × 杜鹃红山茶 F_1 代新品种'夏日粉黛'（ C. azalea × C. azalea F_1 new hybrid 'Xiari Fendai'）。

回交苗编号： DFD-No.67。

性状： 冷凉季节花朵红色，炎热季节花朵粉红色，瓣面中部呈粉白色，单瓣型，中型花，花径9.2～10.0cm，花瓣7枚，呈松散排列，狭长倒卵形，瓣面可见清晰红色脉纹，边缘两侧外翻，雄蕊基部略连生。叶片浓绿色，叶背面灰绿色，革质，长椭圆形，长9.0～9.8cm，宽2.3～3.4cm，基部楔形，叶缘具浅齿。植株紧凑，枝叶稠密，生长旺盛。夏初始花，秋、冬季盛花，春季零星开花。

> **Abstract:** Flowers red in cold season and pink in hot season, white pink at the mid-part of petal surfaces, single form, medium size, 9.2-10.0cm across with 7 narrow and long obovate petals arranged loosely, stamens slightly united at the base. Leaves dark green, leathery, cuneate at the base, margins shallowly serrate. Plant compact, branches dense and growth vigorous. Starts to bloom from early-summer, fully blooms from autumn to winter and sporadically in spring.

HAR-03-8. 八月飘雪
Bayue Piaoxue (August's Snowing)

回交组合： 杜鹃红山茶 × 杜鹃红山茶 F_1 代新品种'夏日粉黛'（ *C. azalea* × *C. azalea* F_1 new hybrid 'Xiari Fendai'）。

回交苗编号： DFD-No.100。

性状： 花朵淡粉红色，瓣面可见无数小白点，单瓣型，中型花，花径8.0～8.5cm，花瓣5～6枚，倒卵形，外翻，先端近圆，雄蕊基部略连生，开花稠密。叶片浓绿色，叶背面灰绿色，革质，椭圆形，长6.5～7.5cm，宽3.0～3.5cm，基部楔形，叶缘光滑。植株紧凑，枝叶稠密，生长旺盛。夏初始花，秋、冬季盛花，春季零星开花。

> **Abstract:** Flowers light pink with countless tiny white spots on petal surfaces, single form, medium size, 8.0-8.5cm across with 5-6 obovate petals, petals rolled outward at the top, stamens slightly united at the base, bloom dense. Leaves dark green, leathery, elliptic, cuneate at the base, margins smooth. Plant compact, branches dense and growth vigorous. Starts to bloom from early-summer, fully blooms from autumn to winter and sporadically in spring.

⊙ 回交组合 HAR-04. 杜鹃红山茶 × 杜鹃红山茶 F_1 代新品种 '夏日红霞'
HAR-04. *C. azalea* × *C. azalea* F_1 new hybrid 'Xiari Hongxia'

本回交组合将介绍 6 个新品种。

'夏日红霞'是从杜鹃红山茶 × 红山茶品种'霍伯'杂交组合中得到的（见本书 HA-35-5）。

本杂交组合获得的新品种在花色上倾向于父本'夏日红霞',而植株、叶片和开花期的性状则倾向于母本杜鹃红山茶。

Six new hybrids will be introduced in this backcross combination.

'Xiari Hongxia' was obtained from the cross-combination between *C. azalea* and *C. japonica* cultivar 'Bob Hope' (See HA-35-5, this book).

The flower color of the new hybrids obtained from this backcross combination tends to their male parent 'Xiari Hongxia', and the characteristics of plants, leaves and the blooming period of the backcross hybrids all tend to their female parent *C. azalea*.

HAR-04-1. 红红火火
Honghong Huohuo (Prosperity & Jollification)

回交组合：杜鹃红山茶 × 杜鹃红山茶 F₁ 代新品种'夏日红霞'（ *C. azalea* × *C. azalea* F₁ new hybrid 'Xiari Hongxia'）。

回交苗编号：DHX-No.1。

性状：花朵艳红色，花芯小花瓣偶有白条纹，半重瓣型，中型花，花径 8.0～9.0cm，花朵厚 4.8cm，大花瓣 16～19 枚，呈 2～3 轮排列，阔倒卵形，外翻，花芯偶有瓣化雄蕊瓣，花丝红色，开花稠密。叶片浓绿色，叶背面灰绿色，厚革质，阔椭圆形，长 9.0～9.5cm，宽 4.0～4.5cm，上部边缘叶齿浅。植株紧凑，生长旺盛。夏初始花，秋、冬季盛花，春季零星开花。

Abstract: Flowers bright red, occasionally with white stripes on the small petals at the center, semi-double form, medium size, 8.0-9.0cm across and 4.8cm deep with 16-19 large petals arranged in 2-3 rows, petals broad obovate, rolled outward, occasionally some petaloids at the center, filaments red, bloom dense. Leaves dark green, heavy leathery, broad elliptic, margins shallowly serrate at the upper part. Plant compact and growth vigorous. Starts to bloom from early-summer, fully blooms from autumn to winter and sporadically in spring.

HAR-04-2. 精致绣品
Jingzhi Xiupin (Delicate Embroidery)

回交组合： 杜鹃红山茶 × 杜鹃红山茶 F_1 代新品种'夏日红霞'（ C. azalea × C. azalea F_1 new hybrid 'Xiari Hongxia'）。

回交苗编号： DHX-No.5。

性状： 花朵红色至栗红色，花芯小花瓣偶有白斑，半重瓣型至玫瑰重瓣型，夏季多为小型至中型花，花径 7.0～9.0cm，花朵厚 4.5cm，冷凉季节也有大型花，花径可达 12.0cm，外部大花瓣 15～18 枚，呈 2～3 轮排列，阔倒卵形，小花瓣半直立，略扭曲，雄蕊簇生，花丝红色。叶片浓绿色，叶背面灰绿色，厚革质，椭圆形，长 8.5～9.0cm，宽 3.0～3.5cm，基部楔形，上部边缘叶齿浅。植株紧凑，生长旺盛。夏初始花，秋、冬季盛花，春季零星开花。

Abstract: Flowers red to maroon red, occasionally with white markings on the small petals at the center, semi-double to rose-double form, small to medium size with 7.0-9.0cm across and 4.5cm deep in summer, however, also large size with 12.0cm across in cold season, 15-18 exterior large petals arranged in 2-3 rows, stamens clustered, filaments red. Leaves dark green, heavy leathery, elliptic, cuneate at the base, margins shallowly serrate at the upper part. Plant compact and growth vigorous. Starts to bloom from early-summer, fully blooms from autumn to winter and sporadically in spring.

HAR-04-3. 长辫姑娘
Changbian Guniang (Long Braid Girl)

回交组合：杜鹃红山茶 × 杜鹃红山茶 F_1 代新品种'夏日红霞'（ *C. azalea* × *C. azalea* F_1 new hybrid 'Xiari Hongxia'）。

回交苗编号：DHX-No.6。

性状：花朵粉红色至深粉红色，渐向花瓣先端呈淡粉红色，半重瓣型，中型至大型花，花径 9.5～12.0cm，花朵厚 4.6cm，花瓣 20～22 枚，呈 3 轮松散排列，花瓣长倒卵形，边缘内弯，呈勺状，中部偶有雄蕊瓣，花丝散射。叶片浓绿色，叶背面灰绿色，硬革质，长披针形，长 10.0～11.7cm，宽 3.0～3.5cm，基部急楔形，边缘全缘。植株立性，生长旺盛。夏初始花，秋、冬季盛花，春季零星开花。

Abstract: Flowers pink to deep pink, fading to light pink at the top of the petals, semi-double form, medium to large size, 9.5-12.0cm across and 4.6cm deep with 20-22 petals arranged loosely in 3 rows, petals long obovate, spoon shaped, filaments scattered. Leaves dark green, hard leathery, long lanceolate, urgently cuneate at the base, margins smooth. Plant upright and growth vigorous. Starts to bloom from early-summer, fully blooms from autumn to winter and sporadically in spring.

HAR-04-4. 粉色田野
Fense Tianye (Pink Field)

回交组合：杜鹃红山茶 × 杜鹃红山茶 F_1 代新品种'夏日红霞'（ *C. azalea* × *C. azalea* F_1 new hybrid 'Xiari Hongxia'）。

回交苗编号：DHX-No.15。

性状：花朵粉红色，瓣面可见纵向深红色脉纹，半重瓣型，中型至大型花，花径 9.5～11.5cm，花朵厚 4.5cm，花瓣 13～15 枚，呈 2 轮紧实排列，阔倒卵形，外翻，雄蕊近离生，散射。叶片浓绿色，叶背面灰绿色，硬革质，长椭圆形，长 10.2～11.5cm，宽 4.5～5.0cm，基部楔形，边缘齿浅。植株立性，矮灌状，生长旺盛。夏初始花，秋、冬季盛花，春季零星开花。

Abstract: Flowers pink, lengthwise deep red veins visible on the petal surfaces, semi-double form, medium to large size, 9.5-11.5cm across and 4.5cm deep with 13-15 petals arranged tightly in 2 rows, petals broad obovate, rolled outward, stamens nearly separated. Leaves dark green, hard leathery, long elliptic, margins shallowly serrate. Plant upright, bushy and growth vigorous. Starts to bloom from early-summer, fully blooms from autumn to winter and sporadically in spring.

HAR-04-5. 浮翠流丹
Fucui Liudan (Floating Green & Flowing Red)

回交组合：杜鹃红山茶 × 杜鹃红山茶 F_1 代新品种'夏日红霞'（ *C. azalea* × *C. azalea* F_1 new hybrid 'Xiari Hongxia'）。

回交苗编号：DHX-No.16。

性状：花朵黑红色，泛蜡光，单瓣型，大型花，花径 12.0～12.8cm，花瓣 8～9 枚，长倒卵形，厚质，略外翻，雄蕊基部连生，花丝红色，开花稠密。叶片浓绿色，叶背面灰绿色，革质，椭圆形，长 9.5cm，宽 4.5cm，边缘近全缘。植株紧凑，生长旺盛。夏初始花，秋、冬季盛花，春季零星开花。

Abstract: Flowers dark red with waxy luster, single form, large size, 12.0-12.8cm across with 8-9 petals, thick texture, stamens united at the base, filaments red, bloom dense. Leaves dark green, leathery, elliptic, margins nearly smooth. Plant compact and growth vigorous. Starts to bloom from early-summer, fully blooms from autumn to winter and sporadically in spring.

HAR-04-6. 六脉神剑
Liumai Shenjian (Six Pulses God Sword)

回交组合： 杜鹃红山茶 × 杜鹃红山茶 F_1 代新品种'夏日红霞'（C. azalea × C. azalea F_1 new hybrid 'Xiari Hongxia'）。

回交苗编号： DHX-No.21。

性状： 花朵艳红色至黑红色，泛紫色调，花瓣先端近白色，完全重瓣型，大型花，花径 12.0～12.5cm，花朵厚5.3cm，花瓣60枚以上，呈10轮以上有序排列，多呈六角形，花瓣两侧略内卷，花芯小球状，花瓣逐片掉落。叶片浓绿色，叶背面灰绿色，革质，狭长椭圆形，长 10.0～11.0cm，宽3.0～4.0cm，基部急楔形，边缘全缘。植株紧凑，灌木状，生长旺盛。夏初始花，秋、冬季盛花，春季零星开花。

Abstract: Flowers bright red to dark red with purple hue, petal apices nearly white, formal double form, large size, 12.0-12.5cm across and 5.3cm deep with more than 60 petals arranged orderly in over 10 rows which shaped as a hexagon, both sides of the petals slightly rolled inward, flower core ball shaped, petals fall one by one. Leaves dark green, leathery, narrow long elliptic, margins smooth, urgently cuneate at the base. Plant compact, bushy and growth vigorous. Starts to bloom from early-summer, fully blooms from autumn to winter and sporadically in spring.

⦿ 回交组合 HAR-05. 杜鹃红山茶 × 杜鹃红山茶 F₁ 代新品种 '夏日七心'

HAR-05. *C. azalea* × *C. azalea* F₁ new hybrid 'Xiari Qixin'

本回交组合将介绍 2 个新品种。

'夏日七心'是从杜鹃红山茶 × 云南山茶品种'帕克斯先生'杂交组合中得到的（见第一部书第 266～267 页）。

母本和父本的花朵都是艳红色的，而在获得的 2 个回交新品种中却出现了一个花朵是粉红色的，这可能与'夏日七心'品种的杂合性有关。

Two new hybrids will be introduced in this backcross combination.

'Xiari Qixin' was obtained from the cross-combination between *C. azalea* and *C. reticulata* cultivar 'Dr Cliiford Parks' (See p.266-267, our first book).

Both female parent and male parent are bright red in their flower color, but one of the backcross new hybrids opens pink flowers. It might be related to the heterozygosity of hybrid 'Xiari Qixin'.

HAR-05-1. 清风胧月
Qingfeng Longyue (Gentle Breeze & Hazy Moon)

回交组合： 杜鹃红山茶 × 杜鹃红山茶 F_1 代新品种'夏日七心'（ C. azalea × C. azalea F_1 new hybrid 'Xiari Qixin'）。

回交苗编号： DQX-No.4。

性状： 花朵柔和的粉红色，随着花朵开放，粉色逐渐变淡，半重瓣型，小型至中型花，花径 7.0～10.0cm，花朵厚 4.5cm，花瓣约 20 枚，呈 3 轮松散排列，阔倒卵形，雄蕊基部略连生。叶片浓绿色，叶背面灰绿色，革质，阔椭圆形，长 8.5～9.5cm，宽 4.5～5.0cm，边缘具浅齿。植株紧凑，灌丛状，生长旺盛。夏初始花，秋、冬季盛花，春季零星开花。

Abstract: Flowers soft pink, as flowers opening, the pink color gradually changes into light pink, semi-double form, small to medium size, 7.0-10.0cm across and 4.5cm deep with about 20 petals arranged loosely in 3 rows, petals broad obovate, stamens slightly united at the base. Leaves dark green, leathery, broad elliptic, margins shallowly serrate. Plant compact, bushy and growth vigorous. Starts to bloom from early-summer, fully blooms from autumn to winter and sporadically in spring.

HAR-05-2. 艳红沐夏
Yanhong Muxia (Bright Red Bathed Summer)

回交组合：杜鹃红山茶 × 杜鹃红山茶 F_1 代新品种'夏日七心'（*C. azalea* × *C. azalea* F_1 new hybrid 'Xiari Qixin'）。

回交苗编号：DQX-No.11。

性状：花朵艳红色，内部花瓣偶有白斑，半重瓣型，中型花，花径 9.5～10.0cm，花朵厚 4.8cm，花瓣 30～32 枚，呈 5 轮排列，倒卵形，雄蕊基部略连生，呈散射状，花丝淡红色。嫩叶泛红色，成熟叶浓绿色，叶背面灰绿色，厚革质，长椭圆形，长 10.5～11.0cm，宽 3.5～4.0cm，边缘近光滑。植株立性，紧凑，生长旺盛。夏初始花，秋、冬季盛花，春季零星开花。

> **Abstract:** Flowers bright red, inner petals occasionally with white markings, semi-double form, medium size, 9.5-10.0cm across and 4.8cm deep with 30-32 petals arranged in 5 rows, petals obovate, stamens slightly united at the base, diffuse, filaments pale red. Tender leaves reddish, mature leaves dark green, heavy leathery, long elliptic, margins nearly smooth. Plant upright, compact and growth vigorous. Starts to bloom from early-summer, fully blooms from autumn to winter and sporadically in spring.

四季茶花杂交新品种彩色图集（第二部）

◉ 回交组合 HAR-06. 杜鹃红山茶 × 杜鹃红山茶 F_1 代新品种 '夏日台阁'

HAR-06. *C. azalea* × *C. azalea* F_1 new hybrid 'Xiari Taige'

本回交组合将介绍 9 个新品种。

'夏日台阁'是从杜鹃红山茶 × 云南山茶品种'帕克斯先生'杂交组合中得到的（见第一部书第 268～269 页）。本杂交组合获得的新品种中，有 7 个是多瓣型的，占品种总数的 77.8%，由此可以初步判定，本回交组合是培育多瓣品种的良好组合。

Nine new hybrids will be introduced in this backcross combination.

'Xiari Taige' was obtained from the cross-combination between *C. azalea* and *C. reticulata* cultivar 'Dr Clifford Parks' (See p.268-269, our first book). In the backcross new hybrids, seven hybrids are multi-petals which accounts for 77.8% of the total number of the hybrids. Therefore, it can be preliminarily concluded that this backcross combination is a good one for breeding multi-petals hybrids.

HAR-06-1. 花样年华
Huayang Nianhua (Diversity of Life)

回交组合： 杜鹃红山茶 × 杜鹃红山茶 F_1 代新品种'夏日台阁'（ *C. azalea* × *C. azalea* F_1 new hybrid 'Xiari Taige'）。

回交苗编号： DTG-No.4。

性状： 花朵桃红色至红色，瓣面有模糊的小白点，半重瓣型至牡丹型，小型至中型花，花径 6.8～8.0cm，花朵厚 4.5cm，大花瓣 18 枚左右，呈 3 轮松散排列，花瓣阔倒卵形，外翻，先端圆，中部少数几枚花瓣直立，雄蕊基部连生，开花稠密。叶片浓绿色，具光泽，叶背面灰绿色，革质，长椭圆形，长 8.5～9.0cm，宽 3.0～3.5cm，叶缘近光滑。植株紧凑，生长中等。夏初始花，秋、冬季盛花，春季零星开花。

Abstract: Flowers peach red to red with some fuzzy white dots, semi-double to peony form, small to medium size, 6.8-8.0cm across and 4.5cm deep with about 18 large petals arranged loosely in 3 rows, a few of central petals erected, bloom dense. Leaves dark green, leathery, long elliptic, margins nearly smooth. Plant compact and growth normal. Starts to bloom from early-summer, fully blooms from autumn to winter and sporadically in spring.

HAR-06-2. 群蜂纷飞
Qunfeng Fenfei (A Swarm of Bees Flying)

回交组合：杜鹃红山茶 × 杜鹃红山茶 F_1 代新品种'夏日台阁'(*C. azalea* × *C. azalea* F_1 new hybrid 'Xiari Taige')。

回交苗编号：DTG-No.6。

性状：花朵深粉红色，中部小花瓣偶有一枚白斑，单瓣型，小型至中型花，花径6.5～8.0cm，多数花朵呈半开放状态，簇拥在叶腋间，花瓣5～6枚，阔倒卵形，厚质，扭曲，雄蕊基部连生，呈短筒状，开花稠密。叶片浓绿色，叶背面灰绿色，稠密，半直立，革质，长椭圆形，长9.0～9.5cm，宽3.0～3.5cm，叶缘近光滑。植株紧凑，立性，生长旺盛。夏初始花，秋、冬季盛花，春季零星开花。

Abstract: Flowers deep pink, occasionally with a white marking on a small petal at the center, single form, small to medium size, 6.5-8.0cm across, most of the flowers in semi-opening, 5-6 petals, broad obovate, thick texture and twisted, stamens short tube-like, bloom dense. Leaves dark green, dense, semi-erected, leathery, long elliptic, margins nearly smooth. Plant compact, upright and growth vigorous. Starts to bloom from early-summer, fully blooms from autumn to winter and sporadically in spring.

HAR-06-3. 大家闺秀
Dajia Guixiu (Young Girl from Respectable Family)

回交组合： 杜鹃红山茶 × 杜鹃红山茶 F_1 代新品种'夏日台阁'（ *C. azalea* × *C. azalea* F_1 new hybrid 'Xiari Taige'）。

回交苗编号： DTG-No.7。

性状： 花朵刚开放时为桃红色至红色，瓣面可见无数小白点，而后渐变为粉红色，花芯偶有白条纹，半重瓣型至牡丹型，中型至大型花，花径9.8～11.3cm，花朵厚4.5cm，外部大花瓣19～22枚，呈3轮排列，花瓣倒卵形，中部小花瓣半直立，间生于雄蕊之中，雄蕊簇生，少数雄蕊瓣化，开花稠密。叶片浓绿色，叶背面灰绿色，厚革质，椭圆形，长9.5～10.0cm，宽3.0～3.5cm，叶缘近光滑。植株紧凑，生长较缓慢。夏初始花，秋、冬季盛花，翌年春季零星开花。

Abstract: Flowers peach red to red with many tiny white dots visible on the petal surfaces when just opening, and then gradually change into pink, occasionally with a few of white stripes on the central petals, semi-double to peony form, medium to large size, 9.8-11.3cm across and 4.5cm deep with 19-22 exterior large petals arranged in 3 rows, central small petals semi-erected, stamens clustered. Leaves dark green, heavy leathery, elliptic, margins nearly smooth. Plant compact and growth slower. Starts to bloom from early-summer, fully blooms from autumn to winter and sporadically in spring.

HAR-06-4. 红粉舞会
Hongfen Wuhui（Red-pink Dancing Party）

回交组合：杜鹃红山茶 × 杜鹃红山茶 F_1 代新品种'夏日台阁'（*C. azalea* × *C. azalea* F_1 new hybrid 'Xiari Taige'）。

回交苗编号：DTG-No.10。

性状：花朵深粉红色至红色，具白色斑块或条纹，随着花朵开放，逐渐变为桃红色，半重瓣型至玫瑰重瓣型，中型至大型花，花径8.5～11.0cm，花朵厚5.3cm，花瓣30～35枚，呈3轮松散排列，狭长倒卵形，略外卷，中部花瓣直立，雄蕊簇生，开花与抽梢同时进行，开花稠密。叶片浓绿色，叶背面灰绿色，厚革质，长椭圆形，长8.2cm，宽3.0cm，基部急楔形，叶缘光滑。植株紧凑，枝叶稠密，生长旺盛。夏初始花，秋、冬季盛花，翌年春季零星开花。

> **Abstract:** Flowers deep pink to red with white markings or white stripes, as flowers opening, the flower color gradually change into peach red, semi-double to rose-double form, medium to large size, 8.5-11.0cm across and 5.3cm deep with 30-35 petals arranged loosely in 3 rows, petals narrow long obovate, stamens clustered. Bloom and shoot simultaneous, bloom dense. Leaves dark green, heavy leathery, long elliptic, urgently cuneate at the base, margins smooth. Plant compact, branches and leaves dense and growth vigorous. Starts to bloom from early-summer, fully blooms from autumn to winter and sporadically in spring.

HAR-06-5. 粉浪迎秋
Fenlang Yingqiu (Pink Waves Welcome Autumn)

回交组合：杜鹃红山茶 × 杜鹃红山茶 F_1 代新品种'夏日台阁'（C. azalea × C. azalea F_1 new hybrid 'Xiari Taige'）。

回交苗编号：DTG-No.19。

性状：花朵粉红色至深粉红色，夏季花色较浅，冬季花色较深，具白色斑点和条纹，玫瑰重瓣型至完全重瓣型，中型至大型花，花径 8.5～11.0cm，花朵厚 4.7cm，花瓣 40 枚以上，呈 5～10 轮有序排列，倒卵形，中部少量小花瓣直立，全开放后花芯处偶见少量散生短雄蕊，开花稠密。叶片浓绿色，具光泽，叶背面灰绿色，厚革质，椭圆形，长 8.0～9.5cm，宽 3.0～3.5cm，叶缘光滑。植株紧凑，立性，生长旺盛。夏初始花，秋、冬季盛花，翌年春季零星开花。

Abstract: Flowers pink to deep pink, the colors of flowers are lighter in summer and deeper in winter with white spots and stripes, rose-double to formal double form, medium to large size, 8.5-11.0cm across, 4.7cm deep with over 40 petals arranged orderly in 5-10 rows, a few of small and erected petals at the center, stamens short and diffuse, bloom dense. Leaves dark green, heavy leathery, elliptic, margins smooth. Plant compact, upright and growth vigorous. Starts to bloom from early-summer, fully blooms from autumn to winter and sporadically in spring.

HAR-06-6. 红铃报喜
Hongling Baoxi（Happy News from Red Bells）

回交组合：杜鹃红山茶 × 杜鹃红山茶 F_1 代新品种'夏日台阁'（ C. azalea × C. azalea F_1 new hybrid 'Xiari Taige'）。

回交苗编号：DTG-No.27。

性状：花朵艳红色至深红色，单瓣型，小型至中型花，花径 6.5～8.0cm，花瓣 5 枚，阔倒卵形，多呈半开放状态，波浪状，雄蕊基部连生，开花稠密，似串铃。叶片浓绿色，叶背面灰绿色，稠密，半直立，厚革质，椭圆形，长 7.5～8.5cm，宽 3.5～4.0cm，基部楔形，叶缘近全缘。植株紧凑，枝软，生长旺盛。夏初始花，秋、冬季盛花，春季零星开花。

> **Abstract:** Flowers bright red to deep red, single form, small to medium size, 6.5-8.0cm across with 5 petals, most of which are semi-opened and wavy, stamens united at the base, bloom dense which is like a string of bells. Leaves dark green, dense, semi-erected, heavy leathery, elliptic, cuneate at the base, margins nearly smooth. Plant compact, branches soft and growth vigorous. Starts to bloom from early-summer, fully blooms from autumn to winter and sporadically in spring.

HAR-06-7. 六六大顺
Liuliu Dashun (Double Sixs Making Everything Smooth)

回交组合： 杜鹃红山茶 × 杜鹃红山茶 F_1 代新品种'夏日台阁'（ *C. azalea* × *C. azalea* F_1 new hybrid 'Xiari Taige'）。

回交苗编号： DTG-No.32。

性状： 花朵深红色，渐向中部变为浅粉红色，玫瑰重瓣型至完全重瓣型，中型至大型花，花径 8.5～11.5cm，花朵厚 5.3cm，花瓣 60～80 枚，6～8 轮呈六角形、有序、紧实排列，花瓣长倒卵形，开花稠密。叶片浓绿色，具光泽，叶背面灰绿色，厚革质，长椭圆形，长 7.5～8.0 cm，宽 2.3～2.7cm，基部楔形，叶缘全缘。植株紧凑，立性，生长旺盛。夏初始花，秋、冬季盛花，春季零星开花。

> **Abstract:** Flowers deep red, fading to light pink at the center, rose-double to formal double form, medium to large size, 8.5-11.5cm across and 5.3cm deep with 60-80 petals arranged ordely and tightly in 6-8 rows that looked like a hexagon in shape, bloom dense. Leaves dark green, heavy leathery, long elliptic, cuneate at the base, margins smooth. Plant compact, upright and growth vigorous. Starts to bloom from early-summer, fully blooms from autumn to winter and sporadically in spring.

HAR-06-8. 冬夏出彩
Dongxia Chucai (Colorfull in Winter & Summer)

回交组合： 杜鹃红山茶 × 杜鹃红山茶 F$_1$ 代新品种'夏日台阁'（ C. azalea × C. azalea F$_1$ new hybrid 'Xiari Taige' ）。

回交苗编号： DTG-No.48。

性状： 花朵深粉红色至紫粉红色，渐向花瓣上部粉色变淡，半重瓣型至玫瑰重瓣型，中型花，花径 9.3～9.8cm，花朵厚 4.8cm，花瓣 35 枚以上，呈 3 轮松散排列，长倒卵形，中部花瓣半直立，雄蕊簇生，开花稠密。叶片浓绿色，叶背面灰绿色，厚革质，椭圆形，长 8.2cm，宽 3.2cm，叶缘光滑。植株紧凑，生长旺盛。夏初始花，秋、冬季盛花，翌年春季零星开花。

Abstract: Flowers deep pink to purple pink, gradually fading to light-pink at the petals' upper parts, semi-double to rose-double form, medium size, 9.3-9.8cm across and 4.8cm deep with over 35 petals arranged loosely in 3 rows, central petals semi-erected, stamens clustered, bloom dense. Leaves dark green, heavy leathery, elliptic, margins smooth. Plant compact and growth vigorous. Starts to bloom from early-summer, fully blooms from autumn to winter and sporadically in spring.

HAR-06-9. 获奖喜悦
Huojiang Xiyue (The Joy of Winning an Award)

回交组合： 杜鹃红山茶 × 杜鹃红山茶 F₁ 代新品种'夏日台阁'（*C. azalea* × *C. azalea* F₁ new hybrid 'Xiari Taige'）。

回交苗编号： DTG-No.49。

性状： 花朵红色，随着花朵开放，渐渐出现紫色调，瓣面偶有白斑块，半重瓣型至牡丹型，中型至大型花，花径 9.5～12.0cm，花朵厚 5.0cm，大花瓣 18～23 枚，呈 3 轮排列，花瓣长倒卵形，中部小花瓣半直立，与雄蕊混生。叶片浓绿色，叶背面灰绿色，厚革质，椭圆形，长 9.0～10.5cm，宽 3.5～4.0cm，基部急楔形，叶缘近光滑。植株紧凑，生长缓慢。夏初始花，秋、冬季盛花，翌年春季零星开花。

Abstract: Flowers red, as flowers opening, gradually appear some purple hue, occasionally with a few white markings on the petal surfaces, semi-double to peony form, medium to large size, 9.5-12.0cm across and 5.0cm deep with 18-23 large petals arranged in 3 rows, petals long obovare, central small petals semi-erected and mixed with stamens. Leaves dark green, heavy leathery, elliptic, urgently cuneate at the base, margins nearly smooth. Plant compact and growth slower. Starts to bloom from early-summer, fully blooms from autumn to winter and sporadically in spring.

⦿ 回交组合 HAR-07. 杜鹃红山茶 × 杜鹃红山茶 F$_1$ 代新品种 '夏日探戈'

HAR-07. *C. azalea* × *C. azalea* F$_1$ new hybrid 'Xiari Tange'

本回交组合将介绍 2 个新品种。

'夏日探戈'是从红山茶品种'媚丽'× 杜鹃红山茶的杂交组合中得到的（见第一部书第 137 ～ 138 页）。

回交获得的新品种比其父本的花色更加柔和，花朵更加漂亮。

Two new hybrid will be introduced in this backcross combination.

'Xiari Tange' was obtained from the cross-combination between *C. japonica* cultivar 'Tama Beauty' and *C. azalea* (See p.137-138, our first book).

Comparing with the new hybrids' male parent, the flower color of the hybrids are softer and the flowers are more beautiful.

HAR-07-1. 深山九妹
Shenshan Jiumei (Ninth Sister in Remote Mountains)

回交组合： 杜鹃红山茶 × 杜鹃红山茶 F_1 代新品种'夏日探戈'（*C. azalea* × *C. azalea* F_1 new hybrid 'Xiari Tange'）。

回交苗编号： DXRTG-No.1。

性状： 刚开放时，花朵淡红色，然后逐渐变为深粉红色，有时具少量白斑，渐向边缘花色变淡，半重瓣型至牡丹型，中型花，花径 8.0～9.0cm，花朵厚 4.0～4.5cm，花瓣 16～18 枚，呈 3～4 轮排列，倒卵形。叶片浓绿色，叶背面灰绿色，厚革质，倒披针形，长 8.5～10.5cm，宽 3.5～4.0cm，基部急楔形，边缘光滑。植株立性，生长旺盛。夏初始花，秋、冬季盛花，春季零星开花。

Abstract: Flowers light red when just opening and gradually change into deep pink, sometimes with a few white markings, the color fading to light at the edges, semi-double to peony form, medium size, 8.0-9.0cm across and 4.0-4.5cm deep with 16-18 petals arranged in 3-4 rows, obovate. Leaves dark green, heavy leathery, oblanceolate, urgently cuneate at the base, margins smooth. Plant upright and growth vigorous. Starts to bloom from early-summer, fully blooms from autumn to winter and sporadically in spring.

HAR-07-2. 明星风范
Mingxing Fengfan (Star Manner)

回交组合： 杜鹃红山茶 × 杜鹃红山茶 F_1 代新品种'夏日探戈'（ *C. azalea* × *C. azalea* F_1 new hybrid 'Xiari Tange'）。

回交苗编号： DXRTG-No.5。

性状： 花朵淡桃红色至橘粉红色，中部小花瓣具白条纹，半重瓣型至牡丹型，中型至大型花，花径9.5～11.5cm，花朵厚4.8cm，大花瓣18～23枚，呈2～3轮排列，花瓣阔倒卵形，中部小花瓣扭曲，雄蕊基部略连生，花丝散射状。叶片浓绿色，叶背面灰绿色，厚质，椭圆形，长9.0～10.5cm，宽3.5～4.5cm，基部楔形，边缘近全缘。植株紧凑，生长旺盛。夏初始花，秋、冬季盛花，春季零星开花。

Abstract: Flowers light peach red to orange pink, central small petals with white stripes, semi-double to peony form, medium to large size, 9.5-11.5cm across and 4.8cm deep with 18-23 large petals arranged in 2-3 rows, petals broad obovate, central small petals wrinkled, stamens slightly united at the base, filaments scattered. Leaves dark green, leathery, cuneate at the base, margins nearly smooth. Plant compact and growth vigorous. Starts to bloom from early-summer, fully blooms from autumn to winter and sporadically in spring.

⦿ 回交组合 HAR-08. 杜鹃红山茶 × 杜鹃红山茶 F₁ 代新品种 '夏咏国色'

HAR-08. *C. azalea* × *C. azalea* F₁ new hybrid 'Xiayong Guose'

本回交组合将介绍 8 个新品种。

'夏咏国色'是从杜鹃红山茶 × 红山茶品种'花牡丹'杂交组合中得到的（见第一部书第 196～197 页）。

本组合尽管出现了 2 个花朵单瓣类型的品种，但是大部分新品种的花朵是多瓣类型的，这进一步验证了回交新品种在花的性状上倾向于父本'夏咏国色'，在植株和叶片的性状上倾向于母本杜鹃红山茶。

Eight new hybrids will be introduced in this backcross combination.

'Xiayong Guose' was obtained from the cross-combination between *C. azalea* and *C. japonica* cultivar 'Daikagura' (See p.196-197, our first book).

The flowers of most new hybrids in this combination are multi-petals, although two single form hybrids have appeared, which further verifies that the flower characteristics of the backcross hybrids in the cross-combination mainly tend to their male parent 'Xiayong Guose' and the characteristics of both plants and leaves tend to their female parent *C. azalea*.

HAR-08-1. 夏初巧遇
Xiachu Qiaoyu (Early Summer's Chance Encounter)

回交组合： 杜鹃红山茶 × 杜鹃红山茶 F_1 代新品种'夏咏国色'（ C. azalea × C. azalea F_1 new hybrid 'Xiayong Guose' ）。

回交苗编号： DGS-No.4。

性状： 花朵刚开放时为红色，而后渐渐变为深粉红色，偶有白斑，渐向花芯呈粉红色，玫瑰重瓣型，中型至大型花，花径 8.5～12.5cm，花朵厚 6.0～7.0cm，外部大花瓣 30～40 枚，呈 3～4 轮有序排列，倒卵形，外翻，中部花瓣半直立，先端内扣。叶片浓绿色，叶背面灰绿色，半直立，稠密，狭长椭圆形，厚革质，较小，长 6.5～7.0cm，宽 2.2～2.7cm，略波浪状，基部急楔形，边缘几乎全缘。植株紧凑，灌丛状，枝叶稠密，生长旺盛。夏初始花，秋、冬季盛花，春季零星开花。

Abstract: Flowers red when just opening and then gradually change into deep pink. occasionally with white markings, fading to pink at the center, rose-double form, medium to large size, 8.5-12.5cm across and 6.0-7.0cm deep with 30-40 petals arranged orderly in 3-4 rows, petals obovate. Leaves dark green, heavy leathery, smaller, urgently cuneate at the base, margins nearly smooth. Plant compact, bushy, branches dense, and growth vigorous. Starts to bloom from early-summer, fully blooms from autumn to winter and sporadically in spring.

HAR-08-2. 炎夏魔红
Yanxia Mohong (Hot Summer's Magic Red)

回交组合：杜鹃红山茶 × 杜鹃红山茶 F_1 代新品种'夏咏国色'(*C. azalea* × *C. azalea* F_1 new hybrid 'Xiayong Guose')。

回交苗编号：DGS-No.5。

性状：花朵红色，渐渐变为淡红色，中部花瓣可见少量白条纹，玫瑰重瓣型，中型至大型花，花径 9.5～10.5cm，花朵厚 5.0cm，花瓣近 40 枚，呈 4 轮有序排列，先端外翻，长倒卵形，全开放后花芯处可见少量雄蕊。叶片浓绿色，叶背面灰绿色，半直立，稠密，厚革质，长椭圆形，长 8.5～9.0cm，宽 2.4～2.7cm，略波浪状，基部急楔形，边缘几乎全缘。植株紧凑，立性，但矮化，枝叶稠密，生长中等。夏初始花，秋、冬季盛花，春季零星开花。

Abstract: Flowers red, gradually turn to light red, a few white stripes are visible on the central petals, rose-double form, medium to large size, 9.5-10.5cm across and 5.0cm deep with nearly 40 petals arranged orderly in 4 rows, petals turned outward at the top parts. Leaves dark green, dense, heavy leathery, long elliptic, urgently cuneate at the base, margins nearly smooth. Plant compact, upright but dwarf, branches dense and growth normal. Starts to bloom from early-summer, fully blooms from autumn to winter and sporadically in spring.

HAR-08-3. 汉森之悦
Hansen Zhiyue（Mr Waldemar Max Hansen's Delight）

回交组合： 杜鹃红山茶 × 杜鹃红山茶 F_1 代新品种'夏咏国色'（*C. azalea* × *C. azalea* F_1 new hybrid 'Xiayong Guose'）。

回交苗编号： DGS-No.6。

性状： 花朵桃红色，外部花瓣淡红色至粉红色，瓣面具清晰的红脉纹和隐约的白斑，内部花瓣红色较深，随着花朵开放，全朵花逐渐变为粉白色，半重瓣型至玫瑰重瓣型，大型至巨型花，花径12.5～14.8cm，花朵厚4.5cm，外部大花瓣20枚以上，呈3～4轮排列，阔倒卵形，外翻，顶部皱褶，中部花瓣边缘内扣。叶片浓绿色，稠密，叶背面灰绿色，椭圆形，厚革质，长9.0～11.0cm，宽3.0～4.2cm，上部边缘有稀齿，嫩叶片黄绿色，略扭曲。植株紧凑，灌丛状，枝叶稠密，生长旺盛。春末始花，夏、秋、冬季盛花，春季零星开花。

注：本品种是为表彰和纪念国际山茶协会前副主席，德国的沃尔德马·马克斯·汉森（Waldemar Max Hansen）先生对茶花世界的贡献而命名的。

Abstract: Flowers peach red, exterior petals light red to pink with faint white markings, interior petals deep red, semi-double to rose-double form, large to very large size, 12.5-14.8cm across and 4.5cm deep with over 20 large petals arranged in 3-4 rows. Leaves dark green, heavy leathery, margins thinly serrate at the top part. Plant compact and bushy, branches dense, growth vigorous. Starts to bloom from late-spring, fully blooms in summer, autumn and winter and sporadically in spring.

Mark: The hybrid has named after ICS former vice president, Mr. Waldemar Max Hansen, for commending and commemorate his great contribution to the camellia world.

HAR-08-4. 红衣仙女
Hongyi Xiannü (The Fairy Dressing Red)

回交组合：杜鹃红山茶 × 杜鹃红山茶 F_1 代新品种'夏咏国色'（ *C. azalea* × *C. azalea* F_1 new hybrid 'Xiayong Guose'）。

回交苗编号：DGS-No.9。

性状：花朵艳红色，偶有白条纹，炎夏花色变淡，玫瑰重瓣型，冬季有时会出现牡丹型，中型至大型花，花径 8.5～11.5cm，花朵厚 5.0cm，花瓣 35～40 枚，呈 4 轮以上紧实排列，倒卵形。叶片浓绿色，叶背面灰绿色，稠密，厚革质，卵形，长 7.0～9.0cm，宽 4.0～5.0cm，边缘几乎全缘。植株紧凑，矮性，枝叶稠密，生长旺盛。夏初始花，秋、冬季盛花，春季零星开花。

Abstract: Flowers bright red, occasionally with white stripes, fading to light red in hot summer, rose-double form, sometimes peony form can appear in winter, medium to large size, 8.5-11.5cm across and 5.0cm deep with 35-40 petals arranged tightly in more than 4 rows, petals obovate. Leaves dark green, dense, heavy leathery, ovate, margins nearly smooth. Plant compact, dwarf, branches dense and growth vigorous. Starts to bloom from early-summer, fully blooms from autumn to winter and sporadically in spring.

HAR-08-5. 百媚千娇
Baimei Qianjiao（Enchanting Beauty）

回交组合： 杜鹃红山茶 × 杜鹃红山茶 F_1 代新品种'夏咏国色'（ C. azalea × C. azalea F_1 new hybrid 'Xiayong Guose'）。

回交苗编号： DGS-No.11。

性状： 花朵刚开放时为橘红色，2～3天后渐渐变为粉红色，完全重瓣型，大型至巨型花，花径12.0～13.5cm，花朵厚4.8～5.5cm，花瓣50枚以上，呈6～7轮有序排列，长倒卵形，外部花瓣外翻，中部花瓣半直立，内扣，花朵全开放后，偶然可见少量雄蕊。叶片浓绿色，叶背面灰绿色，薄革质，狭长椭圆形，长10.5cm，宽3.0～3.5cm，基部急楔形，边缘全缘。植株紧凑，立性，枝叶繁茂，生长旺盛。夏初始花，秋、冬季盛花，春季零星开花。

Abstract: Flowers orange red when just opening, after 2-3 days, gradually change into pink, formal double form, large to very large size, 12.0-13.5cm across and 4.8-5.5cm deep with over 50 petals arranged orderly in 6-7 rows, petals long obovate. Leaves dark green, thinly leathery, narrow-long elliptic, urgently cuneate at the base, margins smooth. Plant compact, upright, branches dense and growth vigorous. Starts to bloom from early-summer, fully blooms from autumn to winter and sporadically in spring.

HAR-08-6. 七夕礼品
Qixi Lipin (Double-Seventh Day's Gift)

回交组合：杜鹃红山茶 × 杜鹃红山茶 F_1 代新品种'夏咏国色'（*C. azalea* × *C. azalea* F_1 new hybrid 'Xiayong Guose'）。

回交苗编号：DGS-No.13。

性状：花朵橘粉红色，渐向花芯粉色变淡，玫瑰重瓣型，中型至大型花，花径9.5～11.5cm，花朵厚5.2cm，大花瓣29～32枚，呈4轮有序排列，阔倒卵形，中部小花瓣3～5枚，与雄蕊混生，全开放后外部两轮外翻，恰似玫瑰，花瓣逐片掉落。叶片浓绿色，叶背面灰绿色，革质，椭圆形，长8.5～9.5cm，宽3.0～3.5cm，基部楔形，边缘全缘。植株紧凑，灌丛状，枝叶稠密，生长旺盛。夏初始花，秋、冬季盛花，春季零星开花。

Abstract: Flowers orange pink, fading to light color at the center, rose-double form, medium to large size, 9.5-11.5cm across and 5.2cm deep with 29-32 petals arranged orderly in 4 rows, 3-5 small petals at the center mixed with stamens, which are rose-like, petals fall one by one. Leaves dark green, leathery, cuneate at the base, margins smooth. Plant compact, bushy, branches dense and growth vigorous. Starts to bloom from early-summer, fully blooms from autumn to winter and sporadically in spring.

HAR-08-7. 粉色烛影
Fense Zhuying (Shadows of Pink Candle)

回交组合： 杜鹃红山茶 × 杜鹃红山茶 F$_1$ 代新品种'夏咏国色'（ *C. azalea* × *C. azalea* F$_1$ new hybrid 'Xiayong Guose'）。

回交苗编号： DGS-No.14。

性状： 花朵多单生，淡粉红色至深粉红色，单瓣型，花瓣 6～7 枚，多呈半开放状态，略扭曲，雄蕊基部连生，呈筒状，花丝淡粉红色，花药淡黄色。叶片浓绿色，叶背面灰绿色，厚革质，基部急楔形，叶缘光滑。植株开张，生长旺盛。夏初始花，秋、冬季盛花，翌年春季零星开花。

> **Abstract:** Flowers mostly solitary, light pink to deep pink, single form, 6-7 petals which are semi-opened, slightly twisty, stamens united into a column at the base. Leaves dark green, heavy leathery, urgently cuneate at the base, margins smooth. Plant spread and growth vigorous. Starts to bloom from early-summer, fully blooms from autumn to winter and sporadically in spring.

HAR-08-8. 邻家红姐
Linjia Hongjie (Neighbor's Red Sister)

回交组合： 杜鹃红山茶 × 杜鹃红山茶 F_1 代新品种'夏咏国色'（*C. azalea* × *C. azalea* F_1 new hybrid 'Xiayong Guose'）。

回交苗编号： DGS-No.30。

性状： 花朵淡红色，渐渐变为深粉红色，单瓣型，中型至大型花，花径9.5～11.0cm，花瓣8枚，倒卵形，雄蕊基部连生，呈短管状，开花稠密。叶片浓绿色，叶背面灰绿色，革质，狭长椭圆形，长11.0～12.5cm，宽2.5～3.0cm，基部急楔形，先端尖，边缘光滑。植株开张，灌丛状，枝叶稠密，生长旺盛。夏初始花，秋、冬季盛花，春季零星开花。

Abstract: Flowers light red, gradually change into deep pink, single form, medium to large size, 9.5-11.0cm across with 8 petals, petals obovate, stamens united into a short tube at the base, bloom dense. Leaves dark green, back surfaces pale green, leathery, narrow-long elliptic, urgently cuneate at the base, margins smooth. Plant spread, bushy, branches dense and growth vigorous. Starts to bloom from early-summer, fully blooms from autumn to winter and sporadically in spring.

⦿ 回交组合 HAR-09. 杜鹃红山茶 × 杜鹃红山茶 F₁ 代新品种 '茶香飘逸'
HAR-09. *C. azalea* × *C. azalea* F₁ new hybrid 'Chaxiang Piaoyi'

本回交组合将介绍 2 个新品种。

'茶香飘逸'是从杜鹃红山茶 × 非云南山茶杂交种'香漩涡'杂交组合中得到的一个能够多季开花、略有茶香的品种（见第一部书第 336～337 页）。

很遗憾的是，本组所获得的这两个回交品种只有在天气晴朗时才能嗅到淡淡的茶香。尽管如此，该组回交新品种仍然具有多季开花的特性，而且花朵艳红色至黑红色，具光泽，观赏价值颇高，是盆栽和园林美化的好品种。

Two new hybrids will be introduced in this backcross combination.

'Chaxiang Piaoyi' was obtained from the cross-combination between *C. azalea* and Non-*reticulata* hybrid 'Scented Swirl' (See p.336-337, our first book), which has the traits of ever-blooming and slight fragrance.

Unfortunately, the two backcrossed hybrids obtained in the combination only can be smelt the faint aroma of tea in a sunny day. Even so, the new hybrids obtained from the backcrosses still have ever-blooming characteristics and their flowers are bright red to dark red with waxy luster and high ornamental value. They will be good varieties for potting and landscaping.

HAR-09-1. 将军风度
Jiangjun Fengdu (General's Elegant Demeanor)

回交组合：杜鹃红山茶 × 杜鹃红山茶 F_1 代新品种'茶香飘逸'（ *C. azalea* × *C. azalea* F_1 new hybrid 'Chaxiang Piaoyi'）。

回交苗编号：DCXPY-No.1。

性状：花朵黑红色，具蜡光，中部小花瓣偶有白条纹，略有茶香，半重瓣型，大型至巨型花，花径12.0～13.5cm，花朵厚7.0～7.5cm，外部大花瓣30～33枚，呈5轮排列，阔倒卵形，厚质，外翻，中部几枚小花瓣直立，与雄蕊混生，花丝散射状。叶片浓绿色，叶背面灰绿色，革质，椭圆形，长10.0～11.0cm，宽5.0～4.0cm，基部楔形，叶缘上部齿钝，下部光滑。植株立性，生长旺盛。夏初始花，秋、冬季盛花，春季零星开花。

Abstract: Flowers dark red with waxy luster, occasionally with white stripes on the central petals and slightly with tea aroma, semi-double form, large to very large size, 12.0-13.5cm across and 7.0-7.5cm deep with 30-33 exterior large petals arranged in 5 rows, petals broad obovate, thick texture, central several small petals erected and mixed with stamens, filaments diffuse. Leaves dark green, leathery, elliptic, cuneate at the base, margins obtusely serrate at the upper part and smooth at the lower part. Plant upright and growth vigorous. Starts to bloom from early-summer, fully blooms from autumn to winter and sporadically in spring.

HAR-09-2. 红浪滔天
Honglang Taotian (Surge Red Waves)

回交组合：杜鹃红山茶 × 杜鹃红山茶 F_1 代新品种'茶香飘逸'（ C. azalea × C. azalea F_1 new hybrid 'Chaxiang Piaoyi'）。

回交苗编号：DCXPY-No.3。

性状：花朵艳红色，具蜡光，瓣面偶有白条纹，略有茶香，半重瓣型至牡丹型，中型至大型花，花径9.0～11.5cm，花朵厚5.0cm，外部大花瓣16～18枚，呈2轮排列，倒卵形，波浪状，中部小花瓣直立，与雄蕊混生，花丝簇生，开花稠密。叶片浓绿色，叶背面灰绿色，革质，椭圆形，长10.0～12.0cm，宽3.5～4.5cm，基部楔形，叶缘齿钝。植株紧凑，生长旺盛。夏初始花，秋、冬季盛花，春季零星开花。

Abstract: Flowers bright red with waxy luster, occasionally with white stripes on petal surfaces, slightly with tea aroma, semi-double to peony form, medium to large size, 9.0-11.5cm across and 5.0cm deep with 16-18 exterior large petals arranged in 2 rows, petals obovate, wavy, central small petals erected and mixed with stamens, filaments clustered. Leaves dark green, leathery, elliptic, cuneate at the base, margins obtusely serrate. Plant compact and growth vigorous. Starts to bloom from early-summer, fully blooms from autumn to winter and sporadically in spring.

⊙ 回交组合 HAR-10. 杜鹃红山茶 × 杜鹃红山茶 F₁ 代新品种 '红屋积香'

HAR-10. *C. azalea* × *C. azalea* F₁ new hybrid 'Hongwu Jixiang'

×

本回交组合将介绍 10 个新品种。

'红屋积香'是从杜鹃红山茶 × 红山茶品种'克瑞墨大牡丹'杂交组合中得到的一个能够多季开花、具芳香的杂交种（见第一部书第 72～74 页）。

本组所获得的这 10 个回交新品种中，仅有 2 个具芳香。尽管如此，本组的回交新品种仍然具有多季开花的特性，且观赏性很好。

Ten new hybrids will be introduced in this backcross combination.

'Hongwu Jixiang' is a fragrant hybrid with ever-blooming trait that was obtained from cross-combination of *C. azalea* and *C. japonica* cultivar 'Kramer's Supreme' (See p.72-74, our first book).

There are only two new hybrids with fragrant characteristics in the ten new backcross hybrids. Even so, the new hybrids gotten from the backcross still have the ever-blooming characteristic with nice ornamental value.

HAR-10-1. 梦中幻云
Mengzhong Huanyun (Magic Clouds in Dream)

回交组合：杜鹃红山茶 × 杜鹃红山茶 F_1 代新品种'红屋积香'（ C. azalea × C. azalea F_1 new hybrid 'Hongwu Jixiang'）。

回交苗编号：DHW-No.1。

性状：花朵橘红色，花瓣中上部具白斑，略带芳香，半重瓣型，中型至大型花，花径 9.0～11.5cm，花朵厚 4.8cm，大花瓣 15～18 枚，呈 3 轮有序排列，花瓣阔倒卵形，外翻，先端近圆，雄蕊散射状，开花稠密。叶片浓绿色，叶背面灰绿色，厚革质，椭圆形，长 9.0～9.5cm，宽 4.0～4.3cm，基部急楔形，叶缘近光滑。植株立性，生长旺盛。夏初始花，秋、冬季盛花，春季零星开花。

Abstract: Flowers orange red with white markings on the middle-upper parts of petal surfaces, slightly with fragrance, semi-double form, medium to large size, 9.0-11.5cm across and 4.8cm deep with 15-18 large petals arranged orderly in 3 rows, petals broad obovate, rolled outward, apices nearly round, stamens diffused, bloom dense. Leaves dark green, heavy leathery, elliptic, urgently cuneate at the base, margins nearly smooth. Plant upright and growth vigorous. Starts to bloom from early-summer, fully blooms from autumn to winter and sporadically in spring.

HAR-10-2. 十月红透
Shiyue Hongtou（Red Through in October）

回交组合：杜鹃红山茶 × 杜鹃红山茶 F_1 代新品种'红屋积香'（ C. azalea × C. azalea F_1 new hybrid 'Hongwu Jixiang'）。

回交苗编号：DHW-No.3。

性状：花朵红色至深红色，略泛蜡光，半重瓣型，中型花，花径 8.5～10.0cm，花朵厚 5.3cm，花瓣 15～18 枚，呈 2～3 轮排列，花瓣阔倒卵形，边缘内扣，雄蕊基部略连生，偶有白色花药。叶片浓绿色，叶背面灰绿色，革质，狭窄椭圆形，长 6.5～7.5cm，宽 1.5～2.0cm，基部急楔形，叶缘光滑。植株紧凑，矮灌状，生长旺盛。夏初始花，秋、冬季盛花，春季零星开花。

Abstract: Flowers red to deep red, slightly with waxy luster, semi-double form, medium size, 8.5-10.0cm across and 5.3cm deep with 15-18 petals arranged in 2-3 rows, petals broad obovate, edges rolled inward, stamens slightly united at the base, anthers occasionally white. Leaves dark green, leathery, narrow elliptic, urgently cuneate at the base, margins smooth. Plant compact, bushy and growth vigorous. Starts to bloom from early-summer, fully blooms from autumn to winter and sporadically in spring.

HAR-10-3. 红色星空
Hongse Xingkong (Red Starry Sky)

回交组合：杜鹃红山茶 × 杜鹃红山茶 F_1 代新品种'红屋积香'（ C. azalea × C. azalea F_1 new hybrid 'Hongwu Jixiang'）。

回交苗编号：DHW-No.4。

性状：花朵红色，偶有白斑，单瓣型，小型至中型花，花径 7.0～8.5cm，花瓣 5～6 枚，阔倒卵形，外翻或波浪状，雄蕊基部连生，呈柱状，开花稠密。叶片浓绿色，叶背面灰绿色，革质，椭圆形，长 8.5～9.5cm，宽 3.5～4.0cm，基部急楔形，叶缘近光滑。植株紧凑，生长旺盛。夏初始花，秋、冬季盛花，春季零星开花。

> **Abstract:** Flowers red, occasionally with white markings, single form, small to medium size, 7.0-8.5cm across with 5-6 petals, petals broad obovate and rolled outward or wavy, stamens united into a column at the base, bloom dense. Leaves dark green, leathery, elliptic, urgently cuneate at the base, margins nearly smooth. Plant compact and growth vigorous. Starts to bloom from early-summer, fully blooms from autumn to winter and sporadically in spring.

HAR-10-4. 红瓣金心
Hongban Jinxin (Red Petals & Golden Heart)

回交组合：杜鹃红山茶 × 杜鹃红山茶 F_1 代新品种'红屋积香'(*C. azalea* × *C. azalea* F_1 new hybrid 'Hongwu Jixiang')。

回交苗编号：DHW-No.5 + 6 + 12。

性状：花朵艳红色至黑红色，瓣面上部红色逐渐变淡，单瓣型，中型花，花径8.5～10.0cm，花瓣6枚，阔倒卵形，雄蕊基部略连生，呈柱状，花丝红色，开花稠密。叶片浓绿色，叶背面灰绿色，革质，椭圆形，长8.0～9.5cm，宽4.0～4.5cm，基部急楔形，叶缘齿浅钝。植株紧凑，生长旺盛。夏初始花，秋、冬季盛花，春季零星开花。

Abstract: Flowers bright red to dark red and then the red color gradually fading at the upper part of the petal surfaces, single form, medium size, 8.5-10.0cm across with 6 petals, petals broad obovate, stamens slightly united into a column at the base, filaments red, bloom dense. Leaves dark green, leathery, elliptic, urgently cuneate at the base, margins shallowly and obtusely serrate. Plant compact and growth vigorous. Starts to bloom from early-summer, fully blooms from autumn to winter and sporadically in spring

HAR-10-5. 九月惊艳
Jiuyue Jingyan (September's Amazement)

回交组合： 杜鹃红山茶 × 杜鹃红山茶 F_1 代新品种'红屋积香'（ C. azalea × C. azalea F_1 new hybrid 'Hongwu Jixiang'）。

回交苗编号： DHW-No.8。

性状： 花朵桃红色，具少量白条纹，炎热季节偶有模糊的大白斑块，半重瓣型，中型至大型花，花径 8.0～11.2cm，花朵厚 4.8cm，大花瓣 22～25 枚，呈 3 轮有序排列，花瓣长倒卵形，雄蕊基部略连生，散射状。叶片浓绿色，叶背面灰绿色，革质，椭圆形，长 8.0～8.5cm，宽 4.0～5.0cm，基部急楔形，叶缘齿钝。植株紧凑，生长旺盛。夏初始花，秋、冬季盛花，春季零星开花。

Abstract: Flowers peach red with a few white stripes, occasionally with large and hazy white patches in hot season, semi-double form, medium to large size, 8.0-11.2cm across and 4.8cm deep with 22-25 large petals arranged orderly in 3 rows, petals long obovate, stamens slightly united at the base, scattered. Leaves dark green, leathery, elliptic, urgently cuneate at the base, margins obtusely serrate. Plant compact and growth vigorous. Starts to bloom from early-summer, fully blooms from autumn to winter and sporadically in spring.

HAR-10-6. 八月踏浪 Bayue Talang (Walking in Waves in August)

回交组合： 杜鹃红山茶 × 杜鹃红山茶 F_1 代新品种'红屋积香'（*C. azalea* × *C. azalea* F_1 new hybrid 'Hongwu Jixiang'）。

回交苗编号： DHW-No.9。

性状： 花朵红色，玫瑰重瓣型，中型至大型花，花径 9.0～11.5cm，花朵厚 5.5cm，花瓣 28～30 枚，呈 5 轮以上有序排列，阔倒卵形，外翻，花芯常为珠球状，偶有少量雄蕊。叶片浓绿色，叶背面灰绿色，厚革质，椭圆形，长 8.5～9.5cm，宽 4.5～5.0cm，基部急楔形，叶缘近光滑。植株紧凑，生长旺盛。夏初始花，秋、冬季盛花，春季零星开花。

Abstract: Flowers red, rose-double form, medium to large size, 9.0-11.5cm across and 5.5cm deep with 28-30 petals arranged orderly in over 5 rows, petals broad obovate, rolled outward, a small ball usually visible at the center, occasionally with a few stamens. Leaves dark green, heavy leathery, elliptic, urgently cuneate at the base, margins nearly smooth. Plant compact and growth vigorous. Starts to bloom from early-summer, fully blooms from autumn to winter and sporadically in spring.

HAR-10-7. 七月开幕
Qiyue Kaimu (In July Curtain-up)

回交组合： 杜鹃红山茶 × 杜鹃红山茶 F_1 代新品种'红屋积香'（ *C. azalea* × *C. azalea* F_1 new hybrid 'Hongwu Jixiang'）。

回交苗编号： DHW-No.13。

性状： 花朵红色，泛紫色调，略具芳香，半重瓣型，中型至大型花，花径 8.0～11.0cm，花朵厚 5.5cm，大花瓣 15～20 枚，呈 3～4 轮排列，花瓣多呈半直立状态，略波浪状，长倒卵形，雄蕊少量，散射状。叶片浓绿色，叶背面灰绿色，厚革质，椭圆形，长 8.0～9.5cm，宽 4.0～4.5cm，基部急楔形，叶缘近光滑。植株紧凑，生长旺盛。夏初始花，秋、冬季盛花，春季零星开花。

Abstract: Flowers red with purple hue, slightly fragrant, semi-double form, medium to large size, 8.0-11.0cm across and 5.5cm deep with 15-20 large petals arranged in 3-4 rows, petals usually semi-erected, slightly wavy, long obovate, stamens scattered. Leaves dark green, heavy leathery, elliptic, urgently cuneate at the base, margins nearly smooth. Plant compact and growth vigorous. Starts to bloom from early-summer, fully blooms from autumn to winter and sporadically in spring.

HAR-10-8. 刘村彩虹
Liucun Caihong (Liu Village's Rainbow)

回交组合： 杜鹃红山茶 × 杜鹃红山茶 F_1 代新品种'红屋积香'（ C. azalea × C. azalea F_1 new hybrid 'Hongwu Jixiang'）。

回交苗编号： DHW-No.14。

性状： 花朵深橘红色，瓣面上部具白色隐斑，半重瓣型，中型至大型花，花径 8.5～11.5cm，花朵厚 4.8cm，花瓣 12～13 枚，呈 2 轮排列，倒卵形，雄蕊基部略连生，花丝散射，开花稠密。叶片浓绿色，叶背面灰绿色，革质，长披针形，长 10.0～11.5cm，宽 3.5～4.0cm，基部急楔形，叶缘齿浅钝。植株紧凑，生长旺盛。夏初始花，秋、冬季盛花，春季零星开花。

> **Abstract:** Flowers deep orange red with faint white markings on petal surfaces, semi-double form, medium to large size, 8.5-11.5cm across and 4.8cm deep with 12-13 petals arranged in 2 rows, broad obovate, stamens slightly united at the base, filaments scattered, bloom dense. Leaves dark green, leathery, long lanceolate, urgently cuneate at the base, margins shallowly and obtusely serrate. Plant compact and growth vigorous. Starts to bloom from early-summer, fully blooms from autumn to winter and sporadically in spring.

HAR-10-9. 回龙新貌
Huilong Xinmao（Huilong Town's New Look）

回交组合：杜鹃红山茶 × 杜鹃红山茶 F₁ 代新品种'红屋积香'（ C. azalea × C. azalea F₁ new hybrid 'Hongwu Jixiang'）。

回交苗编号：DHW-No.16。

性状：花朵艳红色，泛紫色调，半重瓣型至牡丹型，中型至大型花，花径 9.0～12.5cm，花朵厚 5.0cm，花瓣 18～21 枚，呈 2 轮松散排列，阔倒卵形，雄蕊短柱状，开花稠密。叶片浓绿色，叶背面灰绿色，革质，长椭圆形，长 9.5～11.0cm，宽 4.5～5.0cm，基部急楔形，叶缘齿浅钝。植株开张，生长旺盛。夏初始花，秋、冬季盛花，春季零星开花。

Abstract: Flowers bright red with some purple hue, semi-double to peony form, medium to large size, 9.0-12.5cm across and 5.0cm deep with 18-21 petals that arranged loosely in 2 rows, petals broad obovate, stamens short columnar, bloom dense. Leaves dark green, leathery, long elliptic, urgently cuneate at the base, margins shallowly and obtusely serrate. Plant spread and growth vigorous. Starts to bloom from early-summer, fully blooms from autumn to winter and sporadically in spring.

HAR-10-10. 午夜灯火 Wuye Denghuo（Midnight Lights）

回交组合： 杜鹃红山茶 × 杜鹃红山茶 F_1 代新品种'红屋积香'（*C. azalea* × *C. azalea* F_1 new hybrid 'Hongwu Jixiang'）。

回交苗编号： DHW-No.17 + 15 + 16。

性状： 花朵黑红色，泛蜡光，半重瓣型，中型花，花径8.5～9.0cm，花朵厚4.7cm，花瓣16～18枚，呈3轮排列，倒卵形，雄蕊少量，基部略连生，开花稠密。叶片浓绿色，叶背面灰绿色，革质，阔椭圆形，长7.5～8.0cm，宽4.5～5.0cm，基部楔形，叶缘齿浅。植株紧凑，生长旺盛。夏初始花，秋、冬季盛花，春季零星开花。

Abstract: Flowers dark red with waxy luster, semi-double form, medium size, 8.5-9.0cm across and 4.7cm deep with 16-18 petals arranged in 3 rows, petals obovate, a few stamens, slightly united at the base, bloom dense. Leaves dark green, back surfaces pale green, leathery, broad elliptic, cuneate at the base, margins shallowly serrate. Plant compact and growth vigorous. Starts to bloom from early-summer, fully blooms from autumn to winter and sporadically in spring.

◉ 回交组合 HAR-11. 杜鹃红山茶 × 杜鹃红山茶 F₁ 代新品种 '香夏红娇'

HAR-11. *C. azalea* × *C. azalea* F₁ new hybrid 'Xiangxia Hongjiao'

本回交组合将介绍 3 个新品种。

'香夏红娇'是从杜鹃红山茶 × 非云南山茶杂交种'香漩涡'杂交组合中得到的（见第一部书第 338～339 页）。

虽然所获得的这 3 个回交新品种没有芳香性状的表达，但是，它们仍然具有多季开花的特性，而且花色艳丽，植株紧凑，是盆栽和园林美化的好品种。

Three new hybrids will be introduced in this backcross combination.

'Xiangxia Hongjiao' was obtained from cross-combination of *C. azalea* and Non-*reticulata* hybrid 'Scented Swirl' (See p.338-339, our first book).

The three backcrossed hybrids obtained in the combination do not express any fragrant characteristic, but they still have ever-blooming characteristic, their flowers are beautiful and their plants are compact, making them good varieties for potting and landscaping.

HAR-11-1. 金灿红云
Jincan Honyun (Gold Shining in Red Clouds)

回交组合： 杜鹃红山茶 × 杜鹃红山茶 F_1 代新品种'香夏红娇'（ *C. azalea* × *C. azalea* F_1 new hybrid 'Xiangxia Hongjiao'）。

回交苗编号： DXXHJ-No.3。

性状： 花朵红色，略具蜡光，瓣面显模糊的白斑或白条纹，玫瑰重瓣型，中型花，花径 8.0～9.5cm，花朵厚 5.0cm，外部大花瓣 28～35 枚，呈 4～5 轮排列，花瓣窄倒卵形，外翻，雄蕊短，与中部小花瓣混生，开花稠密。叶片浓绿色，叶背面灰绿色，革质，椭圆形，长 9.0～9.5cm，宽 3.0～3.5cm，基部楔形，叶齿稀而钝。植株紧凑，生长旺盛。夏初始花，秋、冬季盛花，春季零星开花。

Abstract: Flowers red slightly with waxy luster and hazy tiny white patches or stripes, rose-double form, medium size, 8.0-9.5cm across and 5.0cm deep with 28-35 exterior large petals arranged in 4-5 rows, petals narrow obovate, rolled outward, stamens short and mixed with central small petals, bloom dense. Leaves dark green, leathery, elliptic, cuneate at the base, margins sparsely and obtusely serrate. Plant compact and growth vigorous. Starts to bloom from early-summer, fully blooms from autumn to winter and sporadically in spring.

HAR-11-2. 月下瑶台
Yuexia Yaotai (Superb Balcony under the Moon)

回交组合： 杜鹃红山茶 × 杜鹃红山茶 F₁ 代新品种'香夏红娇'（C. azalea × C. azalea F₁ new hybrid 'Xiangxia Hongjiao'）。

回交苗编号： DXXHJ-No.5。

性状： 花朵粉红色至淡桃红色，半重瓣型，中型花，花径 8.5～9.5cm，花朵厚 4.5cm，大花瓣 12～14 枚，呈 2 轮排列，花瓣阔倒卵形，先端微凹，雄蕊散射状，花丝淡红色，花药淡黄色。叶片浓绿色，叶背面灰绿色，革质，长椭圆形，长 8.0～9.0cm，宽 3.0～3.5cm，基部楔形，上部叶齿浅。植株紧凑，生长旺盛。夏初始花，秋、冬季盛花，春季零星开花。

Abstract: Flowers pink to lightly peach red, semi-double form, medium size, 8.5-9.5cm across and 4.5cm deep with 12-14 large petals arranged in 2 rows, petals broad obovate, apices emarginate, stamens radial, filaments pale red, anthers slight yellow. Leaves dark green, leathery, long elliptic, cuneate at the base, margins shallowly serrate at the upper part. Plant compact and growth vigorous. Starts to bloom from early-summer, fully blooms from autumn to winter and sporadically in spring.

HAR-11-3. 幸福时代
Xingfu Shidai (The Happy Era)

回交组合：杜鹃红山茶 × 杜鹃红山茶 F_1 代新品种'香夏红娇'（*C. azalea* × *C. azalea* F_1 new hybrid 'Xiangxia Hongjiao'）。

回交苗编号：DXXHJ-No.7。

性状：花朵红色至暗红色，具少量白斑，中部花瓣淡红色，偶有白条纹，半重瓣型至玫瑰重瓣型，中型至大型花，花径 8.0～10.5cm，花朵厚 5.5cm，外部大花瓣 15 枚，呈 2 轮排列，外翻，中部花瓣波浪状，与雄蕊混生。叶片浓绿色，叶背面灰绿色，革质，椭圆形，长 8.5～9.5cm，宽 3.5～4.5cm，基部楔形，叶齿钝。植株紧凑，生长旺盛。夏初始花，秋、冬季盛花，春季零星开花。

Abstract: Flowers red to dark red with a few white markings, central petals light red occasionally with white stripes, semi-double to rose-double form, medium to large size, 8.0-10.5cm across and 5.5cm deep with 15 exterior large petals arranged in 2 rows, petals rolled outward, central petals wavy and mixed with stamens. Leaves dark green, back surfaces pale green, leathery, elliptic, cuneate at the base, margins obtusely serrate. Plant compact and growth vigorous. Starts to bloom from early-summer, fully blooms from autumn to winter and sporadically in spring.

⊙ 回交组合 HAR-12. 杜鹃红山茶 × 杜鹃红山茶 F_1 代新品种'夏梦华林'

HAR-12. *C. azalea* × *C. azalea* F_1 new hybrid 'Xiameng Hualin'

本回交组合将介绍8个新品种。

'夏梦华林'是从红山茶品种'都鸟'×杜鹃红山茶这一杂交组合中得到的（见第一部书第222～224页）。

本回交组合试图使回交获得的新品种的杜鹃红山茶基因频率达到75%。事实上，我们获得的这些回交种不仅像'夏梦华林'品种一样花大、色艳，而且开花期比'夏梦华林'更长了。

Eight new hybrids will be introduced in this backcross combination.

'Xiameng Hualin' was obtained from the cross-combination between *C. japonica* cultivar 'Miyakodori' and *C. azalea* (See p.222-224, our first book).

The backcross combination attempted to make *C. azalea*'s gene frequency of the backcross hybrids increase to 75%. In fact, the backcross hybrids obtained in the combination not only can open large and beautiful flowers which are similar to the hybrid 'Xiameng Hualin', but also have a longer blooming period than 'Xiameng Hualin'.

HAR-12-1. 迷你瑶姬
Mini Yaoji (Miniature Pretty Yaoji)

回交组合： 杜鹃红山茶 × 杜鹃红山茶 F_1 代新品种'夏梦华林'（ C. azalea × C. azalea F_1 new hybrid 'Xiameng Hualin'）。

回交苗编号： DHL-No.1。

性状： 花朵淡粉红色至粉红色，渐向花芯呈粉白色，完全重瓣型，微型至小型花，花径 5.0～6.3cm，花朵厚 4.2～4.5cm，花瓣 45～55 枚，呈 10 轮以上紧实排列，花瓣长倒卵形，开花稠密。叶片浓绿色，叶背面灰绿色，厚革质，阔倒卵圆形，稠密，节间短，长 8.0～8.5cm，宽 4.0～4.5cm，先端近圆形，颇像杜鹃红山茶，基部急楔形，边缘光滑。植株紧凑，灌丛状，生长旺盛。夏初始花，秋、冬季盛花，春季零星开花。

Abstract: Flowers light pink to pink, fading to pink-white at the center, formal double form, miniature to small size, 5.0-6.3cm across and 4.2-4.5cm deep with 45-55 petals arranged tightly in over 10 rows, petals long obovate, bloom dense. Leaves dark green, heavy leathery, broad obovate, dense, leaf internode short, apices nearly round which are like *C. azalea*, urgently cuneate at the base, margins smooth. Plant compact, bushy and growth vigorous. Starts to bloom from early-summer, fully blooms from autumn to winter and sporadically in spring.

HAR-12-2. 淡云阁雨
Danyun Geyu (Pale Cloud & Pavilion Rain)

回交组合： 杜鹃红山茶 × 杜鹃红山茶 F_1 代新品种'夏梦华林'（ C. azalea × C. azalea F_1 new hybrid 'Xiameng Hualin'）。

回交苗编号： DHL-No.3。

性状： 花朵淡粉红色，渐向花瓣上部呈粉白色，花瓣基部淡橘红色，玫瑰重瓣型至牡丹型，大型至巨型花，花径 12.5～13.5cm，花朵厚 5.5cm，大花瓣 35～45 枚，呈 6～7 轮层叠排列，花瓣长倒卵形，外翻，雄蕊簇状，偶有少量雄蕊瓣。叶片浓绿色，叶背面灰绿色，厚革质，狭长椭圆形，长 10.5～11.0cm，宽 3.0～3.5cm，基部急楔形，先端钝尖，边缘光滑。植株紧凑，灌丛状，生长旺盛。夏初始花，秋、冬季盛花，春季零星开花。

Abstract: Flowers pale pink, fading to pink-white at the top part of petals, light orange red at the base, rose-double to peony form, large to very large size, 12.5-13.5cm across and 5.5cm deep with 35-45 large petals arranged overlappingly in 6-7 rows, petals long obovate, turned outward, stamens clustered, occasionally with a few petaloids. Leaves dark green, heavy leathery, narrow long elliptic, urgently cuneate at the base, margins smooth. Plant compact, bushy and growth vigorous. Starts to bloom from early-summer, fully blooms from autumn to winter and sporadically in spring.

HAR-12-3. 卓越风姿
Zhuoyue Fengzi (Remarkable Posture)

回交组合： 杜鹃红山茶 × 杜鹃红山茶 F_1 代新品种'夏梦华林'（ C. azalea × C. azalea F_1 new hybrid 'Xiameng Hualin'）。

回交苗编号： DHL-No.4。

性状： 花朵艳红色至深红色，略泛紫色调，渐向花瓣上部呈深粉红色，偶有白斑块或白条纹，半重瓣型，大型至巨型花，花径 11.0～13.5cm，花朵厚 4.8cm，大花瓣 15～20 枚，呈 2～3 轮松散排列，花瓣长倒卵形，中部小花瓣扭曲，雄蕊簇状。叶片浓绿色，叶背面灰绿色，革质，椭圆形，长 9.0～11.0cm，宽 4.0～4.5cm，基部急楔形，先端钝尖，边缘光滑。植株紧凑，灌丛状，生长旺盛。夏初始花，秋、冬季盛花，春季零星开花。

Abstract: Flowers bright red to deep red, slightly with purple tone, fading to deep pink at the top part of petals, occasionally with white markings or stripes, semi-double form, large to very large size, 11.0-13.5cm across and 4.8cm deep with 15-20 large petals arranged loosely in 2-3 rows, petals long obovate, central small petals wrinkled, stamens clustered. Leaves dark green, back surfaces pale green, leathery, urgently cuneate at the base, margins smooth. Plant compact, bushy and growth vigorous. Starts to bloom from early-summer, fully blooms from autumn to winter and sporadically in spring.

HAR-12-4. 雨后暮景
Yuhou Mujing（Evening Scene after Rain）

回交组合： 杜鹃红山茶 × 杜鹃红山茶 F₁ 代杂交新品种'夏梦华林'（*C. azalea* × *C. azalea* F₁ new hybrid 'Xiameng Hualin'）。

回交苗编号： DHL-No.7。

性状： 花朵深橘红色，通常花瓣具白斑，牡丹型，大型至巨型花，花径 11.5～15.5cm，花朵厚 6.0cm，大花瓣 18～20 枚，呈 2～3 轮排列，花瓣长倒卵形，先端"V"形缺刻，中部无数瓣化小雄蕊瓣，与雄蕊混生，偶有几枚直立、扭曲的大花瓣，花丝短，辐射状。叶片浓绿色，叶背面灰绿色，革质，椭圆形，长 9.0～11.2cm，宽 3.0～4.0cm，基部急楔形，边缘光滑。植株紧凑，矮灌状，生长旺盛。仲夏始花，秋、冬季盛花，春季零星开花。

Abstract: Flowers deep orange red, usually with some white markings on the petals, peony form, large to very large size, 11.5-15.5cm across and 6.0cm deep with 18-20 large petals arranged in 2-3 rows, petals long obovate, apices with a "V" shaped notch, central countless small petaloids mixed with stamens, occasionally some large, erected and twisted petals at the center, filaments short and radial. Leaves dark green, leathery, elliptic, urgently cuneate at the base, margins smooth. Plant compact, bushy and growth vigorous. Starts to bloom from mid-summer, fully blooms from autumn to winter and sporadically in spring.

HAR-12-5. 风华正茂
Fenghua Zhengmao (Prime of Life)

回交组合： 杜鹃红山茶 × 杜鹃红山茶 F_1 代杂交新品种'夏梦华林'（*C. azalea* × *C. azalea* F_1 new hybrid 'Xiameng Hualin'）。

回交苗编号： DHL-No.11。

性状： 花朵深桃红色，花芯雄蕊瓣偶有白条纹，半重瓣型，大型至巨型花，花径12.0～13.5cm，花朵厚5.0cm，大花瓣18～21枚，呈3轮排列，花瓣阔倒卵形，完全开放后花瓣外翻，雄蕊辐射状。叶片浓绿色，叶背面灰绿色，革质，椭圆形，长8.0～9.0cm，宽4.0～4.5cm，基部楔形，上半部边缘齿浅，下半部边缘光滑。植株立性，生长旺盛。夏初始花，秋、冬季盛花，春季零星开花。

Abstract: Flowers deep peach red, occasionally with some white stripes on the central petaloids, semi-double form, large to very large size, 12.0-13.5cm across and 5.0cm deep with 18-21 large petals arranged in 3 rows, petals broad obovate, rolled outward when fully opening, stamens radial. Leaves dark green, leathery, elliptic, cuneate at the base, margins shallowly serrate at the upper part and smooth at the lower part. Plant upright and growth vigorous. Starts to bloom from early-summer, fully blooms from autumn to winter and sporadically in spring.

HAR-12-6. 橘红云路
Juhong Yunlu (Orange Red Cloud Road)

回交组合： 杜鹃红山茶 × 杜鹃红山茶 F_1 代杂交新品种'夏梦华林'(C. azalea × C. azalea F_1 new hybrid 'Xiameng Hualin')。

回交苗编号： DHL-No.13。

性状： 花朵橘红色，渐向花瓣上部略呈粉白色，花瓣基部和中部深橘红色，玫瑰重瓣型至牡丹型，大型至巨型花，花径 12.5～14.0cm，花朵厚 6.0cm，大花瓣 20～25 枚，呈 3～4 轮排列，中部雄蕊瓣 8～10 枚，具纵向白条纹，与雄蕊混生。叶片浓绿色，叶背面灰绿色，厚革质，长椭圆形，长 11.0～11.5cm，宽 3.5～4.0cm，基部急楔形，边缘光滑。植株紧凑，灌丛状，生长旺盛。仲夏始花，秋、冬季盛花，春季零星开花。

Abstract: Flowers orange red, slightly fading to pink-white at the top part of petals, deep orange red at the base and the center of the petals, rose-double to peony form, large to very large size, 12.5-14.0cm across and 6.0cm deep with 20-25 large petals arranged in 3-4 rows, 8-10 central petaloids with longitudinal white stripes and mixed with stamens. Leaves dark green, heavy leathery, long elliptic, urgently cuneate at the base, margins smooth. Plant compact, bushy and growth vigorous. Starts to bloom from mid-summer, fully blooms from autumn to winter and sporadically in spring.

HAR-12-7. 乡村色彩
Xiangcun Secai (Rural Color)

回交组合： 杜鹃红山茶 × 杜鹃红山茶 F_1 代杂交新品种'夏梦华林'（*C. azalea* × *C. azalea* F_1 new hybrid 'Xiameng Hualin'）。

回交苗编号： DHL-No.14。

性状： 花朵红色，花瓣偶见白斑，松散的牡丹型，中型至大型花，花径 9.0～12.0cm，花朵厚 6.0cm，大花瓣 18～21 枚，呈 3 轮排列，花瓣长倒卵形，花芯处几枚雄蕊瓣与雄蕊混生，花丝散生。叶片浓绿色，叶背面灰绿色，革质，椭圆形，长 8.5～9.0cm，宽 4.0～4.5cm，基部急楔形，边缘近光滑。植株立性，生长旺盛。夏初始花，秋、冬季盛花，春季零星开花。

> **Abstract:** Flowers red, occasionally with some white markings, loose peony form, medium to large size, 9.0-12.0cm across and 6.0cm deep with 18-21 large petals arranged in 3 rows, long obovate, some petaloids mixed with stamens at the center, filaments scattered. Leaves dark green, leathery, elliptic, urgently cuneate at the base, margins nearly smooth. Plant upright and growth vigorous. Starts to bloom from early-summer, fully blooms from autumn to winter and sporadically in spring.

HAR-12-8. 繁华世界
Fanhua Shijie (Bustling World)

回交组合：杜鹃红山茶 × 杜鹃红山茶 F$_1$ 代杂交新品种'夏梦华林'（*C. azalea* × *C. azalea* F$_1$ new hybrid 'Xiameng Hualin'）。

回交苗编号：DHL-No.15。

性状：花朵红色，花瓣中部红色较淡，略泛紫色调，托桂型至松散的牡丹型，大型至巨型花，花径11.0～14.0cm，花朵厚5.5～6.0cm，外部大花瓣10～12枚，呈2轮排列，花瓣长倒卵形，中部小雄蕊瓣15～18枚，松散、簇拥地与雄蕊混生。叶片浓绿色，叶背面灰绿色，厚革质，长椭圆形，长9.0～10.0cm，宽3.0～3.5cm，基部急楔形，边缘光滑。植株紧凑，矮灌状，生长旺盛。夏初始花，秋、冬季盛花，春季零星开花。

Abstract: Flowers red, slightly pale red at the mid-part of petals and with slightly purple hue, anemone to loose peony form, large to very large size, 11.0-14.0cm across and 5.5-6.0cm deep with 10-12 exterior large petals arranged in 2 rows, petals long obovate, 15-18 small petaloids loosely and clustered mixed with stamens at the center. Leaves dark green, leathery, elliptic, urgently cuneate at the base, margins smooth. Plant compact, bushy and growth vigorous. Starts to bloom from early-summer, fully blooms from autumn to winter and sporadically in spring.

◉ 回交组合 HAR-13. 杜鹃红山茶 × 杜鹃红山茶 F₁ 代新品种 '夏梦小旋'

HAR-13. *C. azalea* × *C. azalea* F₁ new hybrid 'Xiameng Xiaoxuan'

本回交组合将介绍4个新品种。

'夏梦小旋'是从杜鹃红山茶 × 红山茶品种'花牡丹'杂交组合中得到的（见第一部书第194～196页）。

这4个回交新品种的花型全是重瓣类型的，因此，我们有理由认为，多瓣类型的父本'夏梦小旋'品种决定了回交种花型的表达。

Four new hybrids will be introduced in this backcross combination.

'Xiameng Xiaoxuan' was obtained from the cross-combination between *C. azalea* and *C. japonica* cultivar 'Daikagura' (See p.194-196, our first book).

The four backcross hybrids are all double in their flower form, therefore, we have reason to believe that the male parent 'Xiameng Xiaoxuan', with multi-petals, decides the expression of the backcross hybrids in flower form.

HAR-13-1. 超级叠粉
Chaoji Diefen (Overlapped Pink Supreme)

回交组合：杜鹃红山茶 × 杜鹃红山茶 F_1 代新品种'夏梦小旋'（ *C. azalea* × *C. azalea* F_1 new hybrid 'Xiameng Xiaoxuan'）。

回交苗编号：DXX-No.2。

性状：花朵粉红色至深粉红色，随着花朵开放，粉色将逐渐变淡，玫瑰重瓣型至完全重瓣型，大型至巨型花，花径10.5～14.0cm，花朵厚5.4cm，花瓣70枚以上，呈7轮以上覆瓦状排列，花瓣倒卵形，花朵中部偶有半直立的小花瓣，开花稠密。叶片浓绿色，叶背面灰绿色，革质，近倒卵形，长9.5～11.0cm，宽4.8～5.7cm，基部急楔形，叶缘近全缘。植株开张，矮性，生长旺盛。夏初始花，秋、冬季盛花，翌年春季零星开花。

Abstract: Flowers pink to deep pink, as the flowers opening, the flower color is gradually fading to light pink, rose-double to formal double form, large to very large size, 10.5-14.0cm across and 5.4cm deep with over 70 petals imbricated in more than 7 rows, petals obovate, bloom dense. Leaves dark green, leathery, nearly obovate, urgently cuneate at the base, margins nearly smooth. Plant spread, dwarf and growth vigorous. Starts to bloom from early-summer, fully blooms from autumn to winter and sporadically in spring.

HAR-13-2. 今日喜儿
Jinri Xier (Today's Miss Xier)

回交组合： 杜鹃红山茶 × 杜鹃红山茶 F_1 代新品种'夏梦小旋'（ C. azalea × C. azalea F_1 new hybrid 'Xiameng Xiaoxuan'）。

回交苗编号： DXX-No.3。

性状： 花朵淡粉红色至深粉红色，渐向花芯粉色变淡，中部花瓣具白条纹，玫瑰重瓣型至完全重瓣型，小型至中型花，花径7.0～9.0cm，花朵厚4.0cm，大花瓣45枚以上，呈9轮有序排列，中部少量雄蕊与小花瓣混生，花瓣倒卵形。叶片浓绿色，叶背面灰绿色，稠密，革质，椭圆形，长7.0～8.0cm，宽3.0～3.5cm，叶缘全缘。植株紧凑，矮性，生长旺盛。夏初始花，秋、冬季盛花，翌年春季零星开花。

Abstract: Flowers light pink to deep pink, fading to very light pink at the center, central petals with white stripes, rose-double to formal double form, small to medium size, 7.0-9.0cm across and 4.0cm deep with over 45 petals arranged orderly in 9 rows, central small petals mixed with a few stamens, petals obovate. Leaves dark green, leathery, elliptic, margins smooth. Plant compact, dwarf and growth vigorous. Starts to bloom from early-summer, fully blooms from autumn to winter and sporadically in spring.

HAR-13-3. 童年回忆
Tongnian Huiyi (Childhood Memories)

回交组合：杜鹃红山茶 × 杜鹃红山茶 F_1 代新品种'夏梦小旋'（ C. azalea × C. azalea F_1 new hybrid 'Xiameng Xiaoxuan'）。

回交苗编号：DXX-No.4。

性状：花朵桃红色至红色，渐向花瓣上部呈淡粉白色，偶有白条纹，瓣面可见深红色脉纹，单瓣型至半重瓣型，中型花，花径 8.5～9.5cm，花朵厚 5.0cm，大花瓣 12～13 枚，呈 1～2 轮螺旋状排列，花瓣阔倒卵形，上部略外翻，雄蕊基部略连生，花丝淡红色，花药黄色。叶片浓绿色，叶背面灰绿色，革质，长椭圆形，长 8.5～10.0cm，宽 3.5～4.5cm，叶缘近全缘。植株紧凑，生长旺盛。夏初始花，秋、冬季盛花，翌年春季零星开花。

Abstract: Flowers peach red to red, fading to light pink-white at the upper part, occasionally with white stripes, deep veins visible on petal surfaces, single to semi-double form, medium size, 8.5-9.5cm across and 5.0cm deep with 12-13 large petals arranged spirally in 1-2 rows, petals broad obovate, slightly curved outward at the top part, stamens slightly united at the base, filaments pale red, anthers yellow. Leaves dark green, leathery, long elliptic, margins smooth. Plant compact and growth vigorous. Starts to bloom from early-summer, fully blooms from autumn to winter and sporadically in spring.

HAR-13-4. 乔之千金
Qiaozhi Qianjin（Dr Georg Ziemes' Beloved Daughter）

回交组合： 杜鹃红山茶 × 杜鹃红山茶 F_1 代新品种'夏梦小旋'（C. azalea × C. azalea F_1 new hybrid 'Xiameng Xiaoxuan'）。

回交苗编号： DXX-No.5。

性状： 花朵红色至深红色，具蜡质感，随着花朵开放，中部花瓣逐渐变为粉红色，玫瑰重瓣型至完全重瓣型，中型至大型花，花径9.0～11.5cm，花朵厚4.8cm，花瓣70枚以上，呈9～10轮覆瓦状排列，有时中部花瓣很多，花瓣倒卵形，内扣，开花稠密。叶片浓绿色，叶背面灰绿色，革质，椭圆形，长10.0～10.5cm，宽3.8～4.3cm，叶缘近全缘。植株开张，矮性，生长旺盛。夏初始花，秋、冬季盛花，春季零星开花。

注：德国的乔治·齐默斯（Georg Ziemes）博士是我们的好朋友，为了鼓励和推动他对茶花的酷爱和追求，本品种特以他的名字命名。

Abstract: Flowers red to deep red with waxy luster, as the flowers opening, central petals gradually change into pink, rose-double to formal double form, medium to large size, 9.0-11.5cm across and 4.8cm deep with over 70 petals arranged imbricately in 9-10 rows, sometimes central petals innumerable, petals obovate, bloom dense. Leaves dark green, leathery, elliptic, margins nearly smooth. Plant spread, dwarf and growth vigorous. Starts to bloom from early-summer, fully blooms from autumn to winter and sporadically in spring.

Mark: Dr. Georg Ziemes, our Germany friend, in order to encourage and promote his passion and pursuit of camellias, this hybrid is specially named after him.

回交组合 HAR-14. 杜鹃红山茶 × 杜鹃红山茶 F₁ 代新品种 '夏日广场'
HAR-14. *C. azalea* × *C. azalea* F₁ new hybrid 'Xiari Guangchang'

本回交组合将介绍24个新品种。

'夏日广场'是从杜鹃红山茶 × 云南山茶品种'帕克斯先生'这一杂交组合中得到的（见第一部书第263～264页）。在第一部书中本组曾介绍过1个叫作'桃园结义'的回交品种（见第一部书第383～384页）。至此，本组总计获得的杂交新品种已达25个。

从这些回交种性状上看，回交种的花色和花型较多，如粉红色、桃红色、艳红色，单瓣型、半重瓣型、牡丹型、玫瑰重瓣型均有出现。应该提及的是，这些回交种虽然花径不大，但其开花更稠密，植株明显矮性，叶片更像杜鹃红山茶。这说明作为回交父本的'夏日广场'品种的基因杂合性较为复杂。本组回交新品种更适合于美化家庭和园林环境。

×

Twenty four new hybrids will be introduced in this backcross combination.

'Xiari Guangchang' was obtained from the cross-combination between *C. azalea* and *C. reticulata* cultivar 'Dr Clifford Parks' (See p.263-264, our first book). One backcross hybrid named 'Taoyuan Jieyi' was introduced in this combination in our first book (See p.383-384, our first book). We have obtained twenty five new hybrids in this backcross combination until now.

From the traits of these new backcross hybrids, there are more flower colors and more flower forms, such as pink, peach-red, bright-red and single, semi-double, peony and rose-double forms, all appear in the combination. It should be mentioned that the flower size of these backcrossed hybrids are not large, but their flowers are more dense, plants are obviously dwarf and leaves are more similar to *C. azalea*'s. It has been shown that the heterozygosity of the femal parent, 'Xiari Guangchang' is very complex, and the new backcross hybrids are more suitable for beautifying family and garden environments.

HAR-14-1. 九天瑶池
Jiutian Yaochi (Jade Pool at the Ninth Heaven)

回交组合：杜鹃红山茶 × 杜鹃红山茶 F_1 代新品种'夏日广场'（ *C. azalea* × *C. azalea* F_1 new hybrid 'Xiari Guangchang'）。

回交苗编号：DGC-No.1。

性状：花朵红色，略泛蜡光，半重瓣型至玫瑰重瓣型，中型至大型花，花径 9.5～12.5cm，花朵厚 6.5cm，大花瓣 30～35 枚，呈 4 轮排列，花瓣阔倒卵形，边缘略内卷，花芯处少量小花瓣与雄蕊混生，雄蕊散射状，开花稠密。叶片浓绿色，叶背面灰绿色，革质，阔椭圆形，长 9.0～10.5cm，宽 5.5～6.0cm，基部楔形，边缘齿浅。植株立性，生长旺盛。夏初始花，秋、冬季盛花，春季零星开花。

Abstract: Flowers red, slightly with waxy luster, semi-double to rose-double form, medium to large size, 9.5-12.5cm across and 6.5cm deep with 30-35 large petals arranged in 4 rows, petals broad obovate, edges slightly rolled inward, a few of small petals mixed with stamens at the center, stamens diffused, bloom dense. Leaves dark green, leathery, broad elliptic, cuneate at the base, margins shallowly serrate. Plant upright and growth vigorous. Starts to bloom from early-summer, fully blooms from autumn to winter and sporadically in spring.

HAR-14-2. 成功之喜
Chenggong Zhixi (Exultation of Success)

回交组合：杜鹃红山茶 × 杜鹃红山茶 F_1 代新品种'夏日广场'（*C. azalea* × *C. azalea* F_1 new hybrid 'Xiari Guangchang'）。

回交苗编号：DGC-No.5。

性状：花朵艳红色至黑红色，半重瓣型至牡丹型，中型至大型花，花径9.5～12.0cm，花朵厚5.7cm，大花瓣24～26枚，呈4轮排列，花瓣倒卵形，中部小花瓣15～19枚，半直立，与雄蕊混生。叶片浓绿色，叶背面灰绿色，稠密，厚革质，长椭圆形，长9.5～11.0cm，宽4.9～5.0cm，基部楔形，边缘光滑。植株紧凑，立性，枝条稠密，生长旺盛。夏初始花，秋、冬季盛花，春季零星开花。

Abstract: Flowers bright red to dark red, semi-double to peony form, medium to large size, 9.5-12.0cm across and 5.7cm deep with 24-26 large petals arranged in 4 rows, petals obovate, 15-19 central small petals semi-erected and mixed with stamens. Leaves dark green, heavy leathery, long elliptic, cuneate at the base, margins smooth. Plant compact, upright, branches dense and growth vigorous. Starts to bloom from early-summer, fully blooms from autumn to winter and sporadically in spring.

HAR-14-3. 月月玫红
Yueyue Meihong (Monthly Rose Red)

回交组合：杜鹃红山茶 × 杜鹃红山茶 F_1 代新品种'夏日广场'(C. azalea × C. azalea F_1 new hybrid 'Xiari Guangchang')。

回交苗编号：DGC-No.12。

性状：花朵艳玫瑰红色，偶有白斑，花朵凋谢前2天，红色渐变为深粉红色，半重瓣型至牡丹型，中型花，花径8.5～10.0cm，花朵厚5.5cm，外部大花瓣25～30枚，呈3轮有序排列，大花瓣倒卵形，外翻，红色脉纹清晰可见，中部花瓣半直立，略皱折，雄蕊簇生，基部略连生，与中部花瓣混生。叶片浓绿色，叶背面灰绿色，半直立，稠密，椭圆形，厚革质，长7.5～8.5cm，宽3.0～3.5cm，略波浪状，基部楔形，边缘光滑。植株紧凑，枝叶稠密，灌丛状，生长旺盛。夏初始花，秋、冬季盛花，春季零星开花。

Abstract: Flowers bright rose red, occasionally with white markings on petals, the red color changes into deep pink 2 days before the flowers wither, semi-double to peony form, medium size, 8.5-10.0cm across and 5.5cm deep with 25-30 exterior large petals arranged orderly in 3 rows, large petals rolled outward, stamens clustered and mixed with the central petals. Leaves dark green, elliptic, heavy leathery, cuneate at the base, margins smooth. Plant compact, branches dense, bushy and growth vigorous. Starts to bloom from early-summer, fully blooms from autumn to winter and sporadically in spring.

HAR-14-4. 美丽盛夏
Meili Shengxia (Beautiful Mid-Summer)

回交组合： 杜鹃红山茶 × 杜鹃红山茶 F_1 代新品种'夏日广场'（ *C. azalea* × *C. azalea* F_1 new hybrid 'Xiari Guangchang'）。

回交苗编号： DGC-No.13。

性状： 花朵粉红色，泛紫色调，半重瓣型，中型花，花径 7.5～9.5cm，花朵厚 4.0cm，外部大花瓣 27～32 枚，呈 3 轮排列，花瓣狭长倒卵形，瓣面平铺，中部偶有少量半直立的小花瓣，雄蕊基部簇生，花丝细而蓬松，开花稠密。叶色和叶形与杜鹃红山茶极为相似，浓绿色，稠密，倒卵形，厚革质，长 7.0～7.5cm，宽 2.5～3.5cm，基部楔形，边缘光滑。植株紧凑，枝叶稠密，灌丛状，生长旺盛。夏初始花，秋、冬季盛花，春季零星开花。

Abstract: Flowers pink with purple hue, semi-double form, medium size, 7.5-9.5cm across and 4.0cm deep with 27-32 large petals arranged in 3 rows, petals narrow obovate with flat surfaces, occasionally a few of semi-erected and small petals in the center, stamens thin and fluffy, bloom dense. Leaves are very similar to those of *C. azalea*, dark green, obovate, heavy leathery, cuneate at the base, margins smooth. Plant compact, branches dense, bushy and growth vigorous. Starts to bloom from early-summer, fully blooms from autumn to winter and sporadically in spring.

HAR-14-5. 温馨感觉
Wenxin Ganjue (Warm Feeling)

回交组合：杜鹃红山茶 × 杜鹃红山茶 F_1 代新品种'夏日广场'(*C. azalea* × *C. azalea* F_1 new hybrid 'Xiari Guangchang')。

回交苗编号：DGC-No.16。

性状：花朵粉红色至深粉红色，随着花朵开放，变为淡粉红色，半重瓣型，中型至大型花，花径9.0～11.5cm，花朵厚4.8cm，花瓣30枚以上，呈3～4轮松散排列，花瓣阔倒卵形，雄蕊基部略连生。叶片浓绿色，叶背面灰绿色，半直立，稠密，厚革质，椭圆形，长7.0～8.0cm，宽3.0～3.5cm，略波浪状，基部楔形，边缘光滑。植株紧凑，枝叶稠密，灌丛状，生长旺盛。仲夏始花，秋、冬季盛花，春季零星开花。

Abstract: Flowers pink to deep pink, as the flowers opening, gradually change into light pink, semi-double form, medium to large size, 9.0-11.5cm across and 4.8cm deep with over 30 petals arranged loosely in 3-4 rows, stamens slightly united at the base. Leaves dark green, semi-erected, dense, heavy leathery, elliptic, slightly wavy, cuneate at the base, margins smooth. Plant compact, branches dense, bushy and growth vigorous. Starts to bloom from mid-summer, fully blooms from autumn to winter and sporadically in spring.

HAR-14-6. 儿童乐园
Ertong Leyuan (Children's Playground)

回交组合： 杜鹃红山茶 × 杜鹃红山茶 F₁ 代新品种'夏日广场'（*C. azalea* × *C. azalea* F₁ new hybrid 'Xiari Guangchang'）。

回交苗编号： DGC-No.19。

性状： 花朵深粉红色至红色，单瓣型，中型花，花径 8.5～9.5cm，花瓣 5～6 枚，狭长倒卵形，略波浪状，雄蕊基部连生，开花稠密。叶片浓绿色，叶背面灰绿色，革质，小倒卵形，长 6.0～7.0cm，宽 3.0～3.5cm，波浪状，基部楔形，边缘光滑，很像杜鹃红山茶。植株紧凑，枝叶稠密，灌丛状，生长旺盛。夏初始花，秋、冬季盛花，春季零星开花。

> **Abstract:** Flowers deep pink to red, single form, medium size, 8.5-9.5cm across with 5-6 petals, petals narrow-long obovate, slightly wavy, stamens united at the base, bloom dense. Leaves dark green, leathery, small obovate, wavy, cuneate at the base, margins smooth which is very similar to those of *C. azalea*. Plant compact, branches dense, bushy and growth vigorous. Starts to bloom from early-summer, fully blooms from autumn to winter and sporadically in spring.

HAR-14-7. 暑期红艳
Shuqi Hongyan (Summer Holidays' Brilliant Red)

回交组合： 杜鹃红山茶 × 杜鹃红山茶 F_1 代新品种'夏日广场'（*C. azalea* × *C. azalea* F_1 new hybrid 'Xiari Guangchang'）。

回交苗编号： DGC-No.20。

性状： 花朵深粉红色至红色，中部花瓣有少量白条纹，随着花朵开放，花色渐渐变淡，半重瓣型，中型花，花径 8.7～9.5cm，花朵厚 4.5cm，外部大花瓣约 28 枚，呈 3 轮排列，长倒卵形，中部花瓣半直立，与雄蕊混生，开花稠密。叶片浓绿色，叶背面灰绿色，半直立状态，稠密，厚革质，长椭圆形，长 8.0～9.5cm，宽 2.5～3.0cm，基部楔形，边缘光滑。植株紧凑，立性，枝叶稠密，生长旺盛。春末始花，夏、秋、冬季盛花，春季零星开花。

Abstract: Flowers deep pink to red, some white stripes on the central petals, as the flowers opening, the color gradually changes into light color, semi-double form, medium size, 8.7-9.5cm across and 4.5cm deep with about 28 exterior large petals arranged in 3 rows, central petals semi-erected and mixed with stamens, bloom dense. Leaves dark green, heavy leathery, long elliptic, cuneate at the base, margins smooth. Plant compact, upright, branches dense and growth vigorous. Starts to bloom from late-spring, fully blooms in summer, autumn and winter and sporadically in spring.

HAR-14-8. 高朋满座
Gaopeng Manzuo (Distinguished Friends Party)

回交组合：杜鹃红山茶 × 杜鹃红山茶 F_1 代新品种'夏日广场'（*C. azalea* × *C. azalea* F_1 new hybrid 'Xiari Guangchang'）。

回交苗编号：DGC-No.21 + 22。

性状：花朵艳红色至黑红色，泛蜡光，单瓣型，小型至中型花，花径6.5～8.0cm，花瓣6～7枚，阔倒卵形，边缘内扣，多数花朵呈半开放状态，雄蕊基部连生，开花稠密。叶片浓绿色，叶背面灰绿色，半直立，稠密，厚革质，小椭圆形，长7.0～8.0cm，宽3.0～3.5cm，基部楔形，边缘光滑。植株紧凑，枝叶稠密，灌丛状，生长旺盛。夏初始花，秋、冬季盛花，春季零星开花。

Abstract: Flowers bright red to dark red with waxy luster, single form, small to medium size, 6.5-8.0cm across with 6-7 petals, petals broad obovate, edges turned inward, most of flowers are semi-opened, stamens united at the base, bloom dense. Leaves dark green, semi-erected, heavy leathery, small elliptic, cuneate at the base, margins smooth. Plant compact, branches dense, bushy and growth vigorous. Starts to bloom from early-summer, fully blooms from autumn to winter and sporadically in spring.

HAR-14-9. 波光粼粼
Boguang Linlin (Sparkling Ripples)

回交组合：杜鹃红山茶 × 杜鹃红山茶 F_1 代新品种'夏日广场'（*C. azalea* × *C. azalea* F_1 new hybrid 'Xiari Guangchang'）。

回交苗编号：DGC-No.26。

性状：花朵艳红色，略泛蜡光，单瓣型，呈五角状，微型至小型花，花径 5.8～7.2cm，花瓣 5～6 枚，长倒卵形，半直立，两侧内卷，边缘波浪状，雄蕊基部连生，开花稠密。叶片浓绿色，叶背面灰绿色，革质，狭长椭圆形，长 8.5～10.0cm，宽 2.0～3.0cm，基部急楔形，略扭曲，边缘光滑。植株紧凑，灌丛状，矮性，生长旺盛。仲夏始花，秋、冬季盛花，春季零星开花。

Abstract: Flowers bright red slightly with waxy luster, single form which is pentagonal, miniature to small size, 5.8-7.2cm across with 5-6 petals, petals obovate, semi-erected, both sides incurved, wavy, stamens united at the base, bloom dense. Leaves dark green, leathery, narrow-long elliptic, urgently cuneate at the base, slightly twisted, margins smooth. Plant compact, bushy, dwarf and growth vigorous. Starts to bloom from mid-summer, fully blooms both in autumn and winter and sporadically in spring.

HAR-14-10. 小鸟依人
Xiaoniao Yiren (Sweet & Helpless Birds)

回交组合：杜鹃红山茶 × 杜鹃红山茶 F_1 代新品种'夏日广场'（ *C. azalea* × *C. azalea* F_1 new hybrid 'Xiari Guangchang'）。

回交苗编号：DGC-No.28 + 27。

性状：夏秋季节花朵为柔和的淡橘红色，冬季花朵为红色，半重瓣型，中型花，花径 8.5～9.5cm，冬季有时会出现大型花，花径可达12.0cm，花朵厚4.0～5.0cm，花瓣15～18枚，呈2～3轮排列，倒卵形，边缘内扣，雄蕊基部略连生，花瓣逐片掉落。叶片浓绿色，叶背面灰绿色，厚革质，阔椭圆形，长9.5～11.0cm，宽5.0～6.0cm，基部楔形，边缘叶齿稀钝。植株紧凑，枝叶稠密，灌丛状，生长旺盛。春末始花，夏、秋、冬季盛花，春季零星开花。

Abstract: Flowers soft pale orange red in summer and autumn, red in winter, semi-double form, medium size, 8.5-9.5cm across, sometimes large size with 12.0cm across appeared in winter, 4.0-5.0cm deep with 15-18 petals arranged in 2-3 rows, petals obovate, edges rolled inward, stamens slightly united at the base, petals fall one by one. Leaves dark green, heavy leathery, broad elliptic, cuneate at the base, margins sparsely and obtusely serrate. Plant compact, branches dense, bushy and growth vigorous. Starts to bloom from late-spring, fully blooms in summer, autumn and winter, and sporadically in spring.

HAR-14-11. 红漫金山
Hongman Jinshan (Red Covering Golden Mountains)

回交组合： 杜鹃红山茶 × 杜鹃红山茶 F_1 代新品种'夏日广场'（ C. azalea × C. azalea F_1 new hybrid 'Xiari Guangchang'）。

回交苗编号： DGC-No.29。

性状： 花朵艳红色，半重瓣型，中型至大型花，花径9.0～12.0cm，花朵厚5.0cm，花瓣18～20枚，呈3轮松散排列，长倒卵形，雄蕊基部连生，呈柱状，偶有雄蕊瓣。叶片浓绿色，叶背面灰绿色，革质，长椭圆形，长9.0～12.0cm，宽3.5～4.0cm，基部急楔形，边缘光滑。植株紧凑，灌丛状，生长旺盛。夏初始花，秋、冬季盛花，春季零星开花。

Abstract: Flowers bright red, semi-double form, medium to large size, 9.0-12.0cm across and 5.0cm deep with 18-20 petals arranged loosely in 3 rows, petals long obovate, stamens united into a column at the base, occasionally with petaloids. Leaves dark green, leathery, long elliptic, urgently cuneate at the base, margins smooth. Plant compact, bushy and growth vigorous. Starts to bloom from early-summer, fully blooms in autumn and winter and sporadically in spring.

HAR-14-12. 烂漫礼花 Lanman Lihua (Brilliant Fireworks)

回交组合：杜鹃红山茶 × 杜鹃红山茶 F_1 代新品种'夏日广场'（ C. azalea × C. azalea F_1 new hybrid 'Xiari Guangchang'）。

回交苗编号：DGC-No.33。

性状：花朵艳红色，瓣面偶具白点或白斑，半重瓣型，中型至大型花，花径 8.5～11.5cm，花朵厚 5.0cm，花瓣 18～20 枚，呈 3 轮松散排列，长倒卵形，半开放状态时花瓣半直立，全开放后花瓣外翻，雄蕊簇生。叶片浓绿色，叶背面灰绿色，厚革质，阔椭圆形，长 8.5～9.5cm，宽 5.5～6.0cm，边缘近光滑。植株立性，生长旺盛。春末始花，夏、秋、冬季盛花，春季零星开花。

Abstract: Flowers bright red, occasionally with some white dots or markings on the petal surfaces, semi-double form, medium to large size, 8.5-11.5cm across and 5.0cm deep with 18-20 petals arranged loosely in 3 rows, petals long obovate, semi-erected when just opening and rolled outward when fully opening, stamens clustered. Leaves dark green, heavy leathery, broad elliptic, margins nearly smooth. Plant upright and growth vigorous. Starts to bloom from late-spring, fully blooms in summer, autumn and winter and sporadically in spring.

HAR-14-13. 淡雅柔粉
Danya Roufen (Quietly Elegant & Soft Pink)

回交组合： 杜鹃红山茶 × 杜鹃红山茶 F_1 代新品种'夏日广场'（ C. azalea × C. azalea F_1 new hybrid 'Xiari Guangchang'）。

回交苗编号： DGC-No.34。

性状： 花朵粉红色至橘红色，单瓣型，中型花，花径8.5～9.5cm，花瓣7～8枚，长倒卵形，长6.0cm，宽3.7cm，平铺，厚质，雄蕊基部略连生，呈短管状，开花稠密。叶片浓绿色，叶背面灰绿色，厚革质，狭长椭圆形，长8.5～9.6cm，宽3.0～3.5cm，基部急楔形，边缘光滑。植株开张，灌丛状，生长中等。夏初始花，秋、冬季盛花，春季零星开花。

Abstract: Flowers pink to orange red, single form, medium size, 8.5-9.5cm across with 7-8 petals, petals flat, thick texture, stamens slightly united at the base, bloom dense. Leaves dark green, heavy leathery, urgently cuneate at the base, margins smooth. Plant spread, bushy and growth normal. Starts to bloom from early-summer, fully blooms from autumn to winter and sporadically in spring.

HAR-14-14. 红蜡雕塑
Hongla Diaosu (Sculpture with Red Wax)

回交组合： 杜鹃红山茶 × 杜鹃红山茶 F_1 代新品种'夏日广场'（ C. azalea × C. azalea F_1 new hybrid 'Xiari Guangchang'）。

回交苗编号： DGC-No.40。

性状： 花朵蜡红色，具蜡光，半重瓣型至玫瑰重瓣型，中型至大型花，花径 8.5～12.0cm，花朵厚 6.0cm，大花瓣 25～30 枚，呈 5～6 轮排列，花瓣长倒卵形，边缘内卷呈沟槽状，有时雄蕊散射。叶片浓绿色，叶背面灰绿色，革质，阔椭圆形，长 8.5～10.0cm，宽 4.5～5.0cm，基部楔形，边缘齿浅。植株立性，生长旺盛。夏初始花，秋、冬季盛花，春季零星开花。

Abstract: Flowers waxy red with waxy luster, semi-double to rose-double form, medium to large size, 8.5-12.0cm across and 6.0cm deep with 25-30 large petals arranged in 5-6 rows, petals long obovate, edges rolled inward into groove shape, sometimes stamens radial. Leaves dark green, leathery, broad elliptic, cuneate at the base, margins shallowly serrate. Plant upright and growth vigorous. Starts to bloom from early-summer, fully blooms from autumn to winter and sporadically in spring.

HAR-14-15. 富丽堂皇
Fuli Tanghuang (Magnificence)

回交组合：杜鹃红山茶 × 杜鹃红山茶 F_1 代新品种'夏日广场'(*C. azalea* × *C. azalea* F_1 new hybrid 'Xiari Guangchang')。

回交苗编号：DGC-No.41。

性状：花蕾锥形，萼片淡绿色。花朵红色，略泛紫色调，半重瓣型至玫瑰重瓣型，中型花，花径 9.0～10.0cm，花朵厚 5.0cm，外部大花瓣 23～25 枚，呈 3 轮排列，花瓣倒卵形，瓣长 4.5cm，宽 2.3cm，先端边缘略内扣，微凹，中部偶有少量半直立的小花瓣，雄蕊簇生，花丝淡粉红色，花药淡黄色。叶片浓绿色，叶背面灰绿色，厚革质，阔倒卵形，长 7.0～8.0cm，宽 4.0～4.5cm，基部楔形，主脉黄绿色，侧脉不明显，叶缘齿浅稀。植株立性，枝叶稠密，生长旺盛。夏初始花，秋、冬季盛花，春季零星开花。

Abstract: Flowers red slightly with purple hue, semi-double to rose-double form, medium size, 9.0-10.0cm across and 5.0cm deep with 23-25 exterior large petals arranged in 3 rows, petals obovate, apices slightly turned inward, occasionally a few semi-erected and small petals at the center, stamens clustered. Leaves dark green, heavy leathery, broad obovate, cuneate at the base, midrib yellow green and lateral veins unconspicuous, margins shallowly and sparsely serrate. Plant upright, branches dense and growth vigorous. Starts to bloom from early-summer, fully blooms from autumn to winter and sporadically in spring.

HAR-14-16. 浪漫夏景
Langman Xiajing（Romantic Summer Scenery）

回交组合：杜鹃红山茶 × 杜鹃红山茶 F_1 代新品种'夏日广场'（ C. azalea × C. azalea F_1 new hybrid 'Xiari Guangchang'）。

回交苗编号：DGC-No.42。

性状：花朵深粉红色至桃红色，有少量白色条纹，随着花朵开放，渐渐变为淡橘粉红色，半重瓣型，中型至大型花，花径9.5～11.0cm，花朵厚4.5cm，大花瓣38～45枚，呈3～5轮紧实排列，花瓣倒卵形，边缘内扣，先端尖，雄蕊近簇生。叶片浓绿色，叶背面灰绿色，厚革质，阔椭圆形，长8.5～9.5cm，宽4.0～4.5cm，边缘光滑。植株开张，灌丛状，生长旺盛。夏初始花，秋、冬季盛花，春季零星开花。

Abstract: Flowers deep pink to peach red, with a few white stripes, as the flowers opening, flowers color gradually changes into light-orange pink, semi-double form, medium to large size, 9.5-11.0cm across and 4.5cm deep with 38-45 large petals arranged tightly in 3-5 rows, apices pointed, stamens almost clustered. Leaves dark green, heavy leathery, broad elliptic, margins smooth. Plant spread, bushy and growth vigorous. Starts to bloom from early-summer, fully blooms from autumn to winter and sporadically in spring.

HAR-14-17. 娇柔含羞
Jiaorou Hanxiu (Charming Soft with Pudency)

回交组合：杜鹃红山茶 × 杜鹃红山茶 F_1 代新品种'夏日广场'(*C. azalea* × *C. azalea* F_1 new hybrid 'Xiari Guangchang')。

回交苗编号：DGC-No.46 + 45。

性状：花朵淡粉红色，常有一片花瓣具白斑，单瓣型，微型至小型花，花径5.5～7.0cm，花瓣6～7枚，倒卵形，薄质，边缘内卷或外翻，雄蕊基部连生，呈短柱状，开花稠密，花朵整朵掉落。叶片浓绿色，叶背面灰绿色，革质，长椭圆形，长7.5～8.5cm，宽3.0～4.0cm，基部急楔形，边缘光滑。植株紧凑，生长旺盛。夏初始花，秋、冬季盛花，春季零星开花。

Abstract: Flowers light pink, always a petal with a white marking, single form, miniature to small size, 5.5-7.0cm across with 6-7 petals, petals obovate, thin texture, rolled inward or outward, stamens united into a short column at the base, bloom dense, flowers fall whole. Leaves dark green, leathery, long elliptic, urgently cuneate at the base, margins smooth. Plant compact and growth vigorous. Starts to bloom from early-summer, fully blooms from autumn to winter and sporadically in spring.

HAR-14-18. 后浪奔涌
Houlang Benyong (Back Wavy Surges)

回交组合： 杜鹃红山茶 × 杜鹃红山茶 F$_1$ 代新品种'夏日广场'（ C. azalea × C. azalea F$_1$ new hybrid 'Xiari Guangchang'）。

回交苗编号： DGC-No.48。

性状： 花朵深桃红色，花瓣上部红色略淡，半重瓣型，大型花，花径 10.5～12.5cm，花朵厚 5.5cm，花瓣 21～23 枚，呈 3 轮松散排列，长倒卵形，波浪状，先端略凹，半直立，雄蕊散生。叶片浓绿色，叶背面灰绿色，厚革质，椭圆形，长 10.0～10.5cm，宽 5.0～5.5cm，基部楔形，叶缘近光滑。植株立性，生长旺盛。夏初始花，秋、冬季盛花，春季零星开花。

Abstract: Flowers deeply peach red, slightly light red at the upper part of petals, semi-double form, large size, 10.5-12.5cm across and 5.5cm deep with 21-23 petals arranged loosely in 3 rows, petals long obovate, wavy, apices emarginate, stamens disperse. Leaves dark green, heavy leathery, elliptic, cuneate at the base, margins nearly smooth. Plant upright and growth vigorous. Starts to bloom from early-summer, fully blooms from autumn to winter and sporadically in spring.

HAR-14-19. 青春之歌
Qingchun Zhige (Youth's Song)

回交组合：杜鹃红山茶 × 杜鹃红山茶 F₁ 代新品种'夏日广场'（*C. azalea* × *C. azalea* F₁ new hybrid 'Xiari Guangchang'）。

回交苗编号：DGC-No.49。

性状：花朵红色，瓣面中部红色略淡，半重瓣型，大型花，花径 11.0～12.0cm，花朵厚 7.0cm，花瓣 10～12 枚，呈 2 轮松散排列，花瓣长倒卵形，勺状，先端深凹，外翻，雄蕊基部连生，呈短柱状，花朵整朵掉落。叶片浓绿色，叶背面灰绿色，革质，长椭圆形，长 11.0～11.5cm，宽 4.0～4.5cm，基部急楔形，上部边缘具浅齿。植株立性，生长旺盛。夏初始花，秋、冬季盛花，春季零星开花。

> **Abstract:** Flowers red, slightly light red at the mid-part of petals, semi-double form, large size, 11.0-12.0cm across and 7.0cm deep with 10-12 petals arranged loosely in 2 rows, petals long obovate, spoon-shaped, apices deeply concaved and turned outward, stamens united into a short column at the base, flowers fall whole. Leaves dark green, leathery, long elliptic, urgently cuneate at the base, upper margins shallowly serrate. Plant upright and growth vigorous. Starts to bloom from early-summer, fully blooms from autumn to winter and sporadically in spring.

HAR-14-20. 红院满福
Hongyuan Manfu (Red Courtyard with Full Happiness)

回交组合：杜鹃红山茶 × 杜鹃红山茶 F₁ 代新品种'夏日广场'（ *C. azalea* × *C. azalea* F₁ new hybrid 'Xiari Guangchang'）。

回交苗编号：DGC-No.51。

性状：花朵橘红色至红色，泛蜡光，偶有白斑或白条纹，半重瓣型至玫瑰重瓣型，中型至大型花，花径8.5～11.5cm，花朵厚5.5cm，大花瓣18～20枚，呈3轮排列，花瓣阔倒卵形，雄蕊近簇生，少量雄蕊瓣与雄蕊混生，开花稠密。叶片浓绿色，叶背面灰绿色，狭长椭圆形，厚革质，长8.5～9.5cm，宽3.0～3.5cm，基部急楔形，边缘光滑。植株紧凑，枝叶稠密，灌丛状，生长旺盛。夏初始花，秋、冬季盛花，春季零星开花。

Abstract: Flowers bright orange red to red with waxy luster, occasionally with white markings or white stripes, semi-double to rose-double form, medium to large size, 8.5-11.5cm across and 5.5cm deep with 18-20 large petals arranged in 3 rows, petals broad obovate, stamens nearly clustered and a few of petaloids mixed with stamens, bloom dense. Leaves dark green, narrow-long elliptic, heavy leathery, urgently cuneate at the base, margins smooth. Plant compact, branches dense, bushy and growth vigorous. Starts to bloom from early-summer, fully blooms from autumn to winter and sporadically in spring.

HAR-14-21. 十月烟火
Shiyue Yanhuo (October's Fireworks)

回交组合：杜鹃红山茶 × 杜鹃红山茶 F_1 代新品种'夏日广场'（*C. azalea* × *C. azalea* F_1 new hybrid 'Xiari Guangchang'）。

回交苗编号：DGC-No.52。

性状：花朵粉红色至深粉红色，渐向基部变为淡粉红色，有时有白条纹，花朵凋萎前，变为淡粉红色，半重瓣型，大型至巨型花，花径11.5～13.2cm，花朵厚4.0cm，外部大花瓣40枚以上，呈3～4轮紧实排列，花瓣倒卵形，先端略波浪状，中部花瓣约15枚，半直立，花朵整朵掉落。叶片浓绿色，叶背面灰绿色，厚革质，椭圆形，长8.0～9.0cm，宽3.0～3.4cm，基部楔形，边缘光滑。植株开张，矮灌丛状，生长旺盛。春末始花，夏、秋、冬季盛花，春季零星开花。

Abstract: Flowers pink to deep pink, fading to light pink at the base, sometimes with white stripes, before withering, the flowers become light pink, semi-double form, large to very large size, 11.5-13.2cm across and 4.0cm deep with over 40 exterior petals arranged tightly in 3-4 rows, petals obovate, apices slightly wavy, about 15 erected petals at the center, flowers fall whole. Leaves dark green, heavy leathery, elliptic, cuneate at the base, margins smooth. Plant spread, dwarf, bushy and growth vigorous. Starts to bloom from late-spring, fully blooms in summer, autumn and winter and sporadically in spring.

HAR-14-22. 桃红如春
Taohong Ruchun (Peach Red as Spring)

回交组合： 杜鹃红山茶 × 杜鹃红山茶 F_1 代新品种'夏日广场'（*C. azalea* × *C. azalea* F_1 new hybrid 'Xiari Guangchang'）。

回交苗编号： DGC-No.53。

性状： 花朵淡桃红色，瓣面具少量白色条纹，半重瓣型，中型至大型花，花径 9.5～12.5cm，花朵厚 5.0cm，花瓣 24～25 枚，呈 3～4 轮松散排列，花瓣长倒卵形，外翻，雄蕊基部略连生，呈散射状。叶片浓绿色，叶背面灰绿色，革质，椭圆形，长 9.5～10.0cm，宽 4.0～4.5cm，基部楔形，上部边缘具浅齿。植株立性，生长旺盛。夏初始花，秋、冬季盛花，春季零星开花。

Abstract: Flowers light peach red with a few white stripes on the surfaces, semi-double form, medium to large size, 9.5-12.5cm across and 5.0cm deep with 24-25 petals arranged loosely in 3-4 rows, petals long obovate, stamens slightly united at the base, diffuse. Leaves dark green, leathery, elliptic, cuneate at the base, upper margins shallowly serrate. Plant upright and growth vigorous. Starts to bloom from early-summer, fully blooms from autumn to winter and sporadically in spring.

HAR-14-23. 八月红浪
Bayue Honglang (August's Red Waves)

回交组合：杜鹃红山茶 × 杜鹃红山茶 F_1 代新品种'夏日广场'（C. azalea × C. azalea F_1 new hybrid 'Xiari Guangchang'）。

回交苗编号：DGC-No.54。

性状：花朵艳红色，半重瓣型至牡丹型，中型至大型花，花径 9.0～11.5cm，花朵厚 5.3cm，外部波浪状的大花瓣 35～40 枚，呈 3 轮松散排列，大花瓣阔倒卵形，花芯处有少量半直立的小花瓣，与雄蕊混生，雄蕊散射或簇生，开花稠密。叶片浓绿色，叶背面灰绿色，稠密，厚革质，长椭圆形，长 9.0～9.5cm，宽 3.5～4.0cm，基部楔形，边缘光滑。植株开张，枝叶稠密，生长旺盛。夏初始花，秋、冬季盛花，春季零星开花。

Abstract: Flowers bright red, semi-double to peony form, medium to large size, 9.0-11.5cm across and 5.3cm deep with 35-40 large wavy petals arranged loosely in 3 rows, central small petals semi-erected and mixed with stamens. Leaves dark green, heavy leathery, long elliptic, cuneate at the base, margins smooth. Plant spread, branches dense and growth vigorous. Starts to bloom from early-summer, fully blooms from autumn to winter and sporadically in spring.

HAR-14-24. 似曾相见
Siceng Xiangjian (Seems to Have Met Before)

回交组合： 杜鹃红山茶 × 杜鹃红山茶 F$_1$ 代新品种'夏日广场'（ C. azalea × C. azalea F$_1$ new hybrid 'Xiari Guangchang'）。

回交苗编号： DGC-No.56。

性状： 花朵红色，单瓣型，小型花，花径 7.0～7.5cm，花朵厚 7cm，花瓣 5～6 枚，呈 1 轮排列，多呈半开放状态，花瓣阔倒卵形，半直立，花瓣上部外翻，略扭曲，呈三角状，像玉兰花状，先端微凹，雄蕊基部连生呈柱状，花丝红色，开花与抽梢同时进行，开花稠密，花朵整朵掉落。叶片浓绿色，叶背面灰绿色，革质，椭圆形，长 8.5～9.5cm，宽 3.0～4.0cm，基部楔形，边缘光滑。植株紧凑，生长旺盛。夏初始花，秋、冬季盛花，春季零星开花。本品种与'夏梦玉兰'（见第一部书第 224～225 页）的最大区别是叶片较小，花朵较大，开花更稠密。

Abstract: Flowers red, single form, small size, 7.0-7.5cm across and 7cm deep with 5-6 petals arranged in 1 row, semi-open, petals broad obovate, semi-erected, rolled outward and slightly twisted into triangles which is like magnolia flower shaped, stamens united into a column at the base, filaments red, flowering and budding simultaneously, bloom dense, flowers fall whole. Leaves dark green, leathery, elliptic, cuneate at the base, margins smooth. Plant compact and growth vigorous. Starts to bloom from early-summer, fully blooms from autumn to winter and sporadically in spring. The major difference between this hybrid and 'Xiameng Yulan' (See p.224-225, our first book) is that this hybrid has smaller leaves, larger flowers and denser bloom.

⦿ 回交组合 HAR-15. 杜鹃红山茶 × 杜鹃红山茶 F₁ 代新品种 '夏日粉丽'

HAR-15. *C. azalea* × *C. azalea* F₁ new hybrid 'Xiari Fenli'

本回交组合将介绍 2 个新品种。

'夏日粉丽'是从杜鹃红山茶 × 云南山茶品种'帕克斯先生'杂交组合中得到的（见第一部书第 261 ～ 263 页）。可以看到，所获得的两个回交新品种都是具有大型至巨型花的好品种。

Two new hybrids will be introduced in this backcross combination.

'Xiari Fenli' was obtained from the cross-combination between *C. azalea* and *C. reticulata* cultivar 'Dr Clifford Parks' (See p.261-263, our first book). It can be seen that the two new hybrids we obtained are all nice varieties that can open large to very large flowers.

HAR-15-1. 伟大复兴
Weida Fuxing (Great Renaissance)

回交组合：杜鹃红山茶 × 杜鹃红山茶 F_1 代新品种'夏日粉丽'（*C. azalea* × *C. azalea* F_1 new hybrid 'Xiari Fenli'）。

回交苗编号：DXRFL-No.1。

性状：花朵红色至深红色，牡丹型，大型花，花径 11.0～13.0cm，花朵厚 5.0cm，花瓣 40 枚左右，呈 3～4 轮排列，长倒卵形，上部边缘略波浪状，雄蕊长 1.5cm，基部略连生。叶片浓绿色，叶背面灰绿色，厚革质，椭圆形，长 6.5～7.0cm，宽 2.5～2.7cm，叶缘近光滑。植株紧凑，矮生，生长旺盛。夏初始花，秋、冬季盛花，春季零星开花。

> **Abstract:** Flowers red to deep red, peony form, large size, 11.0-13.0cm across and 5.0cm deep with about 40 petals arranged in 3-4 rows, petals long obovate, slightly wavy at the upper part edges, stamens slightly united at the base. Leaves dark green, heavy leathery, margins nearly smooth. Plant compact, dwarf and growth vigorous. Starts to bloom from early-summer, fully blooms from autumn to winter and sporadically in spring.

HAR-15-2. 日照香炉
Rizhao Xianglu (Sunshine Censer)

回交组合：杜鹃红山茶 × 杜鹃红山茶 F_1 代新品种'夏日粉丽'(C. azalea × C. azalea F_1 new hybrid 'Xiari Fenli')。

回交苗编号：DXRFL-No.5。

性状：花朵紫粉红色至紫红色，半重瓣型，中型至大型花，花径9.0～12.5cm，花朵厚5.0cm，大花瓣15～18枚，呈3轮排列，长倒卵形，花瓣外翻，雄蕊辐射状。叶片浓绿色，叶背面灰绿色，厚革质，椭圆形，长7.5～8.0cm，宽3.0～3.5cm，基部急楔形，叶缘近光滑。植株紧凑，生长旺盛。夏初始花，秋、冬季盛花，春季零星开花。

> **Abstract:** Flowers purple pink to purple red, semi-double form, medium to large size, 9.0-12.5cm across and 5.0cm deep with 15-18 petals arranged in 3 rows, petals long obovate, rolled outward, stamens radial. Leaves dark green, heavy leathery, urgently cuneate at the base, margins nearly smooth. Plant compact and growth vigorous. Starts to bloom from early-summer, fully blooms from autumn to winter and sporadically in spring.

⦿ 回交组合 HAR-16. 杜鹃红山茶 × 杜鹃红山茶 F₁ 代新品种 '满天红星'

HAR-16. *C. azalea* × *C. azalea* F₁ new hybrid 'Mantian Hongxing'

本回交组合将介绍 1 个新品种。

'满天红星'是从红山茶品种'媚丽'× 杜鹃红山茶的杂交组合中得到的（见第一部书第 106～108 页）。

因为本回交组合中两个亲本都是单瓣型的，因此，所获得的回交种也是单瓣型的。尽管'满天红星'品种没有表达'媚丽'品种花瓣白边的性状，但它已具有'媚丽'品种的基因，所以新获得的回交品种花瓣边缘带白色宽边。

One new hybrid will be introduced in this backcross combination.

'Mantian Hongxing' was obtained from the cross-combination between *C. japonica* cultivar 'Tama Beauty' and *C. azalea* (See p.106-108, our first book).

Because both of the cross parents in the backcross combination are single in flower form, the new hybrid obtained is also single form. Although the hybrid 'Mantian Hongxing' does not have any expression of petal white edges as 'Tama Beauty', it has the gene of 'Tama Beauty', so the flowers of the backcross new hybrid have white edges.

HAR-16-1. 闪光风车
Shanguang Fengche (Glittering Windmill)

回交组合： 杜鹃红山茶 × 杜鹃红山茶 F_1 代新品种'满天红星'（*C. azalea* × *C. azalea* F_1 new hybrid 'Mantian Hongxing'）。

回交苗编号： DMTHX-No.7。

性状： 花朵粉红色，渐向花瓣上部变为白色，形成较宽的白边，单瓣型，中型至大型花，花径 9.5～11.0cm，花瓣 7～8 枚，阔倒卵形，外翻，排列有动感，雄蕊基部略连生。叶片浓绿色，叶背面灰绿色，革质，椭圆形，长 10.0～10.5cm，宽 3.7～4.5cm，基部楔形，叶缘近光滑。植株紧凑，生长中等。夏初始花，秋冬季盛花，春季零星开花。

Abstract: Flowers pink, fading to white at the top parts of the petals, which formed a wide white border, single form, medium to large size, 9.5-11.0cm across with 7-8 petals, petals broad obovate, stamens slightly united at the base. Leaves dark green, leathery, elliptic, cuneate at the base, margins nearly smooth. Plant compact and growth normal. Starts to bloom from early-summer, fully blooms from autumn to winter and sporadically in spring.

⦿ 回交组合 HAR-17. 杜鹃红山茶 × 杜鹃红山茶 F₁ 代新品种 '红波涌金'

HAR-17. *C. azalea* × *C. azalea* F₁ new hybrid 'Hongbo Yongjin'

本回交组合将介绍 3 个新品种。

'红波涌金'是从杜鹃红山茶 × 云南山茶品种'帕克斯先生'杂交组合中得到的（见第一部书第 250～251 页）。

可以看到，获得的回交新品种，在花色上已经遗传了父本'红波涌金'的黑红花色，同时也隔代遗传了'帕克斯先生'品种多瓣的性状。从开花期上看，这 3 个回交新品种都是从夏初开始开花，明显具有杜鹃红山茶的性状。

×

Three new hybrids will be introduced in this backcross combination.

'Hongbo Yongjin' was obtained from the cross-combination between *C. azalea* and *C. reticulata* cultivar 'Dr Clifford Parks' (See p.250-251, our first book).

It can be seen that the hybrids have inherited the characteristic of dark-red flower color from the male parent 'Hongbo Yongjin', and also have inherited the characteristic of multi-petals from the cultivar 'Dr Clifford Parks' in atavism. These three backcross new hybrids all start to bloom from early-summer, which abviously inherit the characteristic of *C. azalea*.

HAR-17-1. 酷夏流金
Kuxia Liujin (Flowing Gold in Hot Summer)

回交组合： 杜鹃红山茶 × 杜鹃红山茶 F_1 代新品种'红波涌金'（*C. azalea* × *C. azalea* F_1 new hybrid 'Hongbo Yongjin'）。

回交苗编号： DHBYJ-No.11+9。

性状： 花朵红色至深红色，半重瓣型，大型至巨型花，花径 10.5～14.0cm，花朵厚 4.5cm，花瓣 18～23 枚，呈 3～4 轮排列，花瓣长倒卵形，雄蕊基部略连生，偶有雄蕊瓣。叶片浓绿色，叶背面灰绿色，革质，长椭圆形，长 7.5～8.5cm，宽 3.0～3.5cm，基部急楔形，边缘近光滑。植株立性，枝叶稠密，生长旺盛。夏初始花，秋、冬季盛花，春季零星开花。

> **Abstract:** Flowers red to deep red, semi-double form, large to very large size, 10.5-14.0cm across and 4.5cm deep with 18-23 petals arranged in 3-4 rows, petals long obovate, stamens slightly united at the base, occasionally with some petaloids. Leaves dark green, leathery, long elliptic, urgently cuneate at the base, margins nearly smooth. Plant upright, branches dense and growth vigorous. Starts to bloom from early-summer, fully blooms from autumn to winter and sporadically in spring.

HAR-17-2. 炎夏红伞 Yanxia Hongsan (Red Umbrella in Hot Summer)

回交组合： 杜鹃红山茶 × 杜鹃红山茶 F_1 代新品种'红波涌金'（ C. azalea × C. azalea F_1 new hybrid 'Hongbo Yongjin'）。

回交苗编号： DHBYJ-No.16。

性状： 花朵红色至深红色，泛蜡质光泽，单瓣型，中型至大型花，花径9.0～11.5cm，花瓣8～11枚，花瓣长倒卵形，厚质，先端凹，上部外翻，似雨伞状，花丝淡红色，花药黄色。叶片浓绿色，叶背面灰绿色，厚革质，椭圆形，长7.5～8.2cm，宽3.0～3.5cm，基部楔形，边缘近全缘。植株紧凑，矮性，生长旺盛。夏初始花，秋、冬季盛花，春季零星开花。

Abstract: Flowers red to deep red, slightly with waxy luster, single form, medium to large size, 9.0-11.5cm across with 8-11 petals, petals long obovate, thick texture, apices emarginate, rolled outward at the upper part which shaped as an umbrella, filaments pale red, anthers yellow. Leaves dark green, heavy leathery, elliptic, cuneate at the base, margins nearly smooth. Plant compact, dwarf and growth vigorous. Starts to bloom from early-summer, fully blooms from autumn to winter and sporadically in spring.

HAR-17-3. 黑火喷金
Heihuo Penjin（Black Fire Spraying Gold）

回交组合：杜鹃红山茶 × 杜鹃红山茶 F_1 代新品种'红波涌金'（ C. azalea × C. azalea F_1 new hybrid 'Hongbo Yongjin'）。

回交苗编号：DHBYJ-No.17。

性状：花朵黑红色，泛蜡质光泽，有时会出现小白斑，单瓣型，小型花，花径6.5～7.0cm，花瓣7枚，长倒卵形，厚肉质，先端凹，花丝红色，开花极稠密。叶片浓绿色，叶背面灰绿色，厚革质，小椭圆形，长6.5～7.2cm，宽3.0～3.3cm，基部楔形，边缘近全缘。植株紧凑，枝叶微稠密，生长旺盛。夏初始花，秋、冬季盛花，春季零星开花。

Abstract: Flowers dark red with waxy luster, sometimes, small white markings appear, single form, small size, 6.5-7.0cm across with 7 petals, petals long obovate, thick fleshy, filaments red, bloom very dense. Leaves dark green, heavy leathery, small elliptic, cuneate at the base, margins nearly smooth. Plant compact, branches faintly dense and growth vigorous. Starts to bloom from early-summer, fully blooms from autumn to winter and sporadically in spring.

⦿ 回交组合 HAR-18. 杜鹃红山茶 × 杜鹃红山茶 F_1 代新品种 '夏梦文清'
HAR-18. *C. azalea* × *C. azalea* F_1 new hybrid 'Xiameng Wenqing'

×

本回交组合将介绍 15 个新品种。

'夏梦文清'是从杜鹃红山茶 × 云南山茶品种'帕克斯先生'杂交组合中得到的（见第一部书第 260～261 页）。

可以看到，本组获得的这些回交新品种，花朵红色，大型至巨型，花瓣排列松散，遗传了父本'夏梦文清'的性状，而其开花期、叶片和植株性状则倾向于母本杜鹃红山茶。

Fifteen new hybrids will be introduced in this backcross combination.

'Xiameng Wenqing' was obtained from the cross-combination between *C. azalea* and *C. reticulata* cultivar 'Dr Clifford Parks' (See p.260-261, our first book).

It can be seen that the flowers of the backcross hybrids obtained are red color, large to very large size, petals are loosely arranged, which have inherited the characteristics from its male parent 'Xiameng Wenqing', while the blooming period, leaf and plant characteristics of the backcross hybrids tend to its female parent *C. azalea*.

HAR-18-1. 精彩无限
Jingcai Wuxian (Splendid Infinity)

回交组合： 杜鹃红山茶 × 杜鹃红山茶 F_1 代新品种'夏梦文清'（ C. azalea × C. azalea F_1 new hybrid 'Xiameng Wenqing'）。

回交苗编号： DWQ-No.3。

性状： 花朵外部花瓣粉白色，中部花瓣红色至黑红色，具白斑，半重瓣型至牡丹型，中型至大型花，花径9.0～11.0cm，花朵厚5.0cm，大花瓣15～17枚，呈3轮排列，略皱褶，中部花瓣直立，与雄蕊混生。叶片浓绿色，叶背面灰绿色，厚革质，椭圆形，长8.0～8.5cm，宽3.5～4.0cm，基部急楔形，边缘近光滑。植株立性，枝叶稠密，生长旺盛。夏初始花，秋、冬季盛花，春季零星开花。

Abstract: Exterior petals of the flowers pinkish white, central petals red to dark-red, with white patches, semi-double to peony form, medium to large size, 9.0-11.0cm across and 5.0cm deep with 15-17 large petals arranged in 3 rows, petal surfaces slightly wrinkled, central petals erected and mixed with stamens. Leaves dark green, heavy leathery, elliptic, urgently cuneate at the base, margins nearly smooth. Plant upright, branches dense and growth vigorous. Starts to bloom from early-summer, fully blooms from autumn to winter and sporadically in spring.

HAR-18-2. 层叠美景
Cengdie Meijing (Layering Beautiful Scenery)

回交组合： 杜鹃红山茶 × 杜鹃红山茶 F$_1$ 代新品种'夏梦文清'（ C. azalea × C. azalea F$_1$ new hybrid 'Xiameng Wenqing'）。

回交苗编号： DWQ-No.6。

性状： 花朵红色，外部花瓣渐向上部呈粉白色，玫瑰重瓣型至完全重瓣型，中型至大型花，花径 9.0～10.5cm，花朵厚 4.5～5.0cm，花瓣 25～30 枚，呈 4～5 轮有序排列，中部花瓣可见隐约的白条纹，完全开放后偶现少量雄蕊。叶片浓绿色，叶背面灰绿色，厚革质，长椭圆形，长 8.3～9.5cm，宽 3.5～4.0cm，基部急楔形，边缘近光滑。植株立性，枝叶稠密，生长旺盛。夏初始花，秋、冬季盛花，春季零星开花。

Abstract: Flowers red, fading to pink-white at the top part of the exterior petals, rose-double to formal double form, medium to large size, 9.0-10.5cm across and 4.5-5.0cm deep with 25-30 petals arranged orderly in 4-5 rows, some faint white stripes visible on the central petals, occasionally a few stamens appear in the center when fully opening. Leaves dark green, heavy leathery, long elliptic, urgently cuneate at the base, margins nearly smooth. Plant upright, branches dense and growth vigorous. Starts to bloom from early-summer, fully blooms from autumn to winter and sporadically in spring.

HAR-18-3. 貌美绝俗 Maomei Juesu (The Beautiful Looks that Never Seen)

回交组合：杜鹃红山茶 × 杜鹃红山茶 F_1 代新品种'夏梦文清'(*C. azalea* × *C. azalea* F_1 new hybrid 'Xiameng Wenqing')。

回交苗编号：DWQ-No.11。

性状：花朵桃红色至红色，随着花朵开放，花色逐渐变浅，玫瑰重瓣型至牡丹型，大型至巨型花，花径 11.5～14.0cm，花朵厚 5.7～6.0cm，大花瓣 20～25 枚，呈 4 轮有序排列，花瓣阔倒卵形，花芯处半直立的小花瓣和少量雄蕊混生，花瓣逐片掉落。叶片浓绿色，叶背面灰绿色，厚革质，中脉凸起，侧脉模糊，狭长椭圆形，长 10.0～10.5cm，宽 3.5～4.0cm，基部急楔形，边缘近光滑。植株立性，枝叶稠密，生长旺盛。夏初始花，秋、冬季盛花，春季零星开花。

Abstract: Flowers peach red to red, as flowers opening, the flower color gradually become lighter, rose-double to peony form, large to very large size, 11.5-14.0cm across and 5.7-6.0cm deep with 20-25 large petals arranged orderly in 4 rows, petals broad obovate, semi-elected small petals mixed with stamens at the center, petals fall one by one. Leaves dark green, heavy leathery, narrow-long elliptic, midrib raised, lateral veins blurry, urgently cuneate at the base, margins nearly smooth. Plant upright, branches dense and growth vigorous. Starts to bloom from early-summer, fully blooms from autumn to winter and sporadically in spring.

HAR-18-4. 姿容皆美
Zirong Jiemei (Pose & Face All Pretty)

回交组合：杜鹃红山茶 × 杜鹃红山茶 F_1 代新品种'夏梦文清'（ C. azalea × C. azalea F_1 new hybrid 'Xiameng Wenqing'）。

回交苗编号：DWQ-No.12。

性状：花朵深粉红色至桃红色，渐向基部呈红色，花瓣上部淡粉红色，玫瑰重瓣型，大型花，花径 12.0～12.5cm，花朵厚 5.8～6.3cm，大花瓣 25～30 枚，呈 3 轮排列，花瓣长倒卵形，中部花瓣半直立，与雄蕊混生。叶片浓绿色，叶背面灰绿色，厚革质，长椭圆形，长 10.0～11.0cm，宽 3.5～4.0cm，中脉略凹，侧脉不明显，基部楔形，边缘齿浅。植株立性，枝叶稠密，生长旺盛。夏初始花，秋、冬季盛花，春季零星开花。

Abstract: Flowers deep pink to peach red, becoming red at the base of petals, the upper part of petals blush pink, rose-double form, large size, 12.0-12.5cm across and 5.8-6.3cm deep with 25-30 large petals arranged in 3 rows, petals long obovate, central petals semi-erected and mixed with stamens. Leaves dark green, heavy leathery, long elliptic, midrib slightly sunken, lateral veins blurry, cuneate at the base, margins shallowly serrate. Plant upright, branches dense and growth vigorous. Starts to bloom from early-summer, fully blooms from autumn to winter and sporadically in spring.

HAR-18-5. 堆金积玉 Duijin Jiyu (Store up Gold & Accumulate Jade)

回交组合：杜鹃红山茶 × 杜鹃红山茶 F_1 代新品种'夏梦文清'（*C. azalea* × *C. azalea* F_1 new hybrid 'Xiameng Wenqing'）。

回交苗编号：DWQ-No.14。

性状：花朵深粉红色至红色，渐向花瓣上部呈粉白色，半重瓣型至玫瑰重瓣型，大型花，花径 11.5～12.5cm，花朵厚 5.7～6.0cm，大花瓣 20～25 枚，呈 3 轮排列，花瓣倒卵形，边缘外翻，中部雄蕊瓣与雄蕊混生。叶片浓绿色，叶背面灰绿色，厚革质，阔椭圆形，长 9.0～10.0cm，宽 4.0～4.5cm，中脉略凹，侧脉不明显，基部楔形，边缘光滑。植株立性，枝叶稠密，生长旺盛。夏初始花，秋、冬季盛花，春季零星开花。

Abstract: Flowers deep pink to red, fading to white pink at the upper part of the petals, semi-double to rose-double form, large size, 11.5-12.5cm across and 5.7-6.0cm deep with 20-25 large petals arranged in 3 rows, petals obovate, edges rolled outward, central petaloids mixed with stamens. Leaves dark green, heavy leathery, broad elliptic, midrib slightly sunken, lateral veins blurry, cuneate at the base, margins smooth. Plant upright, branches dense and growth vigorous. Starts to bloom from early-summer, fully blooms from autumn to winter and sporadically in spring.

HAR-18-6. 幸福之家
Xingfu Zhijia（Happy Family）

回交组合：杜鹃红山茶 × 杜鹃红山茶 F_1 代新品种'夏梦文清'（*C. azalea* × *C. azalea* F_1 new hybrid 'Xiameng Wenqing'）。

回交苗编号：DWQ-No.15。

性状：花朵红色至深红色，渐向花瓣中部呈淡红色或带白条纹，半重瓣型至玫瑰重瓣型，大型至巨型花，花径12.5～13.5cm，花朵厚6.0～6.5cm，外部大花瓣18～20枚，呈3～4轮有序排列，花瓣长倒卵形，中部雄蕊瓣10～15枚，直立，与雄蕊混生。叶片浓绿色，叶背面灰绿色，厚革质，阔椭圆形，长10.0～12.0cm，宽4.0～4.5cm，基部楔形，边缘近光滑。植株立性，生长旺盛。夏初始花，秋、冬季盛花，春季零星开花。

Abstract: Flowers red to deep red, fading to light red or with white stripes at the central part of petals, semi-double to rose-double form, large to very large size, 12.5-13.5cm across and 6.0-6.5cm deep with 18-20 exterior large petals arranged orderly in 3-4 rows, petals long-obovate, 10-15 central petals erected and mixed with stamens. Leaves dark green, heavy leathery, broad elliptic, cuneate at the base, margins nearly smooth. Plant upright and growth vigorous. Starts to bloom from early-summer, fully blooms from autumn to winter and sporadically in spring.

HAR-18-7. 夏红秋丽
Xiahong Qiuli (Red Summer & Beautiful Autumn)

回交组合：杜鹃红山茶 × 杜鹃红山茶 F_1 代新品种'夏梦文清'（ *C. azalea* × *C. azalea* F_1 new hybrid 'Xiameng Wenqing'）。

回交苗编号：DWQ-No.16。

性状：花朵红色，略泛橘红色调，偶有白斑，半重瓣型至牡丹型，大型花，花径 12.0～13.0cm，花朵厚 5.0～5.5cm，大花瓣 26～32 枚，呈 3～4 轮松散排列，花瓣长倒卵形，边缘小波浪状，先端外卷，中部沟槽状，雄蕊簇状，偶有瓣化的小花瓣。叶片浓绿色，叶背面灰绿色，厚革质，长椭圆形，长 8.5～9.0cm，宽 3.8～4.2cm，叶脉不清晰，基部楔形，上端边缘齿浅。植株开张，生长旺盛。夏初始花，秋、冬季盛花，春季零星开花。

> **Abstract:** Flowers red slightly with orange hue, occasionally with white markings, semi-double to peony form, large size, 12.0-13.0cm across and 5.0-5.5cm deep with 26-32 large petals arranged loosely in 3-4 rows, petals long-obovate with wavy edges and apices rolled outward, stamens clustered occasionally with small petaloids. Leaves dark green, heavy leathery, long elliptic, veins blurry, cuneate at the base, margins shallowly serrate at the upper part. Plant spread and growth vigorous. Starts to bloom from early-summer, fully blooms from autumn to winter and sporadically in spring.

HAR-18-8. 夏谷灵感
Xiagu Linggan (Summer Valley's Inspiration)

回交组合： 杜鹃红山茶 × 杜鹃红山茶 F₁ 代新品种'夏梦文清'（ *C. azalea* × *C. azalea* F₁ new hybrid 'Xiameng Wenqing' ）。

回交苗编号： DWQ-No.17。

性状： 花朵橘红色，花瓣中部红色较淡，半重瓣型至牡丹型，中型花，花径 8.5～10.0cm，花朵厚 4.3～4.8cm，大花瓣 18～22 枚，呈 2～3 轮排列，花瓣阔倒卵形，边缘内卷，波浪状，而后边缘外翻，雄蕊散射，花芯处偶有少量雄蕊瓣。叶片浓绿色，叶背面灰绿色，厚革质，长椭圆形，长 7.5～8.0cm，宽 3.5～4.0cm，基部楔形，边缘近光滑。植株紧凑，枝叶稠密，生长旺盛。夏初始花，秋、冬季盛花，春季零星开花。

Abstract: Flowers orange red, fading to light color at the mid-part of the petals, semi-double to peony form, medium size, 8.5-10.0cm across and 4.3-4.8cm deep with 18-22 large petals arranged in 2-3 rows, petals broad obovate, petal edges rolled inward, wavy and then rolled outward, stamens diffused occasionally with a few petaloids. Leaves dark green, heavy leathery, long elliptic, cuneate at the base, margins nearly smooth. Plant compact, branches dense and growth vigorous. Starts to bloom from early-summer, fully blooms from autumn to winter and sporadically in spring.

HAR-18-9. 宝石流霞
Baoshi Liuxia (Gem Bathed in Flowing Rosy Clouds)

回交组合： 杜鹃红山茶 × 杜鹃红山茶 F_1 代新品种'夏梦文清'（ *C. azalea* × *C. azalea* F_1 new hybrid 'Xiameng Wenqing'）。

回交苗编号： DWQ-No.18。

性状： 花朵桃红色至红色，外部花瓣渐变粉白色，具模糊的白条纹，牡丹型至玫瑰重瓣型，中型至大型花，花径 9.5～11.5cm，花朵厚 6.0～6.5cm，外部大花瓣 20～30 枚，呈 3 轮排列，花瓣阔倒卵形，中部花瓣 30 枚以上，直立，与雄蕊混生。叶片浓绿色，叶背面灰绿色，厚革质，阔倒卵形，长 11.0～11.5cm，宽 4.0～4.5cm，基部急楔形，边缘近光滑。植株紧凑，立性，枝叶稠密，生长旺盛。夏初始花，秋、冬季盛花，春季零星开花。

Abstract: Flowers peach red to red, fading to pink-white at the exterior petals with hazy white stripes, peony form to rose-double form, medium to large size, 9.5-11.5cm across and 6.0-6.5cm deep with 20-30 exterior large petals arranged in 3 rows, petals broad obovate, over 30 central small petals erected and mixed with stamens. Leaves dark green, heavy leathery, broad obovate, urgently cuneate at the base, margins nearly smooth. Plant compact, upright, branches dense and growth vigorous. Starts to bloom from early-summer, fully blooms from autumn to winter and sporadically in spring.

HAR-18-10. 多季红冠
Duoji Hongguan (Multiseasonal Red Crown)

回交组合： 杜鹃红山茶 × 杜鹃红山茶 F₁ 代新品种'夏梦文清'（*C. azalea* × *C. azalea* F₁ new hybrid 'Xiameng Wenqing'）。

回交苗编号： DWQ-No.24。

性状： 花朵艳红色，花瓣中部隐约有纵向的白条纹，玫瑰重瓣型，大型至巨型花，花径 11.0～13.5cm，花朵厚 5.0～5.5cm，大花瓣 28～31 枚，呈 6 轮以上有序排列，花瓣长倒卵形，中部沟槽状，中部小花瓣 12 枚以上，完全开放后与少量雄蕊混生。叶片浓绿色，叶背面灰绿色，厚革质，中脉凸起，侧脉模糊，长椭圆形，长 11.0～12.0cm，宽 4.0～4.2cm，基部楔形，边缘齿浅。植株立性，枝叶稠密，生长旺盛。夏初始花，秋、冬季盛花，春季零星开花，花期可延长至 4 月底。

Abstract: Flowers bright red, some vertical and faint white stripes on the mid-part of the petal surfaces, rose-double form, large to very large size, 11.0-13.5cm across and 5.0-5.5cm deep with 28-31 large petals arranged in more than 6 rows, petals long obovate. Leaves dark green, heavy leathery, long elliptic, midrib raised, lateral veins blurry, cuneate at the base, margins shallowly serrate. Plant upright, branches dense and growth vigorous. Starts to bloom from early-summer, fully blooms from autumn to winter and sporadically in spring.

HAR-18-11. 赤诚之心
Chicheng Zhixin (Sincere Heart)

回交组合： 杜鹃红山茶 × 杜鹃红山茶 F_1 代新品种'夏梦文清'（ *C. azalea* × *C. azalea* F_1 new hybrid 'Xiameng Wenqing'）。

回交苗编号： DWQ-No.25。

性状： 花朵黑红色，略泛紫色调，具绒质感，单瓣型，中型花，花径 8.0～9.0cm，花瓣 7～8 枚，长倒卵形，长 6.0cm，宽 3.0cm，厚质，上部外卷，先端近圆，雄蕊基部略连生，花丝红色，花药金黄色，开花稠密。叶片浓绿色，叶背面灰绿色，厚革质，长倒卵形，长 6.5～7.5cm，宽 3.5～4.5cm，侧脉不清晰，基部楔形，边缘齿浅。植株立性，枝叶稠密，生长旺盛。夏初始花，秋、冬季盛花，春季零星开花。

> **Abstract:** Flowers dark red with slightly purple hue and velvet texture, single form, medium size, 8.0-9.0cm across with 7-8 petals, petals long obovate, thick, rolled outward at the upper part, apices nearly round, stamens slightly united at the base, filaments red, anthers golden yellow, bloom dense. Leaves dark green, heavy leathery, long obovate, lateral veins blurry, cuneate at the base, margins shallowly serrate. Plant upright, branches dense and growth vigorous. Starts to bloom from early-summer, fully blooms in autumn and winter and sporadically in spring.

HAR-18-12. 橘红淌金
Juhong Tangjin (Gold Trickling Down from Orange Red)

回交组合：杜鹃红山茶 × 杜鹃红山茶 F_1 代新品种'夏梦文清'（*C. azalea* × *C. azalea* F_1 new hybrid 'Xiameng Wenqing'）。

回交苗编号：DWQ-No.33。

性状：花蕾球形，萼片淡绿色。花朵深橘红色，外部花瓣呈淡粉红色，半重瓣型至牡丹型，中型至大型花，花径8.5～11.0cm，花朵厚4.5～5.0cm，花瓣25～30枚，呈5轮排列，花瓣沟槽状，中部雄蕊与花瓣混生，花丝红色，花药金黄色。叶片浓绿色，叶背面灰绿色，革质，阔椭圆形，长8.5～9.5cm，宽3.5～4.5cm，基部楔形，边缘近光滑。植株紧凑，枝叶稠密，生长旺盛。夏初始花，秋、冬季盛花，春季零星开花。

Abstract: Flowers deep orange red, exterior petals light pink, semi-double to peony form, medium to large size, 8.5-11.0cm across and 4.5-5.0cm deep with 25-30 petals arranged in 5 rows, petals groove shaped, central stamens mixed with petals, filaments red, anthers golden yellow. Leaves dark green, back surfaces pale green, leathery, broad elliptic, cuneate at the base, margins nearly smooth. Plant compact, branches dense and growth vigorous. Starts to bloom from early-summer, fully blooms in autumn and winter and sporadically in spring.

HAR-18-13. 意外收获
Yiwai Shouhuo (Windfall)

回交组合：杜鹃红山茶 × 杜鹃红山茶 F_1 代新品种'夏梦文清'（ *C. azalea* × *C. azalea* F_1 new hybrid 'Xiameng Wenqing'）。

回交苗编号：DWQ-No.34。

性状：花朵红色至艳红色，花瓣中上部淡红色，半重瓣型，中型至大型花，花径 8.5～12.0cm，花朵厚 5.0～5.5cm，大花瓣 17～21 枚，呈 3 轮排列，花瓣阔倒卵形，边缘内卷，雄蕊散射，偶有雄蕊瓣。叶片浓绿色，叶背面灰绿色，厚革质，长椭圆形，长 8.5～9.0cm，宽 3.0～3.5cm，基部急楔形，边缘光滑。植株立性，枝叶稠密，生长旺盛。夏初始花，秋、冬季盛花，春季零星开花。

Abstract: Flowers red to bright red, fading to light red at the upper part of the petals, semi-double form, medium to large size, 8.5-12.0cm across and 5.0-5.5cm deep with 17-21 large petals arranged in 3 rows, petals broad obovate, edges rolled inward, stamens diffused and occasionally with some petaloids. Leaves dark green, heavy leathery, long elliptic, urgently cuneate at the base, margins smooth. Plant upright, branches dense and growth vigorous. Starts to bloom from early-summer, fully blooms from autumn to winter and sporadically in spring.

HAR-18-14. 奇妙形色
Qimiao Xingse (Fantastic Shape & Color)

回交组合： 杜鹃红山茶 × 杜鹃红山茶 F_1 代新品种'夏梦文清'（ C. azalea × C. azalea F_1 new hybrid 'Xiameng Wenqing'）。

回交苗编号： DWQ-No.38。

性状： 花朵红色，花瓣背面略泛蜡光，单瓣型，中型至大型花，花径 9.5～12.0cm，波浪状的花瓣 11～12 枚，呈碗状排列，花瓣长倒卵形，长 7.0cm，宽 3.5cm，雄蕊辐射状，花丝淡红色，花药金黄色，开花稠密。叶片浓绿色，叶背面灰绿色，厚革质，长倒卵形，长 7.0～8.5cm，宽 4.0～4.5cm，基部楔形，边缘光滑。植株立性，紧凑，生长旺盛。夏初始花，秋、冬季盛花，春季零星开花。

Abstract: Flowers red with waxy luster at back surfaces of the petals, single form, medium to large size, 9.5-12.0cm across with 11-12 wavy petals arranged as a bowl, petals long obovate, stamens radial, filaments pale red, anthers golden yellow, bloom dense. Leaves dark green, heavy leathery, long obovate, cuneate at the base, margins smooth. Plant upright, compact and growth vigorous. Starts to bloom from early-summer, fully blooms in autumn and winter and sporadically in spring.

HAR-18-15. 前程似锦
Qiancheng Sijin (Bright Future)

回交组合： 杜鹃红山茶 × 杜鹃红山茶 F₁ 代新品种'夏梦文清'（ *C. azalea* × *C. azalea* F₁ new hybrid 'Xiameng Wenqing')。

回交苗编号： DWQ-No.44。

性状： 花朵红色，泛橘红色调，花瓣先端红色较淡，偶有白斑，牡丹型，大型至巨型花，花径 11.0～13.5cm，花朵厚5.5～6.5cm，大花瓣35～45枚，呈5～6轮紧实排列，花瓣长倒卵形，边缘波浪状，中部花瓣簇拥，偶有瓣化的雄蕊瓣。叶片浓绿色，叶背面灰绿色，厚革质，长椭圆形，长8.0～9.0cm，宽3.0～4.0cm，叶脉不清晰，基部楔形，边缘近光滑。植株立性，生长旺盛。夏初始花，秋、冬季盛花，春季零星开花。

Abstract: Flowers red with orange red hue, fading to light red at the top part of the petals, occasionally with white markings, peony form, large to very large size, 11.0-13.5cm across and 5.5-6.5cm deep with 35-45 large petals arranged tightly in 5-6 rows, petals long obovate with wavy edges, central petals clustered, occasionally petaloids appeared. Leaves dark green, heavy leathery, long elliptic, veins blurry, cuneate at the base, margins nearly smooth. Plant upright and growth vigorous. Starts to bloom from early-summer, fully blooms from autumn to winter and sporadically in spring.

⊙ 回交组合 HAR-19. 杜鹃红山茶 × 杜鹃红山茶 F₁ 代新品种 '不知寒暑'

HAR-19. *C. azalea* × *C. azalea* F₁ new hybrid 'Buzhi Hanshu'

本回交组合将介绍1个新品种。

'不知寒暑'是从红山茶品种'白斑康乃馨'×杜鹃红山茶这一杂交组合中得到的（见第一部书第 241～242 页）。

该回交组合获得的这个新品种，其叶形、开花期等性状颇似杜鹃红山茶，而在花性状上则遗传了其父本'不知寒暑'品种的多瓣、具白条纹的性状。这正是我们所期望的。

One new hybrid will be introduced in this backcross combination.

'Buzhi Hanshu' was obtained from the cross-combination between *C. japonica* cultivar 'Ville de Nantes' and *C. azalea* (See p.241-242, our first book).

The new backcross hybrid obtained is very similar to *C. azalea* in the characteristics of leaf shape and the blooming period, and its flower characteristics like multi-petals and white stripes have inherited from its male parent 'Buzhi Hanshu'. These are exactly what we expected.

HAR-19-1. 狂欢舞曲
Kuanghuan Wuqu (Orgiastic Dance Tune)

回交组合： 杜鹃红山茶 × 杜鹃红山茶 F_1 代新品种'不知寒暑'（ *C. azalea* × *C. azalea* F_1 new hybrid 'Buzhi Hanshu'）。

回交苗编号： DBZHS-No.2。

性状： 花朵深橘红色，瓣面具少量白条纹，半重瓣型，中型花，花径 8.5～10.0cm，花朵厚 6.5cm，花瓣 12～15 枚，呈 2～3 轮松散排列，窄长倒卵形，半直立状态，花芯处小花瓣少量，雄蕊基部略连生。叶片浓绿色，叶背面灰绿色，革质，长椭圆形，长 7.5～8.0cm，宽 2.5～3.0cm，主脉凸起，基部楔形，叶缘近光滑。植株紧凑，枝叶稠密，生长旺盛。春末始花，夏、秋、冬季盛花。

Abstract: Flowers deep orange red with some white stripes, semi-double form, medium size, 8.5-10.0cm across and 6.5cm deep with 12-15 petals arranged loosely in 2-3 rows, petals narrow-long obovate, semi-erected, a few central small petals, stamens slightly united at the base. Leaves dark green, leathery, long elliptic, midrib raised, cuneate at the base, margins nearly smooth. Plant compact, branches dense and growth vigorous. Starts to bloom from late-spring, fully blooms in summer, autumn and winter.

四季茶花杂交新品种彩色图集（第二部）

⦿ 回交组合 HAR-20. 杜鹃红山茶 × 杜鹃红山茶 F₁ 代新品种 '夏梦春陵'
HAR-20. *C. azalea* × *C. azalea* F₁ new hybrid 'Xiameng Chunling'

×

本回交组合将介绍2个新品种。

'夏梦春陵'是从红山茶品种'都鸟'×杜鹃红山茶这一杂交组合中得到的（见第一部书第 221～222 页）。

可以看到，该回交组合获得的这两个新品种，花色更加鲜艳，花径大型至巨型，颇似'夏梦春陵'；而叶色更浓绿，始花期提前到春末，这些性状更类似于杜鹃红山茶。

Two new hybrids will be introduced in this backcross combination.

'Xiameng Chunling' was obtained from the cross-combination between *C. japonica* cultivar 'Miyakodori' and *C. azalea* (See p.221-222, our first book).

It can be seen that the flowers of the two new backcross hybrids obtained are brighter and large to very large size which are similar to 'Xiameng Chunling', and their leaves are dark green and their blooming periods are advanced to late-spring, which are more like *C. azalea*.

HAR-20-1. 娇艳欲滴
Jiaoyan Yudi (Delicate & Charming)

回交组合： 杜鹃红山茶 × 杜鹃红山茶 F_1 代新品种'夏梦春陵'（ *C. azalea* × *C. azalea* F_1 new hybrid 'Xiameng Chunling'）。

回交苗编号： DXMCL-No.2。

性状： 花朵桃红色，瓣面渐向上部边缘泛模糊的白色，牡丹型，大型花，花径 10.5～13.0cm，花朵厚 5.8cm，外部大花瓣 24～28 枚，呈 4 轮排列，外翻，阔倒卵形，长 6.0cm，宽 5.5cm，先端微凹，中部花瓣 10～12 枚，半直立状，与少量雄蕊混生，花芯处小花瓣少量。叶片浓绿色，叶背面灰绿色，厚革质，椭圆形，长 9.0～11.0cm，宽 4.0～4.5cm，主脉凸起，基部楔形，叶缘近光滑。植株紧凑，枝叶稠密，生长旺盛。春末始花，夏、秋、冬季盛花。

Abstract: Flowers peach red, fading to blurry white at the top part of the petals, peony form, large size, 10.5-13.0cm across and 5.8cm deep with 24-28 exterior large petals arranged in 4 rows, large petals rolled outward, broad obovate, apices emarginated, central 10-12 petals erected and mixed with a few stamens. Leaves dark green, heavy leathery, elliptic, midrib raised, cuneate at the base, margins nearly smooth. Plant compact, branches dense and growth vigorous. Starts to bloom in late-spring, fully blooms in summer, autumn and winter.

HAR-20-2. 艺术世家
Yishu Shijia（Artistic Family）

回交组合： 杜鹃红山茶 × 杜鹃红山茶 F_1 代新品种'夏梦春陵'（C. azalea × C. azalea F_1 new hybrid 'Xiameng Chunling'）。

回交苗编号： DXMCL-No.12。

性状： 花朵深桃红色，略泛紫色调，瓣面偶见少量白色条纹，半重瓣型，中型至大型花，花径8.5～10.5cm，花朵厚5.0cm，大花瓣15～18枚，呈2～3轮排列，长倒卵形，长6.5cm，宽4.0cm，先端微凹，中部雄蕊瓣少量，与雄蕊混生，花丝淡红色，花药黄色，开花稠密。叶片浓绿色，叶背面灰绿色，厚革质，狭长椭圆形，长8.0～8.5cm，宽3.0～3.5cm，主脉凸起，侧脉不明显，基部急楔形，叶缘光滑。植株紧凑，枝叶稠密，生长旺盛。春末始花，夏、秋、冬季盛花。

Abstract: Flowers deep peach red, slightly with purple hue, occasionally with a few white stripes, semi-double form, medium to large size, 8.5-10.5cm across and 5.0cm deep with 15-18 large petals arranged in 2-3 rows, large petals long obovate, apices emarginate, a few petaloids mixed with stamens at the center, filaments pale red, anthers yellow, bloom dense. Leaves dark green, heavy leathery, narrow-long elliptic, midrib raised, lateral veins unconspicuous, urgently cuneate at the base, margins smooth. Plant compact, branches dense and growth vigorous. Starts to bloom in late-spring, fully blooms in summer, autumn and winter.

◉ 回交组合 HAR-21. 杜鹃红山茶 × 杜鹃红山茶 F₁ 代新品种 '夏蝶群舞'

HAR-21. *C. azalea* × *C. azalea* F₁ new hybrid 'Xiadie Qunwu'

本回交组合将介绍1个新品种。

'夏蝶群舞'是从红山茶品种'媚丽'×杜鹃红山茶这一杂交组合中得到的（见第一部书第122～123页）。

该回交组合获得的这个新品种，花色艳红，虽然是单瓣型，但是开花非常稠密，开花期长，是园林美化的宝贵新品种。

One new hybrid will be introduced in this backcross combination.

'Xiadie Qunwu' was obtained from the cross-combination between *C. japonica* cultivar 'Tama Beauty' and *C. azalea* (See p.122-123, our first book).

The flower color of the new hybrid obtained from the backcross combination is brilliant red, though its flowers are single form, it blooms very densely and has a long blooming period, which is a valuable new hybrid for gardening.

HAR-21-1. 吉星高照
Jixing Gaozhao（Lucky Stars Shining Brightly）

回交组合： 杜鹃红山茶 × 杜鹃红山茶 F$_1$ 代新品种'夏蝶群舞'（C. azalea × C. azalea F$_1$ hybrid 'Xiadie Qunwu'）。

回交苗编号： DXDQW-No.7+38。

性状： 花朵红色至艳红色，单瓣型，中型花，花径 8.5～10.0cm，花瓣 7 枚，阔倒卵形，略内扣，先端略凹缺，雄蕊 100 枚以上，呈辐射状，基部略连生，花丝淡红色，开花稠密。叶片浓绿色，叶背面灰绿色，椭圆形，厚革质，长 9.3～9.8cm，宽 3.0～3.5cm，叶面略扭曲，叶缘近光滑。植株开张，生长旺盛。夏初始花，秋、冬季盛花，春季零星开花。

> **Abstract:** Flowers red to bright red, single form, medium size, 8.5-10.0cm across with 7 petals, petals broad obovate, apices emarginate, stamens more than 100, radical, bloom dense. Leaves dark green, heavy leathery, margins nearly smooth. Plants spread and growth vigorous. Starts to bloom in early-summer, fully blooms from autumn to winter and sporadically in spring.

⦿ 回交组合 HAR-22. 杜鹃红山茶 × 杜鹃红山茶 F₁ 代新品种 '夏日红绢'

HAR-22. *C. azalea* × *C. azalea* F₁ new hybrid 'Xiari Hongjuan'

本回交组合将介绍 5 个新品种。

'夏日红绢'是从杜鹃红山茶 × 红山茶品种'超级南天武士'这一杂交组合中得到的一个花朵红色、单瓣型，能够多季开花的杂交种（见第一部书第 155 ～ 156 页）。

本回交组合所获得的 5 个杂交新品种都具有两个亲本的特性，而且出现了 2 个多瓣的回交新品种，这说明父本'夏日红绢'已经具有了'超级南天武士'品种的多瓣性状基因，并遗传给了本组的回交新品种。

×

Five new hybrids will be introduced in this backcross combination.

'Xiari Hongjuan' is a hybrid with the characteristics of red flowers, single form and ever-blooming which was obtained from the cross-combination between *C. azalea* and *C. japonica* cultivar 'Dixie knight supreme'(See p.155-156, our first book).

The five backcross hybrids obtained from this backcross combination all have the characteristics of their two parents and two of them are multi-petals which shows that the male-parent 'Xiari Hongjuan' has had the multi-petals genes from 'Dixie Knight Supreme' and has inherited to the new backcross hybrids.

HAR-22-1. 坦荡胸怀
Tandang Xionghuai (Magnanimous Mind)

回交组合： 杜鹃红山茶 × 杜鹃红山茶 F$_1$ 代新品种'夏日红绢'(C. azalea × C. azalea F$_1$ new hybrid 'Xiari Hongjuan')。

杂交苗编号： DHJ-No.1。

性状： 花朵粉红色至深粉红色，单瓣型，巨型花，花径 13.5～14.0cm，花瓣 6 枚，近圆形，长 7.2cm，宽 7.0cm，上部略外弯，瓣面可见明显的纵向脉纹，雄蕊基部连生，呈短柱状，花丝紧实，花药黄色，开花稠密，花朵整朵掉落。叶片浓绿色，叶背面灰绿色，厚革质，阔椭圆形，长 8.5～9.0cm，宽 4.5～5.0cm，边缘上部具浅齿，下部边缘光滑。植株立性，生长旺盛。夏末始花，秋、冬季盛花，春季零星开花。

Abstract: Flowers pink to deep pink, single form, very large size, 13.5-14.0cm across with 6 petals, petals slightly round with 7.2cm long and 7.0cm wide, the upper part of the petals slightly curved outward, longitudinal veins visible on the petal surfaces, stamens united into a short column, filaments tightly, anthers yellow, bloom dense, flowers fall whole. Leaves dark green, heavy leathery, broad elliptic, margins shallowly serrate at the top part. Plant upright and growth vigorous. Starts to bloom from late-summer, fully blooms from autumn to winter and sporadically in spring.

HAR-22-2. 红盘托金
Hongpan Tuojin (Red Plate Holding up Gold)

回交组合： 杜鹃红山茶 × 杜鹃红山茶 F₁ 代新品种'夏日红绢'（*C. azalea* × *C. azalea* F₁ new hybrid 'Xiari Hongjuan'）。

回交苗编号： DHJ-No.2。

性状： 花朵艳红色，具蜡质感，渐向中部花瓣红色变淡，单瓣型，大型至巨型花，花径 12.0～15.0cm，花瓣 6～7 枚，阔倒卵形，平铺，瓣面大波浪状，可见清晰红脉纹，先端凹，雄蕊基部略连生，花丝红色，花朵整朵掉落，开花稠密。叶片浓绿色，叶背面灰绿色，厚革质，长椭圆形，长 9.5～10.5cm，宽 3.5～4.5cm，略扭曲，基部楔形，叶缘近全缘。植株开张，生长旺盛。夏初始花，秋、冬季盛花，春季零星开花。

Abstract: Flowers bright red with waxy luster, fading to light red at the central petals, single form, large to very large size, 12.0-15.0cm across with 6-7 petals, petals broad obovate, flat, big wavy, apices sunken, stamens slightly united at the base, filaments red, bloom dense, flowers fall whole. Leaves dark green, heavy leathery, long elliptic, slightly twisted, cuneate at the base, margins nearly smooth. Plant spread and growth vigorous. Starts to bloom from early-summer, fully blooms from autumn to winter and sporadically in spring.

HAR-22-3. 四季合韵
Siji Heyun（The Rhyme Conforming Four Seasons）

回交组合： 杜鹃红山茶 × 杜鹃红山茶 F_1 代新品种'夏日红绢'（*C. azalea* × *C. azalea* F_1 new hybrid 'Xiari Hongjuan'）。

回交苗编号： DHJ-No.12。

性状： 花朵深粉红色至淡红色，单瓣型，小型至中型花，花径 7.0～8.5cm，花瓣 5～6 枚，阔倒卵形，外翻或波浪状，雄蕊基部连生，开花稠密。叶片浓绿色，叶背面灰绿色，革质，椭圆形，长 7.5～8.5cm，宽 3.0～3.5cm，基部楔形，叶缘光滑。植株立性，生长旺盛。夏初始花，秋、冬季盛花，春季零星开花。

> **Abstract:** Flowers deep pink to light red, single form, small to medium size, 7.0-8.5cm across with 5-6 petals, petals broad obovate, turned outward or wavy, stamens united at the base, bloom dense. Leaves dark green, leathery, elliptic, cuneate at the base, margins smooth. Plant upright and growth vigorous. Starts to bloom from early-summer, fully blooms from autumn to winter and sporadically in spring.

HAR-22-4. 披红俏女
Pihong Qiaonü (The Pretty Draped Red)

回交组合： 杜鹃红山茶 × 杜鹃红山茶 F_1 代新品种'夏日红绢'（ C. azalea × C. azalea F_1 new hybrid 'Xiari Hongjuan'）。

回交苗编号： DHJ-No.15。

性状： 花朵秋冬季节艳红色至黑红色，夏季红色变淡，单瓣型至半重瓣型，中型花，花径9.5～10.0cm，花瓣9～12枚，呈1～2轮松散排列，阔倒卵形，外翻，雄蕊基部连生，花朵整朵掉落，开花稠密。叶片浓绿色，叶背面灰绿色，革质，窄椭圆形，长9.0～9.5cm，宽3.0～3.5cm，基部急楔形，叶缘近全缘。植株开张，生长旺盛。夏初始花，秋、冬季盛花，春季零星开花。

Abstract: Flowers bright red to dark red in autumn and winter, the red color faded in summer, single to semi-double form, medium size, 9.5-10.0cm across with 9-12 petals arranged loosely in 1-2 rows, petals broad obovate, turned outward, stamens united at the base, bloom dense, flowers fall whole. Leaves dark green, leathery, narrow elliptic, urgently cuneate at the base, margins nearly smooth. Plant spread and growth vigorous. Starts to bloom from early-summer, fully blooms from autumn to winter and sporadically in spring.

HAR-22-5. 夏奥赛场
Xiaao Saichang (The Summer Olympic Games' Stadium)

回交组合：杜鹃红山茶 × 杜鹃红山茶 F_1 代新品种'夏日红绢'（*C. azalea* × *C. azalea* F_1 new hybrid 'Xiari Hongjuan'）。

回交苗编号：DHJ-No.21。

性状：花朵红色，略显蜡质感，炎热夏季红色较淡，半重瓣型至牡丹型，大型花，花径 11.0～13.0cm，花朵厚 5.0cm，大花瓣 12～15 枚，呈 2 轮松散排列，花瓣长倒卵形，中部小花瓣半直立，雄蕊簇生，与中部小花瓣混生，开花稠密。嫩叶淡紫红色，成熟叶浓绿色，叶背面灰绿色，革质，长椭圆形，长 8.5～10.0cm，宽 3.5～4.0cm，略扭曲，基部楔形，叶缘近全缘。植株紧凑，枝条稠密，生长旺盛。夏初始花，秋、冬季盛花，春季零星开花。

Abstract: Flowers red, slightly with waxy luster, but the red is lighter in hot summer, semi-double to peony form, large size, 11.0-13.0cm across and 5.0cm deep with 12-15 petals arranged loosely in 2 rows, petals long obovate, central small petals semi-erected, stamens clustered and mixed with the small petals. Tender leaves light purple red, mature leaves dark green, leathery, long elliptic, slightly twisted, cuneate at the base, margins nearly smooth. Plant compact, branches dense and growth vigorous. Starts to bloom from early-summer, fully blooms from autumn to winter and sporadically in spring.

◉ 回交组合 HAR-23. 杜鹃红山茶 × 杜鹃红山茶 F₁ 代新品种'夏日叠星'

HAR-23. *C. azalea* × *C. azalea* F₁ new hybrid 'Xiari Diexing'

本回交组合将介绍 3 个新品种。

'夏日叠星'是从红山茶品种'媚丽'×杜鹃红山茶这一杂交组合中得到的杂交新品种（见第一部书第 130 ～ 131 页），其花朵深粉红色，半重瓣型至牡丹型，且能够多季开花。

本组获得的 3 个新品种都能多季开花，而且花朵稠密，叶片浓绿，叶缘光滑，证明它们已经获得了杜鹃红山茶的基因。另外，新品种的花朵都是多瓣型的，证明新品种已经从其父本'夏日叠星'获得了多瓣花朵的性状。由此可以推定，我们所获得的这 3 个杂交新品种都是真实的杂交种。

Three new hybrids will be introduced in this backcross combination.

'Xiari Diexing' was obtained from the cross-combination between *C. japonica* cultivar 'Tama Beauty' and *C. azalea* (See p.130-131, our first book) and its flowers are deep pink, semi-double to peony form and can bloom year-round.

The three new hybrids obtained in the backcross combination all can bloom year-round and densely, their leaves are dark green and leave margins are smooth, which shows that the new hybrids have inherited the *C. azalea*'s genes. In addition, the flowers of the new hybrids are all multi-petals, which shows that the new hybrids have obtained multi-petal genes from their male parent 'Xiari Diexing'. It can be presumed from above that the three new hybrids we obtained are true.

HAR-23-1. 福寿宝塔
Fushou Baota (Happiness & Longevity Pagoda)

回交组合：杜鹃红山茶 × 杜鹃红山茶 F_1 代新品种'夏日叠星'（*C. azalea* × *C. azalea* F_1 new hybrid 'Xiari Diexing'）。

回交苗编号：DDX-No.1。

性状：夏季花朵红色，秋冬季节花朵黑红色，偶具白斑，紧实的牡丹型至玫瑰重瓣型，大型至巨型花，花径12.0～13.5cm，花朵厚6.5～7.0cm，大花瓣50～60枚，呈8轮以上排列，多呈半开放状态，呈塔状，花瓣长倒卵形，长6.0cm，宽3.0cm，略波浪状，雄蕊基部略连生，花丝淡红色，花药黄色。叶片浓绿色，叶背面灰绿色，革质，椭圆形，长10.0～11.0cm，宽4.0～4.5cm，基部急楔形，叶缘近光滑。植株立性，枝叶稠密，生长旺盛。夏初始花，秋、冬季盛花，春季零星开花。

Abstract: Flowers red in summer and dark red in autumn and winter, occasionally with some white markings, compacted peony form to rose-double form, large to verry large size, 12.0-13.5cm across and 6.5-7.0cm deep with 50-60 large petals arranged in over 8 rows, the flowers usually semi-opened which shaped as a pagoda, petals long obovate, slightly wavy, stamens slightly united at the base, filaments pale red, anthers yellow. Leaves dark green, leathery, elliptic, urgently cuneate at the base, margins nearly smooth. Plant upright, branches dense and growth vigorous. Starts to bloom from early-summer, fully blooms in autumn and winter and sporadically in spring.

HAR-23-2. 高氏佳作
Gaoshi Jiazuo（Mr Gao's Excellent Masterpiece）

回交组合：杜鹃红山茶 × 杜鹃红山茶 F_1 代新品种'夏日叠星'（*C. azalea* × *C. azalea* F_1 new hybrid 'Xiari Diexing'）。

回交苗编号：DDX-No.2。

性状：花朵外部花瓣近白色，略泛淡粉色调，内部花瓣淡橘红色，并显白色条纹，玫瑰重瓣型，中型至大型花，花径 8.0～11.5cm，花朵厚 5.0cm，外部大花瓣 12～15 枚，呈 2 轮排列，中部花瓣 15～18 枚，呈 2～3 轮有序排列，偶有少量雄蕊出现。叶片浓绿色，叶背面灰绿色，革质，阔椭圆形，长 7.5～8.5cm，宽 4.0～4.5cm，基部楔形，叶缘光滑。植株紧凑，生长旺盛。夏初始花，秋、冬季盛花，春季零星开花。

Abstract: Flowers exterior petals white with pale pink hue, the central petals pale orange red with white stripes, rose-double form, medium to large size, 8.0-11.5cm across and 5.0cm deep with 12-15 exterior large petals arranged in 2 rows, 15-18 central petals arranged orderly in 2-3 rows, occasionally a few stamens visible. Leaves dark green, leathery, broad elliptic, cuneate at the base, margins smooth. Plant compact and growth vigorous. Starts to bloom from early-summer, fully blooms in autumn and winter and sporadically in spring.

HAR-23-3. 夏日娇韵
Xiari Jiaoyun (Summer's Beautiful Rhymes)

回交组合： 杜鹃红山茶 × 杜鹃红山茶 F_1 代新品种'夏日叠星'（ C. azalea × C. azalea F_1 new hybrid 'Xiari Diexing'）。

回交苗编号： DDX-No.3。

性状： 花朵橘红色，随着花朵开放，花瓣边缘逐渐变为淡红色，半重瓣型至松散的牡丹型，中型至大型花，花径9.0～11.0cm，花朵厚6.0cm，花瓣20～23枚，呈3轮以上排列，花瓣长倒卵形，长6.0cm，宽3.0cm，厚质，雄蕊散射，基部略连生，花丝淡红色，花药金黄色，开花稠密。叶片浓绿色，叶背面灰绿色，革质，长椭圆形，长12.0～12.5cm，宽3.5～4.0cm，基部急楔形，叶缘光滑。植株开张，枝条坚挺，生长旺盛。夏初始花，秋、冬季盛花，春季零星开花。

Abstract: Flowers orange red, as the flowers opening, petal edges gradually change into light red, semi-double to loose peony form, medium to large size, 9.0-11.0cm across and 6.0cm deep with 20-23 petals arranged in over 3 rows, petals long obovate, thick texture, stamens scattered, slightly united at the base, filaments pale red, anthers golden yellow, bloom dense. Leaves dark green, leathery, long elliptic, urgently cuneate at the base, margins smooth. Plant spread, branches strong and growth vigorous. Starts to bloom from early-summer, fully blooms in autumn and winter and sporadically in spring.

◉ 回交组合 HAR-24. 杜鹃红山茶 × 杜鹃红山茶 F₁ 代新品种 '夏梦可娟'

HAR-24. *C. azalea* × *C. azalea* F₁ new hybrid 'Xiameng Kejuan'

本回交组合将介绍 2 个新品种。

'夏梦可娟'是从杜鹃红山茶 × 云南山茶品种'帕克斯先生'这一杂交组合中得到的（见第一部书第 258～260 页），它花朵艳红色，单瓣型，开花稠密，且能多季开花。

本组所获得的 2 个新品种虽然都是单瓣型，但是一个是粉红花色，另一个是艳红至黑红花色，而且开花稠密，都可以重复开花，叶片浓绿，叶缘基部楔形，遗传了杜鹃红山茶和'夏梦可娟'的性状。由此可以推断，它们是真实的回交种。

Two new hybrids will be introduced in this backcross combination.

'Xiameng Kejuan' was obtained from the cross-combination between *C. azalea* and *C. reticulata* cultivar 'Dr Clifford Parks'(See p.258-260, our first book), its flowers are bright red, single form and bloom densely and year-round.

The two new hybrids obtained in this backcross combination are all single in their flower form, one is pink and the other is dark red, and both of them bloom very densely and can ever-bloom with dark green and cuneate base leaves, which are inherited from *C. azalea* and 'Xiameng Kejuan'. It can be presumed from above that the two new hybrids are real backcross hybrids.

HAR-24-1. 风和日丽
Fenghe Rili (Breezy & Sunny)

回交组合： 杜鹃红山茶 × 杜鹃红山茶 F_1 代新品种'夏梦可娟'（ C. azalea × C. azalea F_1 new hybrid 'Xiameng Kejuan'）。

回交苗编号： DXMKJ-No.2。

性状： 花朵粉红色，略泛橘红色调，花瓣顶部淡粉红色，单瓣型，中型花，花径8.0～9.5cm，花瓣7～8枚，排列松散，长倒卵形，雄蕊基部连生，呈柱状，花丝淡粉红色，花药黄色。叶片浓绿色，叶背面灰绿色，革质，椭圆形，长8.0～9.5cm，宽3.0～4.0cm，基部急楔形，叶缘光滑。植株紧凑，枝叶稠密，生长旺盛。仲夏始花，秋、冬季盛花，春季零星开花。

Abstract: Flowers pink slightly with orange red hue, light pink at the top part of petals, single form, medium size, 8.0-9.5cm across with 7-8 petals arranged loosely, petals long obovate, stamens united into a column at the base, filaments pale pink, anthers yellow. Leaves dark green, leathery, elliptic, urgently cuneate at the base, margins smooth. Plant compact, branches dense and growth vigorous. Starts to bloom from mid-summer, fully blooms in autumn and winter and sporadically in spring.

HAR-24-2. 碧血丹心
Bixue Danxin (Red Blood & Loyal Heart)

回交组合： 杜鹃红山茶 × 杜鹃红山茶 F_1 代新品种'夏梦可娟'（ C. azalea × C. azalea F_1 new hybrid 'Xiameng Kejuan' ）。

回交苗编号： DXMKJ-No.4。

性状： 花朵黑红色，泛蜡光，偶具模糊的白斑，单瓣型，中型花，花径8.5～10.0cm，花瓣6～7枚，倒卵形，长5.0cm，宽3.0cm，边缘波浪状，雄蕊基部连生，花丝淡粉色，花药黄色，开花稠密。叶片浓绿色，叶背面灰绿色，革质，椭圆形，长10.0～11.5cm，宽3.0～4.0cm，基部急楔形，叶缘具浅齿。植株紧凑，枝叶稠密，生长旺盛。仲夏始花，秋、冬季盛花，春季零星开花。

> **Abstract:** Flowers dark red with waxy luster, occasionally with fuzzy white markings, single form, medium size, 8.5-10.0cm across with 6-7 petals, petals obovate, edges wavy, stamens united at the base, filaments pale pink, anthers yellow, bloom dense. Leaves dark green, leathery, elliptic, urgently cuneate at the base, margins shallowly serrate. Plant compact, branches dense and growth vigorous. Starts to bloom from mid-summer, fully blooms in autumn and winter and sporadically in spring.

◉ 回交组合 HAR-25. 杜鹃红山茶 × 杜鹃红山茶 F_1 代新品种 '夏梦玉兰'

HAR-25. *C. azalea* × *C. azalea* F_1 new hybrid 'Xiameng Yulan'

本回交组合将介绍1个新品种。

'夏梦玉兰'是从红山茶品种'都鸟'×杜鹃红山茶这一杂交组合中得到的（见第一部书第224～225页）。该品种花朵淡粉红色、单瓣型、玉兰花状，开花稠密。

根据我们的调查，母本和父本均为单瓣型的杂交组合所获得的杂交种通常也是单瓣型的。而本杂交组合虽然两个亲本都是单瓣型的，但是获得了半重瓣型、大花型的杂交种。究其原因，可能是其父本'夏梦玉兰'具有的'都鸟'品种的多瓣杂合基因在本回交新品种中表达的结果。这给我们一个启示，单瓣的杜鹃红山茶 F_1 代，如果它的母本是多瓣类型的，其回交新品种仍有可能出现多瓣类型的新品种。

One new hybrid will be introduced in this backcross combination.

'Xiamemg Yulan' was obtained from the cross-combination between *C. japonica* cultivar 'Miyakodori' and *C. azalea* (See p.211-212, our first book). The flowers of the hybrid are light pink, single, magnolia shaped and dense.

According to our investigations, the flowers are always single form, if the hybrids obtained from the cross-combination that both female and male parents are single form. Although the two parents of this backcross combination are single form, the flowers of the hybrid obtained are semi-double form and large size. Maybe the reason is that the heterozygous gene of 'Miyakodori' with multi-petals has expressed out in the new backcrossed hybrid of the male parent 'Xiameng Yulan'. It gives us a revelation that if single-petal *C. azalea*'s F_1 has the female parent which is multi-petal, the new hybrids with multi-petals might still appear in the backcross combination.

HAR-25-1. 迷人红裙
Miren Hongqun (Attractive Red Dress)

回交组合： 杜鹃红山茶 × 杜鹃红山茶 F_1 代新品种'夏梦玉兰'（ C. azalea × C. azalea F_1 new hybrid 'Xiameng Yulan'）。

回交苗编号： DYL-No.2。

性状： 花朵桃红色至深桃红色，偶具隐约的白条纹，半重瓣型，大型至巨型花，花径12.0～13.5cm，花朵厚5.5cm，花瓣18～20枚，呈3轮有序排列，花瓣阔倒卵形，长6.5cm，宽5.5cm，先端近圆形，雄蕊基部略连生，花丝淡黄色，花药淡黄色。叶片浓绿色，叶背面灰绿色，厚革质，阔椭圆形，长9.5～10.0cm，宽4.5～5.0cm，基部楔形，上部边缘叶齿稀疏。植株立性，枝叶稠密，生长旺盛。夏初始花，秋、冬季盛花，春季零星开花。

Abstract: Flowers peach red to deep peach red, occasionally with faint white stripes, semi-double form, large to very large size, 12.0-13.5cm across and 5.5cm deep with 18-20 petals arranged orderly in 3 rows, petals broad obovate, apices nearly round, stamens slightly united at the base, filaments pale yellow, anthers pale yellow. Leaves dark green, heavy leathery, broad elliptic, cuneate at the base, the upper margins shallowly and thinly serrate. Plant upright, branches dense and growth vigorous. Starts to bloom from early-summer, fully blooms in autumn and winter and sporadically in spring.

回交组合 HAR-26. 杜鹃红山茶 × 杜鹃红山茶 F₁ 代新品种 '浪漫粉娘'

HAR-26. *C. azalea* × *C. azalea* F$_1$ new hybrid 'Langman Fenniang'

本回交组合将介绍 1 个新品种。

'浪漫粉娘'是从红山茶品种'石达琳'× 杜鹃红山茶这一杂交组合中得到的（见第一部书第 211 ~ 212 页）。该品种花朵淡粉红色，略显橙红色调，单瓣型，开花稠密。

可以看出，该回交组合两个亲本的花型均为单瓣型，因此我们获得的回交新品种也是单瓣型的。然而，回交新品种的花色更浅了，花朵也变小了，其原因有待进一步研究和分析。尽管获得的这个回交品种不太理想，但是它的花朵小而素雅，植株紧凑，开花稠密，而且能夏季开花，适合家庭盆栽。

One new hybrid will be introduced in this backcross combination.

'Langman Fenniang' was obtained from the backcross combination between *C. japonica* cultivar 'Shidalin' and *C. azalea* (See p.211-212, our first book). The flowers of the hybrid are light pink slightly with orange red hue in color, single form and very dense.

It can be seen that the flowers of the two cross parents are all single in flower form, therefore, the backcross new hybrid is single form too. However, the flower color of the hybrid becomes lighter and the flower size becomes smaller, which the reasons need to be further studied and analyzed. Although the backcross hybrid obtained is not ideal, it has small but elegant flowers, compact plant, dense flowers and can bloom in summer, which make it suitable for potted cultivation at home.

HAR-26-1. 粉雅佳境
Fenya Jiajing (Pink Elegant & Wonderful Place)

回交组合：杜鹃红山茶 × 杜鹃红山茶 F_1 代新品种'浪漫粉娘'（ C. azalea × C. azalea F_1 new hybrid 'Langman Fenniang' ）。

回交苗编号：DLMFN-No.3。

性状：花朵淡粉红色，单瓣型，小型花，花径6.0～7.5cm，花瓣5～6枚，阔倒卵形，长4.5cm，宽3.5cm，先端略凹，雄蕊基部略连生，花丝淡粉红色，花药淡黄色，开花稠密。叶片浓绿色，叶背面灰绿色，革质，长椭圆形，长7.5～8.0cm，宽2.5～3.0cm，基部急楔形，上部叶缘有稀疏的浅齿。植株立性，枝叶稠密，生长旺盛。仲夏始花，秋、冬季盛花，春季零星开花。

Abstract: Flowers light pink, single form, small size, 6.0-7.5cm across with 5-6 broad obovate petals, petals apices emarginate, stamens slightly united at the base, filaments pale pink, bloom dense. Leaves dark green, leathery, long elliptic, urgently cuneate at the base, upper margins shallowly and thinly serrate. Plant upright, branches dense and growth vigorous. Starts to bloom from mid-summer, fully blooms in autumn and winter and sporadically in spring.

回交组合 HAR-27. 杜鹃红山茶 × 杜鹃红山茶 F₁ 代新品种 '火红牡丹'
HAR-27. *C. azalea* × *C. azalea* F₁ new hybrid 'Huohong Mudan'

本回交组合将介绍 1 个新品种。

'火红牡丹'是从杜鹃红山茶 × 云南山茶品种'帕克斯先生'这一杂交组合中得到的（见第一部书第 251～252 页）。该品种花朵艳红色、牡丹型、中型花。

虽然本组父本的花朵是牡丹型的，我们获得的新品种却是单瓣型的，这可能是因为父本杂合度较高，其子代就会出现多种花型。

One new hybrid will be introduced in this backcross combination.

'Huohong Mudan' was obtained from the cross-combination between *C. azalea* and *C. reticulata* cultivar 'Dr Clifford Parks'(See p.251-252, our first book). Its flowers are bright red, peony form and medium size.

Although the flowers of the male parent are peony form, the flowers of the new hybrid we obtained are single form, which might due to the male parent has higher heterozygosity making the offspring appear various flower forms.

HAR-27-1. 奔放舞者
Benfang Wuzhe (Unrestrained Dancer)

回交组合：杜鹃红山茶 × 杜鹃红山茶 F_1 代新品种'火红牡丹'（*C. azalea* × *C. azalea* F_1 new hybrid 'Huohong Mudan'）。

回交苗编号：DHHMD-No.1。

性状：花朵淡桃红色，单瓣型，大型至巨型花，花径 10.5～13.5cm，花瓣 6～8 枚，倒卵形，长 6.5cm，宽 3.5cm，花瓣上部外翻或波浪状，先端略凹，雄蕊基部连生，呈柱状，花丝淡粉红色，花药黄色，开花稠密。叶片浓绿色，叶背面灰绿色，革质，椭圆形，长 8.5～9.5cm，宽 3.5～4.0cm，基部急楔形，上部叶缘具浅齿。植株立性，枝叶稠密，生长旺盛。夏初始花，秋、冬季盛花，春季零星开花。

Abstract: Flowers light peach red, large to very large size, 10.5-13.5cm across with 6-8 obovate petals, petals rolled outward or wavy at the upper part and apices emarginate, stamens united into a column at the base, filaments pale pink, bloom dense. Leaves dark green, leathery, elliptic, urgently cuneate at the base, margins shallowly serrate at the upper part. Plant upright, branches dense and growth vigorous. Starts to bloom from early-summer, fully blooms in autumn and winter and sporadically in spring.

四季茶花杂交新品种彩色图集（第二部）

◉ 回交组合 HAR-28. 杜鹃红山茶 F_1 代新品种'满天红星'× 杜鹃红山茶 F_1 代新品种'吉利牡丹'

HAR-28. *C. azalea* F_1 new hybrid 'Mantian Hongxing' × *C. azalea* F_1 new hybrid 'Jili Mudan'

本回交组合将介绍 1 个新品种。

'满天红星'是由红山茶品种'媚丽'× 杜鹃红山茶获得的一个花朵艳红色、单瓣型、中等花径的杂交新品种（见第一部书第 106～108 页），而'吉利牡丹'则是由杜鹃红山茶 × 云南山茶品种'帕克斯先生'获得的一个花朵桃红色至红色、牡丹型、中等花径的杂交新品种（见第一部书第 252～253 页）。由此可见，该回交组合实际上属于两个杜鹃红山茶 F_1 代之间的杂交，理论上其新品种的杜鹃红山茶的基因频率应是 50%。

可以看到，本组所获得的杂交新品种能四季开花，其花朵红色、半重瓣型、整朵掉落，叶片厚，这些性状似乎都与两个亲本的杂合度相关。

One new hybrid will be introduced in this cross-combination.

'Mantian Hongxing' obtained from *C. japonica* cultivar 'Tama Beauty' × *C. azalea* is a new hybrid with bright red, single form and medium size flowers (See p.106-108, our first book) and 'Jili Mudan' obtained from *C. azalea* × *C. reticulata* cultivar 'Dr Clifford Parks' is a new hybrid with peach red to red, peony form and medium size flowers (See p.252-253, our first book). From above, we can see that the cross-combination actually belongs to the cross between two *C. azalea*'s F_1. In theory, the *C. azalea*'s gene frequency of the new hybrid should be 50%.

It can be seen that the new hybrid obtained in this cross-combination can bloom in four seasons, its flowers are red, semi-double form, and fall whole, leaves are thick, these characteristics all seem to be related to the heterozygosity of its parents.

HAR-28-1. 皱瓣金心
Zhouban Jinxin (Wrinkled Petals with Golden Heart)

回交组合：杜鹃红山茶 F_1 代新品种'满天红星'×杜鹃红山茶 F_1 代新品种'吉利牡丹'（*C. azalea* F_1 new hybrid 'Mantian Hongxing'× *C. azalea* F_1 new hybrid 'Jili Mudan'）。

回交苗编号：MTHXJL-No.1。

性状：花朵桃红色至红色，半重瓣型，中型花，花径8.0～9.0cm，花朵厚4.5cm，花瓣12～14枚，呈2轮紧实排列，花瓣皱褶，厚质，阔倒卵形，先端微凹，雄蕊基部略连生，花丝和花药淡黄色，花朵整朵掉落。叶片浓绿色，叶背面灰绿色，厚革质，阔椭圆形，长8.5～9.0cm，宽4.5～5.0cm，基部楔形，上部边缘叶齿稀疏。植株立性，枝叶稠密，生长旺盛。夏末始花，秋、冬季盛花，春季零星开花。

Abstract: Flowers peach red to red, semi-double form, medium size, 8.0-9.0cm across and 4.5cm deep with 12-14 petals arranged tightly in 2 rows, petals wrinkled, thick texture, broad obovate, apices emarginate, stamens slightly united at the base, filaments and anthers pale yellow, flowers fall whole. Leaves dark green, back surfaces pale green, heavy leathery, broad elliptic, cuneate at the base, margins shallowly and thinly serrate at the upper part. Plant upright, branches dense and growth vigorous. Starts to bloom from late-summer, fully blooms in autumn and winter and sporadically in spring.

回交组合 HAR-29. 杜鹃红山茶 F_1 代新品种'水月紫鹃' × 杜鹃红山茶 F_1 代新品种'夏梦文清'

HAR-29. *C. azalea* F_1 new hybrid 'Shuiyue Zijuan' × *C. azalea* F_1 new hybrid 'Xiameng Wenqing'

本回交组合将介绍 1 个新品种。

'水月紫鹃'是从红山茶品种'锯叶椿'×杜鹃红山茶这一杂交组合获得的（见第一部书第 185～186 页）。它虽然是单瓣型的，但是花朵呈紫红色，显得别具一格。

很明显，本回交组合属于杜鹃红山茶 F_1 代姊妹品种之间的杂交，其杂交种的杜鹃红山茶基因频率是 50%。

本回交组合期望获得花色为紫红色，花径像其父本'夏梦文清'一样巨型，且能四季开花的新品种。实际上，所获得的新品种在性状上基本符合了我们的期盼。

One new hybrid will be introduced in this cross-combination.

'Shuiyue Zijuan' was obtained from the cross-combination between *C. japonica* cultivar 'Nokogiriba-tsubaki' and *C. azalea* (See p.185-186, our first book). Although it is single in flower form, its flower color is purple red which is very unique.

Obviously, this cross-combination belongs to the hybridization between sister-hybrids of *C. azalea* F_1, so the *C. azalea*'s gene frequency of its hybrid is 50%.

The backcross combination was to get the new hybrid that flowers are purple-red, as large as its male parent 'Xiameng Wenqing' and could bloom in four seasons. The new hybrid obtained basically conforms to our expectation.

HAR-29-1. 艳紫妖红
Yanzi Yaohong (Brilliant Purple & Peculiar Red)

回交组合： 杜鹃红山茶 F_1 代新品种'水月紫鹃'× 杜鹃红山茶 F_1 代新品种'夏梦文清'（*C. azalea* F_1 new hybrid 'Shuiyue Zijuan' × *C. azalea* F_1 new hybrid 'Xiameng Wenqing'）。

回交苗编号： SYZJWQ-No.7。

性状： 花朵淡紫红色，花芯部分雄蕊瓣具白斑，牡丹型，中型至大型花，花径 8.0～11.0cm，花朵厚 5.0cm，大花瓣 12～15 枚，呈 2 轮排列，花瓣外翻，阔倒卵形，雄蕊基部连生，雄蕊瓣直立，扭曲。叶片浓绿色，叶背面灰绿色，革质，椭圆形，长 8.5～9.5cm，宽 4.0～4.5cm，基部楔形，上部边缘具疏齿。植株立性，枝叶稠密，生长旺盛。仲夏始花，秋、冬季盛花，春季零星开花。

Abstract: Flowers light purple red, central petaloids with white markings, peony form, medium to large size, 8.0-11.0cm across and 5.0cm deep with 12-15 large petals arranged in 2 rows, petals rolled outward, broad obovate, stamens united at the base, petaloids erected and twisted. Leaves dark green, leathery, elliptic, cuneate at the base, margins sparsely serrate at the upper part. Plant upright, branches dense and growth vigorous. Starts to bloom from mid-summer, fully blooms from autumn to winter and sporadically in spring.

回交组合 HAR-30. 杜鹃红山茶 × 杜鹃红山茶 F_1 代新品种 '夏日粉裙'

HAR-30. *C. azalea* × *C. azalea* F_1 new hybrid 'Xiari Fenqun'

本回交组合将介绍1个新品种。

'夏日粉裙'是由红山茶品种'媚丽'×杜鹃红山茶这一杂交组合得到的（见第一部书第131～133页），其花朵为粉红色、半重瓣型、中型花。

本回交组合所获得的新品种，其杜鹃红山茶基因频率高达75%，而常规茶花的基因频率则降为25%。因为'夏日粉裙'粉色花基因的介入，我们所获得的新品种的花色为粉红色。

One new hybrid will be introduced in this backcross combination.

'Xiari Fenqun' was obtained from the cross-combination between *C. japonica* cultivar 'Tama Beauty' and *C. azalea* (See p.131-133, our first book) and its flowers are pink in color, semi-double in form and medium in size.

In the new hybrid obtained in this combination, its gene frequency of *C. azalea* can increase to 75%, while the gene frequency of the normal camellia cultivar can reduce to 25%. Because of the pink flower genes of 'Xiari Fenqun', the flowers of the new hybrid we obtained are pink.

HAR-30-1. 超级粉冠
Chaoji Fenguan (Supreme Pink Crown)

回交组合： 杜鹃红山茶 × 杜鹃红山茶 F_1 代新品种'夏日粉裙'（ C. azalea × C. azalea F_1 new hybrid 'Xiari Fenqun'）。

回交苗编号： DFQ-No.3。

性状： 花朵粉红色至深粉红色，雄蕊瓣具白条纹，松散的牡丹型，巨型花，花径13.5cm以上，最大花径可达18.0cm，花朵厚6.5cm，大花瓣21～25枚，呈3～4轮排列，花瓣阔倒卵形，外翻，花芯处少量小雄蕊瓣与雄蕊混生。叶片浓绿色，叶背面灰绿色，厚革质，阔椭圆形，长11.0～11.5cm，宽4.0～4.5cm，基部楔形，边缘光滑。植株立性，生长旺盛。夏初始花，秋、冬季盛花，春季零星开花。

Abstract: Flowers pink to deep pink, central petaloids with white stripes, loose peony form, very large size, 13.5cm or more across, the largest across can reach to 18.0cm, 6.5cm deep, 21-25 large petals arranged in 3-4 rows, petals broad obovate, rolled outward, a few petaloids mixed with stamens at the center. Leaves dark green, heavy leathery, broad elliptic, cuneate at the base, margins smooth. Plant upright and growth vigorous. Starts to bloom from early-summer, fully blooms from autumn to winter and sporadically in spring.

⦿ 回交组合 HAR-31. 杜鹃红山茶 × 杜鹃红山茶 F₁ 代新品种 '书香之家'
HAR-31. *C. azalea* × *C. azalea* F₁ new hybrid 'Shuxiang Zhijia'

本回交组合将介绍 1 个新品种。

'书香之家'是由红山茶品种'白斑康乃馨'× 杜鹃红山茶 F₁ 代新品种'红屋积香'这一杂交组合得到的（见本书 HAR-36-1）。理论上它具有 25% 的杜鹃红山茶基因和 75% 的常规茶花品种基因。

本回交组合所获得的新品种，理论上其杜鹃红山茶基因有 62.5%，而常规茶花基因则为 37.5%，因此，新品种可以保持较强的多季开花的性状。

One new hybrid will be introduced in this backcross combination.

'Shuxiang Zhijia' was obtained from the backcross combination between *C. japonica* cultivar 'Ville de Nantes' and *C. azalea* F₁ new hybrid 'Hongwu Jixiang' (See HAR-36-1, this book). In theory, it has 25% *C. azalea* genes and 75% normal camellia cultivar genes.

In the new hybrid obtained from this combination, in theory, 62.5% of the genes comes from *C. azalea*, and 37.5% of the genes comes from the normal camellia cultivar, therefore the new hybrid obtained can maintain strong characteristics of multi-seasons flowering.

HAR-31-1. 层层诗意
Cengceng Shiyi (Layers of Poetic Flavor)

回交组合： 杜鹃红山茶 × 杜鹃红山茶 F_1 代新品种'书香之家'（*C. azalea* × *C. azalea* F_1 new hybrid 'Shuxiang Zhijia'）。

回交苗编号： DSXZJ-No.1。

性状： 花朵红色，具白色条纹，半重瓣型，中型至大型花，花径 8.0～11.5cm，花朵厚 5.8cm，大花瓣 16～18 枚，呈 3 轮有序排列，花瓣阔倒卵形，外翻，雄蕊基部略连生，雄蕊瓣少。叶片浓绿色，叶背面灰绿色，革质，长椭圆形，长 9.0～9.5cm，宽 3.0～3.5cm，基部急楔形，叶缘叶齿稀。植株立性，生长旺盛。夏初始花，秋、冬季盛花，春季零星开花。

> **Abstract:** Flowers red with white stripes, semi-double form, medium to large size, 8.0-11.5cm across and 5.8cm deep with 16-18 large petals arranged orderly in 3 rows, petals broad obovate, rolled outward, stamens united at the base. Leaves dark green, leathery, long elliptic, urgently cuneate at the base, margins sparsely serrate. Plant upright and growth vigorous. Starts to bloom from early-summer, fully blooms from autumn to winter and sporadically in spring.

⊙ 回交组合 HAR-32. 杜鹃红山茶 × 杜鹃红山茶 F₁ 代新品种 '粤桂大嫂'

HAR-32. *C. azalea* × *C. azalea* F₁ new hybrid 'Yuegui Dasao'

本回交组合将介绍 1 个新品种。

'粤桂大嫂'品种是由杜鹃红山茶 × 多齿红山茶这一杂交组合获得的（见第一部书第 284 ~ 285 页）。

本回交组合试图让回交新品种保持'粤桂大嫂'品种的黑红花色，同时也保持父本杜鹃红山茶四季开花的性状。

所获得的新品种花朵艳红色至黑红色，大型至巨型花，虽然是半重瓣型，但是花朵中部出现了一些复色瓣化的雄蕊瓣。很明显，这是父本'粤桂大嫂'品种中多齿红山茶基因的表达所致。

One new hybrid will be introduced in this backcross combination.

The hybrid 'Yuegui Dasao' was obtained from the cross-combination between *C. azalea* and *C. polyodonta* (See p. 284-285, our first book).

The backcross combination attempted to make the hybrid keep the dark red flower color of 'Yuegui Dasao', and the ever-blooming trait of *C. azalea*.

The hybrid obtained is bright red to dark red in flower color, large to very large in flower size, although it is semi-double form, some petaloids with multi-color appeared at the center. It is obvious that this is the result of the expression of *C. polyodonta*'s genes in the male parent 'Yuegui Dasao'.

HAR-32-1. 大唐宫灯
Datang Gongdeng（Palace Lantern in Great Tang Dynasty）

回交组合：杜鹃红山茶 × 杜鹃红山茶 F_1 代新品种'粤桂大嫂'（ C. azalea × C. azalea F_1 new hybrid 'Yuegui Dasao'）。

回交苗编号：DYGDS-No.2。

性状：花朵艳红色至黑红色，花瓣略显白条纹，花芯处偶有带白斑的雄蕊瓣，半重瓣型，大型至巨型花，花径 12.0～14.0cm，花朵厚 6.5cm，大花瓣 14～16 枚，呈 2～3 轮排列，花瓣阔倒卵形，厚质，内扣，雄蕊散射，似肥后茶，花丝红色。叶片浓绿色，叶背面灰绿色，厚革质，长椭圆形，长 8.5～9.5cm，宽 3.5～4.0cm，基部楔形，上部边缘叶齿钝。植株立性，枝叶稠密，生长旺盛。夏初始花，秋、冬季盛花，春季零星开花。

Abstract: Flowers bright red to dark red, slightly with white stripes, occasionally with white markings on petaloids at the center, semi-double form, large to very large size, 12.0-14.0cm across and 6.5cm deep with 14-16 petals arranged in 2-3 rows, petals broad obovate, thick texture, rolled inward, stamens radial as Higo camellias, filaments red. Leaves dark green, heavy leathery, long elliptic, cuneate at the base, upper margins obtusely serrate. Plant upright, branches dense and growth vigorous. Starts to bloom from early-summer, fully blooms from autumn to winter and sporadically in spring.

⊙ 回交组合 HAR-33. 杜鹃红山茶 × 杜鹃红山茶 F₁ 代新品种'夏日红杯'

HAR-33. *C. azalea* × *C. azalea* F₁ new hybrid 'Xiari Hongbei'

本回交组合将介绍 1 个新品种。

本组合中，作为父本的'夏日红杯'是由红山茶品种'媚丽'×杜鹃红山茶获得的一个单瓣品种（见第一部书第 135～136 页）。

尽管本回交组合的父本和母本都是单瓣型的，但是获得的杂交新品种偶然也会出现半重瓣型的花朵，而且花瓣变宽了，花期提前了，这说明杜鹃红山茶的基因和'媚丽'品种的基因在起作用。

One new hybrid will be introduced in this backcross combination.

In this combination, the male parent 'Xiari Hongbei' is a single form variety obtained from *C. japonica* cultivar 'Tama Beauty' × *C. azalea* (See p.135-136, our first book).

Although both the male and female parents of this backcross combination are single form, the new hybrid obtained occasionally had semi-double flowers, and the petals became wider, and the flowering time was advanced, which indicates that the genes of *C. azalea* and 'Tama Beauty' cultivar were playing a role.

HAR-33-1. 红瓣竞秀
Hongban Jingxiu (Red Petals' Beauty Competitions)

回交组合：杜鹃红山茶 × 杜鹃红山茶 F_1 代新品种'夏日红杯'（*C. azalea* × *C. azalea* F_1 new hybrid 'Xiari Hongbei'）。

回交苗编号：DXRHB-No.5+12。

性状：花朵红色至深红色，偶有蜡质感，半重瓣型，偶现单瓣型，中型至大型花，花径 9.0～11.5cm，花瓣 12～13 枚，呈 2 轮排列，花瓣阔倒卵形，略波浪状，瓣面可见深色脉纹，雄蕊基部连生，呈短柱状。叶片浓绿色，叶背面灰绿色，厚革质，长椭圆形，长 10.0～11.5cm，宽 3.5～4.0cm，基部急楔形，边缘光滑。植株立性，生长旺盛。夏初始花，秋、冬季盛花，春季零星开花。

Abstract: Flowers red to deep red, occasionally with waxy luster, semi-double form, sometimes single form appeared, medium to large size, 9.0-11.5cm across with 12-13 petals arranged in 2 rows, petals broad obovate, wavy, deep veins visible on the surfaces, stamens united into a column at the base. Leaves dark green, heavy leathery, long elliptic. Starts to bloom from early-summer, fully blooms from autumn to winter and sporadically in spring.

回交组合 HAR-34. 红山茶品种'孔雀玉浦' × 杜鹃红山茶 F₁ 代新品种'红屋积香'

HAR-34. *C. japonica* cultivar 'Tama Peacock' × *C. azalea* F₁ new hybrid 'Hongwu Jixiang'

本回交组合将介绍1个新品种。

正如本书HA-81所述,'孔雀玉浦'品种是一个花瓣边缘具白边、枝条下垂、灌木状的漂亮品种。

本回交组合试图让新品种保持'孔雀玉浦'品种的花色、株形性状,同时也保持父本'红屋积香'品种花朵芳香和四季开花的性状。

所获得的这个杂交新品种虽然花径不大,也不具芳香,但是其花朵淡红色至深粉红色,略泛紫色调,完全重瓣型,非常精致、耐看,且花期较长,遗传了'红屋积香'品种重复开花的性状。此杂交组合所获得的新品种理论上只具有25%的杜鹃红山茶基因,因此,其花期也推迟了。

×

One new hybrid will be introduced in this backcross-combination.

As described in HA-81 of this book, 'Tama Peacock' is a very beautiful cultivar with the characteristics of white bordered petals, hanging branches and bushy plant.

The backcross attempted to make the hybrid keep the flower color and the plant shape of 'Tama Peacock' and keep the characteristics of ever-blooming and fragrance of their male parent 'Hongwu Jixiang'.

The flowers of the new hybrid obtained have small diameters and are not fragrant, but are light red to deep pink with purplish hue, formal double form, which are very delicate and beautiful. Moreover, the hybrid has inherited the ever-blooming characteristics from 'Hongwu Jixiang'. It should be noted that the new hybrid obtained in the combination theoretically has only 25% of the genes of *C. azalea*, so its blooming period is delayed.

HAR-34-1. 孔雀炫丽
Kongque Xuanli (Peacock Flaunting Its Beauty)

回交组合： 红山茶品种'孔雀玉浦'×杜鹃红山茶 F_1 代新品种'红屋积香'（ *C. japonica* cultivar 'Tama Peacock'× *C. azalea* F_1 new hybrid 'Hongwu Jixiang'）。

回交苗编号： KQYPJX-No.3。

性状： 花朵淡红色至深粉红色，渐向花朵中部呈粉红色，完全重瓣型，微型至小型花，花径 5.5～7.0cm，花朵厚 3.5cm，花瓣 50 枚以上，呈 10 轮以上覆瓦状排列，花瓣阔倒卵形，先端微凹。叶片浓绿色，叶背面灰绿色，厚革质，阔椭圆形，长 10.0～10.5cm，宽 4.5～5.0cm，边缘叶齿尖。植株立性，枝叶稠密，生长旺盛。秋季始花，冬季盛花，春季亦能开花。

Abstract: Flowers light red to deep pink, fading to pink at the center of flowers, formal double form, miniature to small size, 5.5-7.0cm across and 3.5cm deep with more than 50 petals arranged imbricately in over 10 rows, petals broad obovate, apices emarginate. Leaves dark green, heavy leathery, broad elliptic, margins sharply serrate. Plant upright, branches dense and growth vigorous. Starts to bloom from autumn, fully blooms in winter and continuously in spring.

⊙ 回交组合 HAR-35. 红山茶品种'玉之浦'× 杜鹃红山茶 F_1 代新品种'红屋积香'

HAR-35. *C. japonica* cultivar 'Tama-no-ura' × *C. azalea* F_1 new hybrid 'Hongwu Jixiang'

 ×

本回交组合将介绍18个新品种。

正如本书 HA-82 所述，母本'玉之浦'花朵红色，花瓣边缘具清晰白边。父本'红屋积香'是从杜鹃红山茶 × 红山茶品种'克瑞墨大牡丹'杂交组合中得到的一个能够多季开花、略带芳香的杂交种（见第一部书第 72～74 页）。

本组合所得到的回交新品种，在理论上杜鹃红山茶基因频率只有25%。即使是这样，回交新品种的整个花期仍可达到8个月，比普通茶花长1个月。这给了我们启示，在培育四季开花的茶花新品种时，杂交新品种中的杜鹃红山茶基因频率仍有进一步降低的空间。

可以看到，本组合的育种目标是获得花朵具白边、有香味、能够多季开花的新品种。本组所获得的18个新品种中，有11个新品种花瓣边缘具白边，占总数的61.1%；有10个新品种具微香，占总数的55.6%，基本达到了我们的预期目标。另外，本组获得的新品种中，大部分花色是黑红色的，花径为微型至小型，这些性状倾向于其母本'玉之浦'。

Eighteen new hybrids will be introduced in this backcross combination.

As described in HA-82 of this book, the female parent, 'Tama-no-ura' can open red flowers with clear white borders. The male parent 'Hongwu Jixiang' with the characteristics of ever-blooming and fragrance was obtained from the cross-combination between *C. azalea* and *C. japonica* cultivar 'Kramer's Supreme'(See p. 72-74, our first book).

The backcross new hybrids obtained in this combination, in theory, only have 25% gene frequency of *C. azalea*. Even so, eight months of the whole blooming period can be reached in the backcross new hybrids, which is one month longer than normal camellias. The inspiration for us is that to breed new camellia hybrids that bloom year-round, there is still room for further reduction of the gene frequency of *C. azalea* in the new hybrids.

It can be seen that the targets of the combination were to obtain those hybrids whose flowers had the characteristics of white edges, fragrance and ever-blooming. There are 11 new hybrids that petals are with white borders, which account for 61.1% of the 18 new hybrids obtained in the combination, and there are ten new hybrids that flowers are fragrant, which account for 55.6 % of the total, therefore, our expected targets have been achieved basically. In addition, in the new hybrids obtained from this combination, most of them are dark red in flower color, miniature to small in flower size, which tend to their female parent 'Tama-no-ura'.

HAR-35-1. 左家娇女
Zuojia Jiaonü (Tender &Cute Girl)

回交组合：红山茶品种'玉之浦'×杜鹃红山茶 F_1 代新品种'红屋积香'（*C. japonica* cultivar 'Tama-no-ura' × *C. azalea* F_1 new hybrid 'Hongwu Jixiang'）。

回交苗编号：YUJX-No.1。

性状：花朵淡红色至桃红色，瓣面具白条纹，边缘具窄白边，随着花朵开放，花色逐渐变为粉红色，春季花朵变为粉白色，半重瓣型至玫瑰重瓣型，小型花，花径7.0～7.5cm，花朵厚3.0～4.0cm，花瓣20～23枚，呈3～4轮有序排列，花瓣倒卵形，雄蕊少量，开花稠密。叶片浓绿色，厚革质，椭圆形，长7.5～8.5cm，宽3.5～4.0cm，基部楔形，上部边缘叶齿尖。植株开张，生长旺盛。秋季始花，冬季盛花，春季可持续开花。

> **Abstract:** Flowers light red to peach red with white stripes on the petal surfaces, bordered white, as flower opening, the flower color gradually changes into pink and even into pink-white in spring, semi-double to rose-double form, small size, 7.0-7.5cm across and 3.0-4.0cm deep with 20-23 petals arranged orderly in 3-4 rows, petals obovate, a few stamens, bloom dense. Leaves dark green, heavy leathery, elliptic, cuneate at the base, margins sharply serrate at the upper part. Plant spread and growth vigorous. Starts to bloom from autumn, fully blooms in winter and continuously in spring.

HAR-35-2. 外婆童谣
Waipo Tongyao (Grandmother's Nursery Rhymes)

回交组合：红山茶品种'玉之浦'×杜鹃红山茶F₁代新品种'红屋积香'（*C. japonica* cultivar 'Tama-no-ura' × *C. azalea* F₁ new hybrid 'Hongwu Jixiang'）。

回交苗编号：YUJX-No.3。

性状：花朵红色至深红色，花芯处雄蕊瓣偶有白斑，略具微香，半重瓣型至牡丹型，小型花，花径6.5～7.5cm，花朵厚4.0cm，大花瓣18～20枚，呈3轮排列，阔倒卵形，厚质，外翻，雄蕊基部连生，花丝红色，少量雄蕊瓣与雄蕊混生，开花稠密。叶片浓绿色，厚革质，阔椭圆形，长8.0～8.5cm，宽5.0～5.5cm，基部楔形，边缘叶齿尖。植株立性，生长旺盛。秋季始花，冬季盛花，春季可持续开花。

Abstract: Flowers red to deep red, central petaloids occasionally with white markings, slightly with fragrance, semi-double to peony form, small size, 6.5-7.5cm across and 4.0cm deep with 18-20 large petals arranged in 3 rows, petals broad obovate, thick texture, stamens united at the base, filaments red, bloom dense. Leaves dark green, heavy leathery, broad elliptic, cuneate at the base, margins sharply serrate. Plant upright and growth vigorous. Starts to bloom from autumn, fully blooms in winter and continuously in spring.

HAR-35-3. 倚天望云
Yitian Wangyun (Look over Cloud against Sky)

回交组合：红山茶品种'玉之浦'× 杜鹃红山茶 F₁ 代新品种'红屋积香'（*C. japonica* cultivar 'Tama-no-ura' × *C. azalea* F₁ new hybrid 'Hongwu Jixiang'）。

回交苗编号：YUJX-No.4。

性状：花朵黑红色，具蜡光，略具清香，瓣缘可见窄白边，开花至春季时雄蕊全部瓣化，花色变为深粉红色，花朵中部显白斑，牡丹型，小型花，花径6.5～7.5cm，花朵厚6.5cm，大花瓣13～15枚，呈2～3轮排列，花瓣阔倒卵形，外翻，先端微凹，小花瓣和大花瓣与雄蕊混生，雄蕊簇生，多为散射状，花药金黄色。叶片浓绿色，革质，椭圆形，长11.0～12.0cm，宽4.0～4.5cm，边缘齿浅。植株立性，紧凑，生长旺盛。秋季始花，冬季盛花，春季可持续开花。

Abstract: Flowers dark red with waxy luster, with slight fragrance, petals with narrow white borders, in the spring all the stamens change into petaloids, flower color changes into deep pink and central white markings visible, peony form, small size, 6.5-7.5cm across and 6.5cm deep with 13-15 large petals arranged in 2-3 rows, petals broad obovate, turned outward, central small and large petals mixed with stamens, stamens clustered, anthers golden yellow. Leaves dark green, leathery, elliptic, margins shallowly serrate. Plant upright, compact and growth vigorous. Starts to bloom from autumn, fully blooms in winter and continuously in spring.

HAR-35-4. 新潮艳口
Xinchao Yankou (Trendy Pretty Mouth)

回交组合：红山茶品种'玉之浦'×杜鹃红山茶F₁代新品种'红屋积香'（*C. japonica* cultivar 'Tama-no-ura' × *C. azalea* F₁ new hybrid 'Hongwu Jixiang'）。

回交苗编号：YUJX-No.5。

性状：花朵淡红色，具白斑或边缘有白边，随着花朵开放，花色逐渐变为深粉红色，春季花朵多为红色，单瓣型，小型花，花径6.5～7.5cm，花瓣5～6枚，阔倒卵形，厚质，雄蕊基部连生，开花稠密。叶片浓绿色，革质，椭圆形，长8.5～9.5cm，宽4.0～4.5cm，基部楔形，边缘叶齿钝。植株立性，生长旺盛。秋季始花，冬季盛花，春季可持续开花。

> **Abstract:** Flowers light red with white markings or bordered white at the edges, as flowers opening, the flower color gradually changes into deep pink, most of flowers red in spring, single form, small size, 6.5-7.5cm across with 5-6 petals, petals broad obovate, thick texture, stamens united at the base, bloom dense. Leaves dark green, leathery, elliptic, cuneate at the base, margins obtusely serrate. Plant upright and growth vigorous. Starts to bloom from autumn, fully blooms in winter and continuously in spring.

HAR-35-5. 时尚小妹
Shishang Xiaomei (Stylish Little Sister)

杂交组合：红山茶品种'玉之浦'×杜鹃红山茶 F₁ 代新品种'红屋积香'（*C. japonica* cultivar 'Tama-no-ura'× *C. azalea* F₁ new hybrid 'Hongwu Jixiang'）。

回交苗编号：YUJX-No.6。

性状：花朵淡桃红色至深桃红色，雄蕊瓣具小白斑，具微香，托桂型，微型花，花径 4.0～5.0cm，花朵厚 3.0cm，外部大花瓣 5～6 枚，呈 1 轮排列，花瓣近圆形，雄蕊全部变为雄蕊瓣，簇拥成球，花丝红色。叶片深绿色，革质，小椭圆形，长 4.0～5.0cm，宽 3.5～4.0cm，基部楔形，边缘叶齿尖。植株立性，紧凑，生长旺盛。秋季始花，冬季盛花，春季可持续开花。

> **Abstract:** Flowers light peach red to deep peach red with small white markings on the petaloids, slightly fragrant, anemone form, miniature size, 4.0-5.0cm across and 3.0cm deep with 5-6 exterior large petals arranged in 1 row, petals nearly round, stamens all changed into petaloids and crowded into a ball, filaments red. Leaves dark green, leathery, small elliptical, cuneate at the base, margins sharply serrate. Plant upright, compact and growth vigorous. Starts to bloom from autumn, fully blooms in winter and continuously in spring.

HAR-35-6. 九九同心
Jiujiu Tongxin (Nine Persons with One Heart)

回交组合：红山茶品种'玉之浦'×杜鹃红山茶F₁代新品种'红屋积香'（*C. japonica* cultivar 'Tama-no-ura'× *C. azalea* F₁ new hybrid 'Hongwu Jixiang'）。

回交苗编号：YUJX-No.7+16。

性状：花朵红色，略泛紫色调，雄蕊瓣具白斑，具淡香，半重瓣型至牡丹型，小型花，花径6.5～7.5cm，花朵厚5.5cm，外部大花瓣约15枚，呈3轮排列，花瓣阔倒卵形，渐向基部变为淡红色，花芯处雄蕊瓣化呈牡丹型，雄蕊簇生。叶片浓绿色，厚革质，椭圆形，长10.0～11.0cm，宽4.0～4.5cm，边缘齿浅。植株立性，紧凑，生长旺盛。秋季始花，冬季盛花，春季可持续开花。

Abstract: Flowers red slightly with purple hue and light scent, petaloids with white markings, semi-double to peony form, small size, 6.5-7.5cm across and 5.5cm deep with about 15 exterior large petals arranged in 3 rows, petals broad obovate, fading to light red at the petal base, when petaloids formed at the center, the flower will become peony form, stamens clustered. Leaves dark green, heavy leathery, elliptic, margins shallowly serrate. Plant upright, compact and growth vigorous. Starts to bloom from autumn, fully blooms in winter and continuously in spring.

HAR-35-7. 红发模特
Hongfa Mote（Red Hair Models）

回交组合：红山茶品种'玉之浦'×杜鹃红山茶 F_1 代新品种'红屋积香'（*C. japonica* cultivar 'Tama-no-ura'× *C. azalea* F_1 new hybrid 'Hongwu Jixiang'）。

回交苗编号：YUJX-No.9。

性状：花朵黑红色，具蜡光和微清香，单瓣型，微型花，花径5.5～6.0cm，花瓣5～6枚，多排列呈半开放状态，花瓣阔倒卵形，雄蕊100多个，花丝红色，基部近离生，花药瓣化。叶片深绿色，光亮，厚革质，椭圆形，长8.0～8.5cm，宽3.5～4.0cm，边缘齿稀浅。植株开张，矮灌状，生长旺盛。秋季始花，冬季盛花，春季可持续开花。

Abstract: Flowers dark red with waxy luster, slightly with fragrance, single form, miniature size, 5.5-6.0cm across with 5-6 petals, mostly semi-opened, petals broad obovate, stamens more than 100, filaments red and nearly free at the base. Leaves deep green, shiny, heavy leathery, elliptic, margins shallowly and thinly serrate. Plants spread, dwarf and growth vigorous. Starts to bloom from autumn, fully blooms in winter and continuously in spring.

HAR-35-8. 戈壁驼铃
Gebi Tuoling (Gobi Camel Bells)

杂交组合：红山茶品种'玉之浦'× 杜鹃红山茶 F₁ 代新品种'红屋积香'（ *C. japonica* cultivar 'Tama-no-ura'× *C. azalea* F$_1$ new hybrid 'Hongwu Jixiang'）。

回交苗编号：YUJX-No.14。

性状：花朵红色至黑红色，泛蜡光，瓣缘偶有窄的白边，单瓣型，小型花，花径 6.5～7.5cm，大花瓣 5 枚，阔倒卵形，内扣，雄蕊基部连生，花丝红色，花药金黄色，开花稠密。叶片浓绿色，厚革质，椭圆形，长 8.5～9.5cm，宽 3.5～4.0cm，基部楔形，边缘叶齿浅。植株立性，高大，生长旺盛。秋季始花，冬季盛花，春季可持续开花。

Abstract: Flowers red to dark red with waxy luster, occasionally with narrow white borders, single form, small size, 6.5-7.5cm across with 5 large petals, petals broad obovate, rolled inward, stamens united at the base, filaments red, anthers golden yellow, bloom dense. Leaves dark green, heavy leathery, elliptic, cuneate at the base, margins shallowly serrate. Plant upright, tall and growth vigorous. Starts to bloom from autumn, fully blooms in winter and continuously in spring.

HAR-35-9. 彩球发结
Caiqiu Fajie (Colorful Ball and Hair Knots)

回交组合：红山茶品种'玉之浦'× 杜鹃红山茶 F_1 代新品种'红屋积香'（ *C. japonica* cultivar 'Tama-no-ura'× *C. azalea* F_1 new hybrid 'Hongwu Jixiang'）。

回交苗编号：YUJX-No.15。

性状：花朵深粉红色至淡红色，偶有白色隐斑，具微香，托桂型至小牡丹型，微型至小型花，花径5.5～7.5cm，花朵厚4.0cm，外轮花瓣12～13枚，呈2轮紧实排列，花瓣近圆形，中部雄蕊瓣达170多枚，勺状，簇拥成团，偶有小白斑，全开放后中部雄蕊微露，开花稠密。叶片深绿色，革质，椭圆形，长8.0～8.5cm，宽3.0～3.5cm，先端钝尖，边缘叶齿尖。植株紧凑，矮性，生长旺盛。秋季始花，冬季盛花，春季可持续开花。

Abstract: Flowers deep pink to light red, occasionally with faint white markings, little fragrance, anemone to small peony form, miniature to small size, 5.5-7.5cm across and 4.0cm deep with 12-13 exterior petals arranged tightly in 2 rows, apices nearly round, more than 170 small petaloids crowded into a ball at the center, occasionally with tiny white dots, bloom dense. Leaves deep green, leathery, elliptic, margins sharply serrate. Plant compact, dwarf and growth vigorous. Starts to bloom from autumn, fully blooms in winter and continuously in spring.

HAR-35-10. 三千墨丽
Sanqian Moli (Three Thousand Ink Works)

回交组合：红山茶品种'玉之浦'×杜鹃红山茶 F_1 代新品种'红屋积香'（ *C. japonica* cultivar 'Tama-no-ura'× *C. azalea* F_1 new hybrid 'Hongwu Jixiang'）。

回交苗编号：YUJX-No.17。

性状：花朵黑红色，泛蜡光，偶有白斑，具微香，托桂型至牡丹型，小型花，花径 6.0～6.5cm，花朵厚 3.0～3.5cm，花瓣 100 枚以上，紧实排列，花瓣小，中部雄蕊瓣短，与花瓣混生，花丝红色，花药黄色。叶片深绿色，革质，椭圆形，长 9.0～9.5cm，宽 3.0～3.5cm，先端尖，边缘叶齿尖。植株紧凑，矮性，生长旺盛。秋季始花，冬季盛花，春季可持续开花。

Abstract: Flowers dark red with waxy luster, occasionally with white markings, little fragrance, anemone to peony form, small size, 6.0-6.5cm across and 3.0-3.5cm deep with over 100 petals tightly arranged, central stamens short and mixed with petals, filaments red, anthers yellow. Leaves deep green, leathery, elliptic, margins sharply serrate. Plant compact, dwarf and growth vigorous. Starts to bloom from autumn, fully blooms in winter and continuously in spring.

HAR-35-11. 红城夜景
Hongcheng Yejing (Red City's Night Scenes)

回交组合：红山茶品种'玉之浦'×杜鹃红山茶 F_1 代新品种'红屋积香'（*C. japonica* cultivar 'Tama-no-ura'× *C. azalea* F_1 new hybrid 'Hongwu Jixiang'）。

回交苗编号：YUJX-No.25。

性状：花朵黑红色，花瓣上具白色斑块，略具芳香，半重瓣型至紧实牡丹型，中型花，花径 8.5～10.0cm，花朵厚 5.0cm，外部花瓣 15～20 枚，呈 2～3 轮紧实排列，花瓣阔倒卵形，中部小花瓣约 10 枚，直立，雄蕊瓣很多，雄蕊基部近离生，散射状，花丝红色，花药金黄色。叶片深绿色，革质，椭圆形，长 9.0～9.5cm，宽 3.5～4.5cm，边缘叶齿浅。植株立性，紧凑，生长旺盛。秋季始花，冬季盛花，春季可持续开花。

> **Abstract:** Flowers dark red with white markings, slightly with fragrance, semi-double to tight peony form, medium size, 8.5-10.0cm across and 5.0cm deep with 15-20 exterior petals arranged tightly in 2-3 rows, petals broad obovate, about 10 small petals and a lot of petaloids at the center, stamens nearly free at the base, filaments scattered and red, anthers golden yellow. Leaves deep green, leathery, elliptic, margins shallowly serrate. Plant upright, compact and growth vigorous. Starts to bloom from autumn, fully blooms in winter and continuously in spring.

HAR-35-12. 黑红精灵
Heihong Jingling（Dark Red Elves）

回交组合： 红山茶品种'玉之浦'×杜鹃红山茶 F_1 代新品种'红屋积香'（*C. japonica* cultivar 'Tama-no-ura' × *C. azalea* F_1 new hybrid 'Hongwu Jixiang'）。

回交苗编号： YUJX-No.27+26。

性状： 花朵黑红色，略泛蜡光，偶显白边，单瓣型，微型花，花径4.0～5.0cm，花瓣5～6枚，阔倒卵形，外翻，雄蕊连生呈柱状，花丝红色，花药黄色。叶片浓绿色，革质，阔椭圆形，长7.0～8.0cm，宽4.0～4.5cm，基部楔形，边缘叶齿浅。植株立性，生长旺盛。秋季始花，冬季盛花，春季可持续开花。

Abstract: Flowers dark red with waxy luster, occasionally bordered with white edges, single form, miniature size, 4.0-5.0cm across with 5-6 petals, petals broad obovate, rolled outward, stamens united into a column, filaments red, anthers yellow. Leaves dark green, leathery, cuneate at the base, broad elliptic, margins shallowly serrate. Plant upright and growth vigorous. Starts to bloom from autumn, fully blooms in winter and continuously in spring.

HAR-35-13. 红城漫金
Hongcheng Manjin (Red City Overflowing Gold)

回交组合： 红山茶品种'玉之浦'× 杜鹃红山茶 F₁ 代新品种'红屋积香'（ *C. japonica* cultivar 'Tama-no-ura'× *C. azalea* F₁ new hybrid 'Hongwu Jixiang'）。

回交苗编号： YUJX-No.29。

性状： 花朵黑红色，中部小花瓣偶有白条纹，略具芳香，托桂型至松散牡丹型，中型至大型花，花径 8.5～11.5cm，花朵厚 5.0cm，外部花瓣 14～15 枚，呈 2 轮松散排列，花瓣倒卵形，先端微凹，中部有大花瓣 5～8 枚和少量雄蕊瓣，雄蕊散射，花丝红色，花药金黄色。叶片浓绿色，革质，椭圆形，长 7.5～8.5cm，宽 4.0～4.5cm，边缘叶齿稀疏。植株立性，紧凑，生长旺盛。秋季始花，冬季盛花，春季可持续开花。

Abstract: Flowers dark red, occasionally with white stripes at central small petals, slightly fragrant, anemone to loose peony form, medium to large size, 8.5-11.5cm across and 5.0cm deep with exterior 14-15 petals arranged loosely in 2 rows, petals obovate, apices emarginate, 5-8 large petals and a few petaloids at the center, stamens scattered, filaments red, anthers golden yellow. Leaves dark green, leathery, elliptic, margins sparsely serrate. Plant upright, compact and growth vigorous. Starts to bloom from autumn, fully blooms in winter and continuously in spring.

HAR-35-14. 田园风光
Tianyuan Fengguang (Rural Scenery)

杂交组合：红山茶品种'玉之浦'×杜鹃红山茶 F₁ 代新品种'红屋积香'（ *C. japonica* cultivar 'Tama-no-ura' × *C. azalea* F₁ new hybrid 'Hongwu Jixiang' ）。

回交苗编号：YUJX-No.31。

性状：花朵鲜艳的淡橘粉红色，雄蕊小花瓣具白条纹，松散的牡丹型，偶然出现单瓣型，小型至中型花，花径 7.0～9.5cm，花朵厚 4.5cm，外轮大花瓣约 6 枚，近圆形，长 5.4cm，宽 5.1cm，外翻，先端微凹，中部几枚小花瓣与簇生雄蕊混生，雌蕊退化，花丝极淡的粉白色，花药金黄色。叶片浓绿色，革质，椭圆形，长 8.0～9.0cm，宽 3.5～4.0cm，基部楔形，边缘叶齿浅而尖。植株立性，生长旺盛。秋季始花，冬季盛花，春季可持续开花。

Abstract: Flowers bright and light orange pink with white stripes, loose peony form, occasionally single form appeared, small to medium size, 7.0-9.5cm across and 4.5cm deep with 6 exterior large petals, exterior petals turned outward, apices emarginate, central some small petals mixed with the clustered stamens, pistil vestigial, filaments very pale pink, anthers golden yellow. Leaves dark green, leathery, elliptic, cuneate at the base, margins sharply and shallowly serrate. Plant upright and growth vigorous. Starts to bloom from autumn, fully blooms in winter and continuously in spring.

HAR-35-15. 福娃新衣
Fuwa Xinyi (Mascots' New Clothes)

回交组合：红山茶品种'玉之浦'× 杜鹃红山茶 F_1 代新品种'红屋积香'（*C. japonica* cultivar 'Tama-no-ura'× *C. azalea* F_1 new hybrid 'Hongwu Jixiang'）。

回交苗编号：YUJX-No.32。

性状：花朵红色至黑红色，具蜡光，边缘有明显的窄白边，单瓣型，中型至大型花，花径9.5～10.5cm，花瓣6枚，阔倒卵形，厚质，外翻，边缘略皱，先端微凹，雄蕊基部连生，呈柱状，花丝红色，花药金黄色。叶片深绿色，叶背面淡绿色，革质，椭圆形，长9.5～10.5cm，宽4.0～5.0cm，边缘叶齿尖。植株立性，枝条开张，生长旺盛。秋季始花，冬季盛花，春季可持续开花。

Abstract: Flowers red to dark red with waxy luster, bordered with obvious narrow white edges, single form, medium to large size, 9.5-10.5cm across with 6 petals, petals broad obovate, thick texture, rolled outward, slightly crinkled at the edges, apices emarginate, stamens united into a column at the base, filaments red, anthers golden yellow. Leaves deep green, leathery, elliptic, margins sharply serrate. Plant upright, branches spread and growth vigorous. Starts to bloom from autumn, fully blooms in winter and continuously in spring.

HAR-35-16. 红绒公主
Hongrong Gongzhu (Red Velvet Princess)

回交组合：红山茶品种'玉之浦'× 杜鹃红山茶 F₁代新品种'红屋积香'（ *C. japonica* cultivar 'Tama-no-ura' × *C. azalea* F₁ new hybrid 'Hongwu Jixiang'）。

回交苗编号：YUJX-No.33。

性状：花朵黑红色，泛蜡光，瓣缘偶有很窄的白边，半重瓣型至牡丹型，小型花，花径 6.0～7.5cm，花朵厚 4.3cm，大花瓣 18～21 枚，呈 3～4 轮排列，阔倒卵形，波浪状，先端微凹，雄蕊簇生，多呈 5 簇，花丝淡红色，花药金黄色，开花稠密。叶片浓绿色，革质，椭圆形，长 8.0～9.0cm，宽 4.5～5.0cm，基部楔形，边缘叶齿浅。植株立性，生长旺盛。秋季始花，冬季盛花，春季可持续开花。

Abstract: Flowers dark red with waxy luster, occasionally bordered with very narrow white edges, semi-double to peony form, small size, 6.0-7.5cm across and 4.3cm deep with 18-21 large petals arranged in 3-4 rows, petals broad obovate, wavy, apices emarginate, stamens always clustered into 5 clusters, filaments pale red, anthers golden yellow, bloom dense. Leaves dark green, leathery, elliptic, cuneate at the base, margins shallowly serrate. Plant upright and growth vigorous. Starts to bloom from autumn, fully blooms in winter and continuously in spring.

HAR-35-17. 红城金库
Hongcheng Jinku (Red City's Gold Vault)

回交组合：红山茶品种'玉之浦'×杜鹃红山茶F_1代新品种'红屋积香'（*C. japonica* cultivar 'Tama-no-ura'× *C. azalea* F_1 new hybrid 'Hongwu Jixiang'）。

回交苗编号：YUJX-No.35。

性状：花朵黑红色至紫红色，具蜡光，具微清香，半重瓣型至牡丹型，微型至小型花，花径5.8～7.0cm，花朵厚3.8cm，外部花瓣6～7枚，阔倒卵形，外翻，先端微凹，有时中部出现小花瓣，与雄蕊混生，直立或扭曲，雄蕊近散射，花丝红色，花药金黄色。叶片浓绿色，光亮，革质，椭圆形，长9.0～10.0cm，宽3.5～5.0cm，边缘叶齿浅。植株立性，紧凑，生长旺盛。秋季始花，冬季盛花，春季可持续开花。

Abstract: Flowers dark red to purple red with waxy luster, slight faint scent, semi-double to peony form, miniature to small size, 5.8-7.0cm across and 3.8cm deep with 6-7 exterior petals, petals broad obovate, rolled outward, apices emarginate, sometimes small petals appear at the center, erected and mixed with stamens, stamens nearly scattered, filaments red, anthers golden yellow. Leaves dark green, leathery, elliptic, margins shallowly serrate. Plant upright, compact and growth vigorous. Starts to bloom from autumn, fully blooms in winter and continuously in spring.

HAR-35-18. 行为艺术
Xingwei Yishu (Performance Art)

回交组合： 红山茶品种'玉之浦'× 杜鹃红山茶 F_1 代新品种'红屋积香'（*C. japonica* cultivar 'Tama-no-ura'× *C. azalea* F_1 new hybrid 'Hongwu Jixiang'）。

回交苗编号： YUJX-No.37。

性状： 花朵艳红色，花瓣下部边缘具窄白边，半重瓣型，小型至中型花，花径 7.0～8.5cm，花朵厚 4.0cm，花瓣 13～15 枚，呈 2 轮排列，长倒卵形，外翻或半直立，雄蕊簇生，直立，花丝淡黄色，花药黄色，开花稠密。叶片浓绿色，革质，阔椭圆形，长 8.0～9.0cm，宽 5.0～5.5cm，基部楔形，边缘叶齿稀。植株立性，生长旺盛。秋季始花，冬季盛花，春季可持续开花。

Abstract: Flowers bright red with narrow white borders at the lower parts of petals, semi-double form, small to medium size, 7.0-8.5cm across and 4.0cm deep with 13-15 petals arranged in 2 rows, petals long obovate, rolled outward or semi-erected, stamens clustered, erected, filaments pale yellow, anthers yellow, bloom dense. Leaves dark green, leathery, broad elliptic, cuneate at the base, margins sparsely serrate. Plant upright and growth vigorous. Starts to bloom from autumn, fully blooms in winter and continuously in spring.

⦿ 回交组合 HAR-36. 红山茶品种'白斑康乃馨' × 杜鹃红山茶 F_1 代新品种'红屋积香'

HAR-36. *C. japonica* cultivar 'Ville de Nantes' × *C. azalea* F_1 new hybrid 'Hongwu Jixiang'

本回交组合将介绍1个新品种。

我们曾介绍过红山茶品种'白斑康乃馨'×杜鹃红山茶这一杂交组合，共获得5个杂交新品种（见第一部书第241～247页），证明'白斑康乃馨'是一个非常好的杂交亲本。

'红屋积香'是从杜鹃红山茶×红山茶品种'克瑞墨大牡丹'杂交组合中获得的一个既具芳香，也具四季开花性状的杂交新品种（见第一部书第72～74页）。从理论上说，'红屋积香'中杜鹃红山茶和'克瑞墨大牡丹'的基因频率各为50%。

本回交组合所得到的新品种中杜鹃红山茶和'克瑞墨大牡丹'的基因频率将分别下降到25%，因此，本回交组合试图探索在杂交亲本的杜鹃红山茶基因频率如此低的情况下，其杂交新品种的四季开花和花朵芳香性状是否依然可以表达。

从获得的这个杂交新品种看，花朵不是太漂亮，但略具芳香，而且四季开花的性状也获得了遗传。

×

One new hybrid will be introduced in this backcross combination.

We have introduced the cross-combination between *C. japonica* cultivar 'Ville de Nantes' and *C. azalea*, and obtained five new hybrids (See p.241-247, our first book), which proved that 'Ville de Nantes' is a very good cross parent.

'Hongwu Jixiang' is a new hybrid with fragrance and ever-blooming traits that obtained from the cross-combination of *C. azalea* × 'Kramer's Supreme' (See p.72-74, our first book). Theoretically, the gene frequencies of *C. azalea* and 'Kramer's Supreme' are 50 % respectively.

The gene frequencies of both *C. azalea* and 'Kramer's Supreme' in the new hybrid obtained from this cross-combination will drop to 25% respectively, therefore, this cross-combination attempts to explore whether the characteristics of ever-blooming and flower fragrance of the new hybrid can still be expressed at such low *C. azalea*'s gene frequencies of the hybrid parents.

The results of the hybridization show that the flowers of the new hybrid are not too pretty, but slight fragrant and the characteristic of ever-blooming is inherited.

HAR-36-1. 书香之家
Shuxiang Zhijia (Literary Family)

回交组合： 红山茶品种'白斑康乃馨'×杜鹃红山茶 F_1 代新品种'红屋积香'（ *C. japonica* cultivar 'Ville de Nantes' × *C. azalea* F_1 new hybrid 'Hongwu Jixiang'）。

回交苗编号： NTJX-No.7。

性状： 花朵红色，略泛紫红色调，微香，半重瓣型，中型花，花径8.0～9.5cm，花瓣10～12枚，呈2轮排列，花瓣倒卵形，波浪状，略皱褶，外翻，先端近圆形，雄蕊基部略连生，呈辐射状，花丝淡粉红色，花药黄色。叶片浓绿色，革质，椭圆形，长8.5～9.5cm，宽3.5～4.0cm，基部楔形。植株紧凑，生长旺盛。秋季始花，冬季盛花，春季可持续开花。

> **Abstract:** Flowers red slightly with purple red hue, slightly fragrant, semi-double form, medium size, 8.0-9.5cm across with 10-12 petals arranged in 2 rows, petals obovate, wavy and slightly wrinkled, apices nearly round, stamens radial, filaments pale pink, anthers yellow. Leaves dark green, leathery, elliptic, cuneate at the base. Plant compact and growth vigorous. Starts to bloom from autumn, fully blooms in winter and continuously in spring.

⊙ 回交组合 HAR-37. 红山茶品种'聚香'× 杜鹃红山茶 F₁ 代新品种'夏梦文清'

HAR-37. *C. japonica* cultivar 'Juxiang' × *C. azalea* F₁ new hybrid 'Xiameng Wenqing'

本回交组合将介绍 4 个新品种。

红山茶品种'聚香'是通过'锯叶椿'品种的机遇苗选育获得的。它的花朵深粉红色，略泛紫色调，虽然是单瓣型，但是具有较浓的香气，因此，我们把它作为母本，与杜鹃红山茶 F₁ 代杂交种'夏梦文清'进行回交，目的不仅是要使杂交种花朵呈紫红色、带芳香，而且也要使杂交种重复开花。

可以看到，所获得的这 4 个回交新品种，花色全都为红色带紫色调，花朵具芳香，而且冬初就可以开花。这进一步证明，即便在杜鹃红山茶的基因频率只有 25% 的情况下，获得的新品种的花性状明显倾向于母本'聚香'品种，而开花期则倾向于父本'夏梦文清'品种。这是一个出色的回交组合。

×

Four new hybrids will be introduced in this backcross combination.

C. japonica cultivar 'Juxiang' (Gathering Fragrance) was obtained from a chance seedling of 'Nokogiriba-tsubaki'. Its flowers are deep pink, slightly with purple hue, single form, but with a strong aroma, therefore, the aim we used it as female parent to cross with *C. azalea*'s hybrid 'Xiameng Wenqing' is to obtain the new hybrids that flowers are purple-red, fragrant and ever-blooming.

It can be seen that the four new hybrids obtained are all red with purple hue and fragrant in flowers and start to bloom in early-winter. It further proves that even in the case of 25% *C. azalea*'s gene frequency, the flower traits of the new hybrids obtained obviously tend to their female parent 'Juxiang' and the blooming periods of them tend to their male parent 'Xiameng Wenqing'. This is an excellent backcross combination.

HAR-37-1. 香菱紫衣
Xiangling Ziyi (Miss Xiangling's Purple Dress)

回交组合： 红山茶品种'聚香'× 杜鹃红山茶 F_1 代新品种'夏梦文清'（*C. japonica* cultivar 'Juxiang' × *C. azalea* F_1 new hybrid 'Xiameng Wenqing'）。

回交苗编号： JXWQ-No.3。

性状： 花朵紫红色，略带芳香，半重瓣型，中型花，花径 8.5～10.0cm，花朵厚 4.5cm，大花瓣 16～18 枚，呈 3 轮排列，花瓣阔倒卵形，波浪状，外翻，瓣面皱褶，中部雄蕊瓣略有白条纹，雄蕊基部略连生，花丝红色，开花稠密。叶片深绿色，革质，椭圆形，长 7.0～8.5cm，宽 3.5～4.0cm，基部楔形，边缘具浅齿。植株紧凑，生长旺盛。冬初始花，一直可持续到翌年春季。

Abstract: Flowers purple red with slight fragrance, semi-double form, medium size, 8.5-10.0cm across and 4.5cm deep with 16-18 large petals arranged in 3 rows, petals broad obovate, wavy, rolled outward, surfaces wrinkled, central petaloids with a few white stripes, stamens slightly united at the base, filaments red, bloom dense. Leaves deep green, leathery, elliptic, cuneate at the base, margins shallowly serrate. Plant compact and growth vigorous. Starts to bloom from early-winter and continues to bloom until the spring of the following year.

HAR-37-2. 古朴清幽
Gupu Qingyou (Ancient Beauty & Tranquility)

回交组合：红山茶品种'聚香' × 杜鹃红山茶 F_1 代新品种'夏梦文清'（ *C. japonica* cultivar 'Juxiang' × *C. azalea* F_1 new hybrid 'Xiameng Wenqing'）。

回交苗编号：JXWQ-No.4+6。

性状：花朵红色，略泛紫色调，偶有白斑，略带芳香，半重瓣型至牡丹型，中型花，花径 8.5～10.0cm，花朵厚 5.2～5.5cm，大花瓣 13～15 枚，呈 3 轮排列，花瓣阔倒卵形，瓣面可见黑红脉纹，中部雄蕊瓣半直立，与雄蕊混生，开花稠密。叶片深绿色，革质，阔椭圆形，长 8.5～9.0cm，宽 3.8～4.0cm，边缘具浅齿。植株紧凑，矮性，生长旺盛。冬初始花，一直可持续到翌年春季。

Abstract: Flowers red, slightly with purple hue, occasionally with white markings, slight fragrant, semi-double to peony form, medium size, 8.5-10.0cm across and 5.2-5.5cm deep with 13-15 large petals arranged in 3 rows, petals broad obovate, dark red veins visible on the surfaces, central petaloids semi-erected and mixed with stamens, bloom dense. Leaves deep green, leathery, broad elliptic, margins shallowly serrate. Plant compact, dwarf and growth vigorous. Starts to bloom from early-winter and continues to bloom until the spring of the following year.

HAR-37-3. 紫墙古宅
Ziqiang Guzhai (Purple Wall Ancient Messuage)

回交组合：红山茶品种'聚香'× 杜鹃红山茶 F_1 代新品种'夏梦文清'（*C. japonica* cultivar 'Juxiang' × *C. azalea* F_1 new hybrid 'Xiameng Wenqing'）。

回交苗编号：JXWQ-No.7。

性状：花朵紫红色，具白条纹，晴天时略带芳香，半重瓣型至牡丹型，中型花，花径9.5～10.0cm，花朵厚5.0～5.5cm，外部大花瓣15～16枚，呈2～3轮松散排列，花瓣阔倒卵形，略皱褶，中部雄蕊瓣10枚以上，直立，与雄蕊混生，花丝淡黄色，花药金黄色，开花稠密。叶片深绿色，厚革质，阔椭圆形，长9.0～10.0cm，宽5.0～5.5cm，边缘具浅齿。植株紧凑，生长旺盛。冬初始花，一直可持续到翌年春季。

Abstract: Flowers purple red with white stripes, slightly with fragrance in fine day, semi-double to peony form, medium size, 9.5-10.0cm across and 5.0-5.5cm deep with 15-16 exterior large petals arranged loosely in 2-3 rows, petals broad obovate, surfaces wrinkled, over 10 central petaloids erected and mixed with the stamens, filaments pale yellow, anthers golden yellow, bloom dense. Leaves deep green, heavy leathery, broad elliptic, margins shallowly serrate. Plant compact and growth vigorous. Starts to bloom from early-winter and continues to bloom until the spring of the following year.

HAR-37-4. 龙章凤姿
Longzhang Fengzi (Dragon's Literate & Phoenix's Posture)

回交组合： 红山茶品种'聚香'× 杜鹃红山茶 F_1 代新品种'夏梦文清'（ C. japonica cultivar 'Juxiang'× C. azalea F_1 new hybrid 'Xiameng Wenqing'）。

回交苗编号： JXWQ-No.9。

性状： 花朵紫红色，具淡清香，偶有隐约的白斑，牡丹型，大型至巨型花，花径11.0～13.2cm，花朵厚6.0cm，大花瓣18～20枚，呈3轮排列，花瓣阔倒卵形，长6.0cm，宽5.0cm，波浪状，外翻，先端凹，中部小花瓣皱褶，与簇生雄蕊混生，花丝淡粉红色，花药淡黄色。叶片浓绿色，厚革质，椭圆形，长9.0～10.5cm，宽5.9～5.5cm，基部楔形。植株紧凑，生长旺盛。冬初始花，一直可持续到翌年春季。

Abstract: Flowers purple red with slight fragrance, occasionally with faint white stripes, peony form, large to very large size, 11.0-13.2cm across and 6.0cm deep with 18-20 large petals arranged in 3 rows, petals broad obovate, wavy, rolled outward, central small petals wrinkled, mixed with clustered stamens, filaments pale pink, anthers pale yellow. Leaves deep green, heavy leathery, elliptic, cuneate at the base. Plant compact and growth vigorous. Starts to bloom from early-winter and continues to bloom until the spring of the following year.

⦿ 回交组合 HAR-38. 越南抱茎茶 × 杜鹃红山茶 F₁ 代新品种'夏日红霞'
HAR-38. *C. amplexicaulis* × *C. azalea* F₁ new hybrid 'Xiari Hongxia'

本回交组合将介绍1个新品种。

本书有越南抱茎茶与常规茶花品种之间的杂交组合（见本书HB杂交组合）。因为本组合有杜鹃红山茶F₁代新品种参与，所以将该组合列入回交组合。

'夏日红霞'是从杜鹃红山茶 × 红山茶品种'霍伯'组合中获得的杂交种，其花色黑红色，花朵牡丹型、花径大，非常特别（见本书HA-35）。

正因为如此，本组获得的新品种尽管花径不大，但是从母本越南抱茎茶那里获得了大叶片、生长旺盛的性状，从父本'夏日红霞'那里获得了黑红花色和四季开花的性状。

One new hybrid will be introduced in this backcross-combination.

There are the cross-combinations of *C. amplexicaulis* and normal camellia cultivars in this book (See HB in this book). Because *C. azalea* F₁ new hybrid was involved in this combination, the combination is categorized into this backcross combination.

'Xiari Hongxia' was obtained from the cross-combination between *C. azalea* and *C. japonica* cultivar 'Bob Hope' and its flowers are dark red, peony form and large size which are very special (See HA-35-5, this book).

Thus, though the flower diameter of the new hybrid obtained is not large, the hybrid has obtained the characteristics of large leaves and vigorous growth from its female parent *C. amplexicaulis*, and the characteristics of dark red flower color and ever-blooming from its male parent 'Xiari Hongxia'.

HAR-38-1. 火山丽景
Huoshan Lijing (Beautiful Scene of Volcano)

回交组合： 越南抱茎茶 × 杜鹃红山茶 F_1 代新品种'夏日红霞'（*C. amplexicaulis* × *C. azalea* F_1 new hybrid 'Xiari Hongxia'）。

回交苗编号： BXRHX-No.4。

性状： 花朵黑红色，泛蜡光，单瓣型，小型至中型花，花径 7.0～8.0cm，花瓣 7～8 枚，花瓣厚质，阔倒卵形，雄蕊基部连生呈柱状，花丝红色，花药金黄色。嫩叶紫红色，成熟叶中等绿色，厚革质，阔椭圆形，长 13.5～14.5cm，宽 5.0～6.5cm，基部楔形，边缘具浅齿。植株紧凑，枝叶稠密，生长旺盛。夏末始花，秋季盛花，冬季零星开花。

Abstract: Flowers dark red with waxy luster, single form, small to medium size, 7.0-8.0cm across with 7-8 petals, petals thick texture, broad obovate, stamens united into a column at the base, filaments red and anthers golden yellow. Tender leaves purple red, mature leaves normal green, heavy leathery, broad elliptic, cuneate at the base, margins shallowly serrate. Plant compact, branches dense and growth vigorous. Starts to bloom from late-summer, fully blooms in autumn and sporadically in winter.

越南抱茎茶杂交新品种育种概况

The Breeding Outline on the *C. amplexicaulis*' New Hybrids

顾名思义，越南抱茎茶原产越南。在山茶属植物分类上它隶属于古茶组，是一个花朵深粉红色至淡紫红色、单瓣型至半重瓣型、花柄长、开花稠密、花期长、叶片硕大、基部抱茎、植株高大、颇具育种价值的山茶原生种（见第一部书第 55～60 页）。自 1997 年该原种引进我国以来，我们便开始了以培育乔木性观赏茶花为目的的育种工作。本章的杂交是越南抱茎茶与常规茶花品种之间的杂交，其杂交方式有 2 种，见下图：

As the name implies, *C. amplexicaulis* is originated in Vietnam. It belongs to Section *Archecamellia* in the taxonomy of Genus *Camellia*. The characteristics of the camellia species are that the flowers are deep pink to violet red, single to semi-double form with long pedicels, bloom is dense, the blooming period is long, leaves are large and amplexicaul at the base and the plant is tall, which make it have higher breeding value (See p.55-60, our first book). Since the species was introduced into China in 1997, we had started breeding ornamental camellias which are arbor. The hybridizations in this chapter refer to the crosses between *C. amplexicaulis* and normal camellia cultivars. There are two patterns of the crossing, as shown in the figures below:

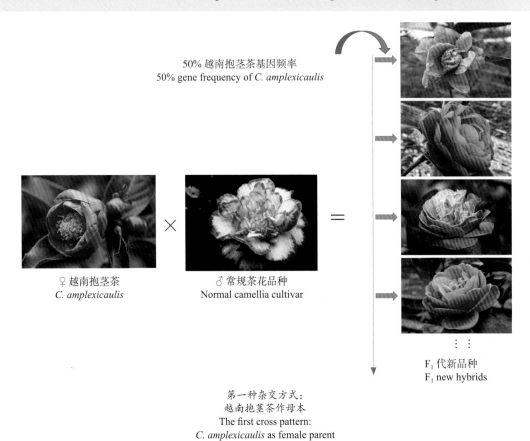

第一种杂交方式：
越南抱茎茶作母本
The first cross pattern:
C. amplexicaulis as female parent

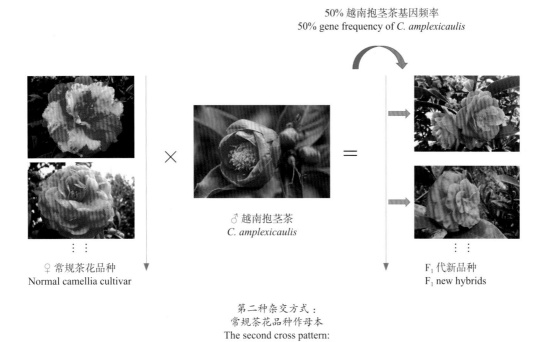

第二种杂交方式：
常规茶花品种作母本
The second cross pattern:
Normal camellia cultivar as female parent

在我们的第一部书中，曾介绍过 55 个越南抱茎茶和常规茶花品种或者山茶物种之间的杂交新品种，它们涉及 20 个杂交组合（见第一部书第 399～484 页）。

本书将介绍 49 个越南抱茎茶杂交新品种。这些新品种共涉及 27 个杂交组合，其中有 6 个杂交组合是第一部书已经有的（见本章 HB-02、HB-08、HB-09、HB-12、HB-14、HB-17）。因为这 6 个组合是不连贯的，因此，它们的组合编号也是不连贯的，而且，这些组合中的新品种的编号将接续第一部书的序列编号，所以，它们的编号都不是从 1 开始的。

如果加上第一部书列出的杂交组合和获得的杂交新品种，到目前为止，越南抱茎茶杂交组合数已达 41 个，获得的杂交新品种总数已达 104 个。

Fifty five new hybrids from cross-combinations between *C. amplexicaulis* and normal camellia cultivars or camellia species were introduced in our first book and the new hybrids were involved to 20 cross-combinations (See p.399-484, our first book).

Forty nine *C. amplexicaulis*' new hybrids will be introduced in this book. These new hybrids are involved to 27 cross-combinations, of these, six hybrid combinations are already present in our first book (See HB-02, HB-08, HB-09, HB-12, HB-14, HB-17 in this chapter). Because these combinations are discontinuous, their combination numbers are also discontinuous, moreover, the new hybrids in these combinations will continue the serial numbers of our first book, so their numbers don't start from No.1.

If the cross-combinations and the new hybrids listed in our first book were added, 41 cross-combinations and 104 new hybrids would be obtained in total until now.

在这些杂交组合中，无论越南抱茎茶是作母本还是作父本，所获得的杂交种，其花朵性状主要倾向于常规茶花品种，而开花期、叶片大小、株形则倾向于越南抱茎茶。另外，在生长态势和抗性方面，杂交新品种明显强于越南抱茎茶。

越南抱茎茶杂交新品种的开花期列表如表3：

In these cross-combinations, whether *C. amplexicaulis* is as female parent or as male parent, the new hybrids obtained mainly tend to normal camellia cultivars in flower characteristics, but tend to *C. amplexicaulis* in the blooming period, leaf size and plant shape. In addition, the growth and resistance of the new hybrids are obviously better than *C. amplexicaulis*.

The blooming period of *C. amplexicaulis*' new hybrids is listed in the following table:

表3 越南抱茎茶及其杂交新品种花期的比较
Table 3　Comparison on the Blooming Periods of *C. amplexicaulis* and Its New Hybrids

月份 Month	01 Jan.	02 Feb.	03 Mar.	04 Apr.	05 May	06 Jun.	07 Jul.	08 Aug.	09 Sept.	10 Oct.	11 Nov.	12 Dec.	全花期/月数 Entire blooming /Mon.	盛花期/月数 Fully blooming /Mon.	无花期/月数 No blooming /Mon.
越南抱茎茶 *C. amplexicaulis*	盛	盛	盛	末	末	始	始	始	始	始	始	无	7	3	5
越南抱茎茶杂交新品种 *C. amplexicaulis*' new hybrids	盛	盛	盛	末	末	始	始	始	始	始	始	盛	8	4	4

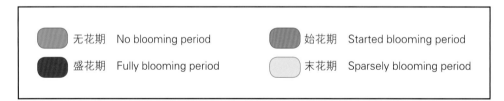

■ 无花期　No blooming period　　■ 始花期　Started blooming period
■ 盛花期　Fully blooming period　　□ 末花期　Sparsely blooming period

从上表可以看出，越南抱茎茶杂交新品种与其亲本越南抱茎茶在开花期上大体一致，但杂交种的盛花期和全花期要长一个月。更为重要的是，越南抱茎茶杂交新品种的花色更艳、花瓣更多、花朵更大、开花更稠密，而且树体高大，非常适合于园林美化。

It can be seen from the above table that the blooming period of the *C. amplexicaulis*' new hybrids is basically as same as what of its parent *C. amplexicaulis*, but the fully blooming period and the entire blooming period of the new hybrids are one month longer than *C. amplexicaulis*' respectively. It is very important that the *C. amplexicaulis*' new hybrids are brighter in flower color, more in petal quantity, larger in flower diameter and denser in blooming and taller in plant shape, which makes the new hybrids very suitable for landscaping.

应该指出的是，在我们的第一部书中下列 6 个杂交组合的品种数分别为：HB-02. 有 5 个（见第一部书第 406～413 页）；HB-08. 有 4 个（见第一部书第 432～437 页）；HB-09. 有 3 个（见第一部书第 438～441 页）；HB-12. 有 1 个（见第一部书第 446～447 页）；HB-14. 有 9 个（见第一部书第 450～462 页）；HB-17. 有 1 个（见第一部书第 472～473 页）。因此，本书中这些组合所获得的新品种的编号将接续第一部书的编号，品种编号不是从 1 开始的。

It should be noted that, in our first book, there are five hybrids in HB-02. (p.406-413); four hybrids in HB-08. (p. 432-437); three hybrids in HB-09. (p.438-441); one hybrid in HB-12. (p.446-447); nine hybrids in HB-14. (p.450-462); one hybrid in HB-17. (p.472-473). Therefore, the new hybrids obtained from these combinations in this book will follow the numbers of our first book, they are do not start from No. 1.

揽月阁
(Lanyuege) HB-27-3

粉玉露
(Fenyulu) HB-26-4

古殿
(Gudian) HB-24-1

粉增秀
(Fenzengxiu) HB-12-2

越南抱茎茶杂交新品种详解
The Detailed Descriptions for the *C. amplexicaulis*' New Hybrids

HB-02-6. 爱心 Aixin (Loving Heart)

杂交组合： 越南抱茎茶 × 非云南山茶杂交种'罗伯特卡特博士'（*C. amplexicaulis* × Non-*reticulata* hybrid 'Dr Robert K. Cutter'）。

杂交苗编号： BLUO-No.8。

性状： 花朵桃红色，略带芳香，牡丹型，大型花，花径 11.0～12.8cm，花朵厚 5.8cm，外部大花瓣 23～25 枚，呈 3～4 轮松散排列，花瓣外翻，内部花瓣皱褶，雄蕊簇生。叶片绿色，厚革质，椭圆形，长 17.8～18.9cm，宽 6.5～7.8cm，叶脉明显，叶齿浅。植株立性，高大，枝条稠密，生长旺盛。冬末至翌年春季开花。

Abstract: Flowers peach red, slightly fragrant, peony form, large size, 11.0-12.8cm across and 5.8cm deep with 23-25 exterior large petals arranged loosely in 3-4 rows, petals rolled outward, central small petals crinkled and mixed with stamens. Leaves green, heavy leathery, elliptic, veins visible, margins shallowly serrate. Plant upright, tall, branches dense and growth vigorous. Blooms from late-winter to the spring of following year.

HB-02-7. 旺财
Wangcai (Flourish Treasures)

杂交组合： 越南抱茎茶 × 非云南山茶杂交种'罗伯特卡特博士'（C. amplexicaulis × Non-reticulata hybrid 'Dr Robert K. Cutter'）。

杂交苗编号： BLUO-No.36。

性状： 花朵红色，泛紫色调，花瓣边缘红色略淡，略带芳香，单瓣型，小型至中型花，花径7.0～8.0cm，花瓣7～8枚，花瓣边缘略皱褶，外翻，雄蕊基部连生呈柱状。叶片绿色，厚革质，长椭圆形，长16.5～17.8cm，宽7.0～8.5cm，叶脉明显，叶齿浅而尖。植株紧凑，高大，枝条稠密，生长旺盛。冬末至翌年春季开花。

Abstract: Flowers red with purple hue, light red at the petal edges, slightly fragrant, single form, small to medium size, 7.0-8.0cm across with 7-8 petals, petals rolled outward, slightly crinkled, stamens united into a column at the base. Leaves green, heavy leathery, long elliptic, veins visible, margins shallowly and sharply serrate. Plant compact, tall, branches dense and growth vigorous. Blooms from late-winter to the spring of following year.

HB-02-8. 淳朴
Chunpu (Unsophistication Simplicity)

杂交组合： 越南抱茎茶 × 非云南山茶杂交种'罗伯特卡特博士'（*C. amplexicaulis* × Non-*reticulata* hybrid 'Dr Robert K. Cutter'）。

杂交苗编号： BLUO-No.37。

性状： 花朵红色，花瓣边缘淡红色，略带芳香，牡丹型，中型至大型花，花径 8.5～11.5cm，花朵厚 5.5cm，外部大花瓣约 18 枚，呈 3 轮排列，花瓣厚质，外翻，中部小花瓣皱褶，与雄蕊混生。叶片绿色，厚革质，长椭圆形，长 15.5～16.8cm，宽 6.0～8.0cm，叶脉明显，叶齿浅而尖。植株立性，高大，枝条稠密，生长旺盛。冬末至翌年春季开花。

Abstract: Flowers red with light red at the petal edges, slightly fragrant, peony form, medium to large size, 8.5-11.5cm across and 5.5cm deep with about 18 exterior large petals arranged in 3 rows, petals thick texture, rolled outward, central small petals crinkled and mixed with stamens. Leaves green, heavy leathery, long elliptic, veins visible, margins shallowly and sharply serrate. Plant upright, tall, branches dense and growth vigorous. Blooms from late-winter to the spring of following year.

HB-02-9. 欢庆
Huanqing (Festivity)

杂交组合： 越南抱茎茶 × 非云南山茶杂交种'罗伯特卡特博士'(C. amplexicaulis × Non-reticulata hybrid 'Dr Robert K. Cutter')。

杂交苗编号： BLUO-No.48。

性状： 花朵红色，中部雄蕊瓣淡红色，略带芳香，牡丹型至托桂型，大型花，花径11.5～13.0cm，花朵厚6.5cm，外部大花瓣23～26枚，呈4～5轮松散排列，花瓣阔倒卵形，中部小花瓣扭曲，与雄蕊混生。叶片绿色，嫩叶紫红色，厚革质，椭圆形，长16.5～17.5cm，宽6.0～7.5cm，叶脉明显，中脉黄色，叶齿浅。植株紧凑，高大，枝条稠密，生长旺盛。冬末至翌年春季开花。

Abstract: Flowers red, central petaloids light red, slightly fragrant, peony to anemone form, large size, 11.5-13.0cm across and 6.5cm deep with 23-26 large petals arranged loosely in 4-5 rows, petals broad obovate, central small petals twisted, mixed with stamens. Leaves green, tender leaves purple red, heavy leathery, elliptic, veins visible, midribs yellow, margins shallowly serrate. Plant compact, tall, branches dense and growth vigorous. Blooms from late-winter to the spring of following year.

HB-08-5. 粉元春
Fenyuanchun (Pink Early Spring)

杂交组合：越南抱茎茶 × 红山茶品种'花妮子'（*C. amplexicaulis* × *C. japonica* cultivar 'Little Bit Var.'）。

杂交苗编号：BHNZ-No.9。

性状：花朵淡粉红色，略显紫色调，花芯小花瓣偶有模糊的白斑，牡丹型，中型至大型花，花径 9.0～12.0cm，花朵厚 5.5cm，大花瓣 20～32 枚，阔倒卵形，长 5.5cm，宽 5.0cm，呈 4～5 轮松散排列，花瓣外翻，波浪状，中部小花瓣与雄蕊混生，瓣面皱褶，雄蕊簇生。叶片中等绿色，厚革质，长椭圆形，长 12.0～13.5cm，宽 5.0～6.0cm，叶脉明显，叶齿尖。植株立性，高大，生长旺盛。冬末至翌年春季开花。

Abstract: Flowers light pink, slightly with purple hue, occasionally some faint white markings at the central small petals, peony form, medium to large size, 9.0-12.0cm across and 5.5cm deep with 20-32 large petals arranged loosely in 4-5 rows, petals rolled outward, wavy, central small petals mixed with stamens, surfaces crinkled. Leaves normal green, heavy leathery, long elliptic, veins obvious, margins sharply serrate. Plant upright, tall and growth vigorous. Blooms from late-winter to the spring of following year.

HB-09-4. 旺德福
Wangdefu (Wonderful)

杂交组合：越南抱茎茶 × 红山茶品种'皇家天鹅绒'（*C. amplexicaulis* × *C. japonica* cultivar 'Royal Velvet'）。

杂交苗编号：BHJ-No.3。

性状：花朵黑红色，单瓣型，中型至大型花，花径9.8～12.0cm，花瓣7～8枚，阔倒卵形，瓣长5.5cm，瓣宽4.0cm，瓣面皱褶或波浪状，先端近圆形，偶有开裂，雄蕊基部连生，呈管状，花丝红色，开花稠密。叶片绿色，嫩叶紫红色，厚革质，阔椭圆形，长13.0～14.0cm，宽7.0～8.0cm，中脉和侧脉凹陷，先端下弯，叶齿细浅。植株立性，高大，枝叶稠密，生长旺盛。冬末至翌年春季开花。

Abstract: Flowers dark red, single form, medium to large size, 9.8-12.0cm across with 7-8 petals, petals broad obovate, wrinkled or wavy, stamens unite into a column at the base, bloom dense. Leaves green, tender leaves purple red, thick leathery, broad elliptic, midribs and lateral veins sunken, margins finely and shallowly serrate. Plant upright, tall and growth vigorous. Blooms from late-winter to the spring of following year.

HB-12-2. 粉增秀
Fenzengxiu（Adding Pink's Beauty）

杂交组合： 越南抱茎茶 × 红山茶品种'雅致'（ *C. amplexicaulis* × *C. japonica* cultivar 'Elegans'）。

杂交苗编号： BYZ-No.3。

性状： 花朵粉红色，渐向花朵中部变为淡粉红色，花芯处偶有白色瓣化雄蕊，玫瑰重瓣型，中型至大型花，花径9.5～11.5cm，花朵厚4.5cm，大花瓣25～30枚，阔倒卵形，呈3～5轮有序排列，偶有花瓣波浪状，开花稠密。叶片中等绿色，厚革质，长椭圆形，长12.5～14.5cm，宽5.5～6.5cm，叶脉明显，叶齿钝。植株开张，枝叶稠密，生长极旺盛。冬末至翌年春季开花。

Abstract: Flowers pink, fading to light pink at the center, occasionally some white petaloids appeared at the center, rose-double form, medium to large size, 9.5-11.5cm across and 4.5cm deep with 25-30 large petals arranged orderly in 3-5 rows, petals occasionally wavy, bloom dense. Leaves normal green, heavy leathery, long elliptic, veins obvious, margins obtusely serrate. Plant spread, branches and leaves dense and growth very vigorous. Blooms from late-winter to the spring of following year.

HB-14-10. 新郎装 Xinlangzhuang (Bridegroom's Clothing)

杂交组合：越南抱茎茶 × 红山茶品种'媚丽'(*C. amplexicaulis* × *C. japonica* cultivar 'Tama Beauty')。

杂交苗编号：BML-No.37。

性状：花朵深粉红色，泛紫色调，中部花瓣边缘具白斑，半重瓣型至玫瑰重瓣型，中型花，花径9.0～10.0cm，花朵厚5.5cm，大花瓣18～20枚，阔倒卵形，呈3轮排列，花瓣边缘外卷，瓣面略皱褶，雄蕊基部连生，开花稠密。叶片中等绿色，厚革质，长椭圆形，长13.5～14.5cm，宽5.0～5.5cm，叶脉明显，叶齿尖。植株立性，枝条稠密，生长旺盛。冬初至翌年春季开花。

Abstract: Flowers deep pink with purple hue and central petals with white markings at petal edges, semi-double to rose-double form, medium size, 9.0-10.0cm across and 5.5cm deep with 18-20 large petals arranged in 3 rows, petals turned outward and surfaces slightly wrinkled, stamens united into a short column at the base, bloom dense. Leaves heavy leathery, long elliptic, veins obvious and margins sharply serrate. Plant upright, branches dense and growth vigorous. Blooms from early-winter to the spring of following year.

HB-17-2. 白富美
Baifumei (White, Rich & Beautiful Girl)

杂交组合：博白大果油茶 × 越南抱茎茶（*C. gigantocarpa* × *C. amplexicaulis*）。

杂交苗编号：BOB-No.35。

性状：花朵白色，渐向花瓣基部呈嫩粉色，略带芳香，单瓣型，小型至中型花，花径7.0～8.0cm，花瓣7～9枚，阔倒卵形，边缘略皱，雄蕊基部连生，花朵整朵掉落。叶片绿色，厚革质，长椭圆形，长16.0～17.5cm，宽7.0～8.5cm，叶脉凹，叶齿密而浅。植株立性，高大，枝叶繁茂，生长旺盛。冬末至翌年春季开花。

Abstract: Flowers white, fading to tender pink at the base of petals, slightly fragrant, single form, small to medium size, 7.0-8.0cm across with 7-9 petals, petals broad obovate, edges slightly crinkled, stamens united at the base, flowers fall whole. Leaves green, heavy leathery, long elliptic, veins sunken, margins densely and shallowly serrate. Plant upright, tall, branches dense and growth vigorous. Blooms from late-winter to the spring of following year.

HB-17-3. 粉富美
Fenfumei (Pink, Rich & Beautiful Girl)

杂交组合： 博白大果油茶 × 越南抱茎茶（*C. gigantocarpa* × *C. amplexicaulis*）。

杂交苗编号： BOB-No.58。

性状： 花朵淡粉红色，渐向瓣缘变为白色，单瓣型，小型至中型花，花径6.0～8.0cm，花瓣6～7枚，阔倒卵形，长4.5cm，宽3.0cm，瓣面皱褶，雄蕊基部连生，花朵整朵掉落。嫩叶黑紫红色，成熟叶绿色，厚革质，长椭圆形，长15.0～16.5cm，宽6.0～7.5cm，叶脉明显，叶齿密而浅。植株立性，高大，枝叶繁茂，生长旺盛。冬末至翌年春季开花。

> **Abstract:** Flowers light pink, fading to white at the petal edges, single form, small to medium size, 6.0-8.0cm across with 6-7 petals, petals broad obovate, surfaces crinkled, stamens united at the base, flowers fall whole. Tender leaves dark purple red, mature leaves green, heavy leathery, long elliptic, veins obvious, margins densely and shallowly serrate. Plant upright, tall, branches dense and growth vigorous. Blooms from late-winter to the spring of following year.

HB-21-1. 梦归
Menggui (Realized Dream)

杂交组合： 红山茶品种'香神'×越南抱茎茶（*C. japonica* cultivar 'Scentsation'× *C. amplexicaulis*）。

杂交苗编号： XSB-No.3。

性状： 花朵深粉红色，具清香，半重瓣型，大型花，花径10.0～11.3cm，花朵厚4.0～4.5cm，花瓣25～30枚，呈4轮排列，花瓣阔倒卵形，外翻，先端近圆形，雄蕊基部略连生，开花稠密。叶片绿色，嫩叶紫红色，厚革质，阔椭圆形，长13.5～15.0cm，宽7.0～8.5cm，弓背状，中脉和侧脉明显，先端下弯，叶齿细浅。植株立性，高大，枝叶稠密，生长旺盛。冬末至翌年春季开花。

Abstract: Flowers deep pink with a faint scent, semi-double form, large size, 10.0-11.3cm across and 4.0-4.5cm deep with 25-30 petals arranged in 4 rows, petals broad obovate, apices nearly round, stamens united at the base, bloom dense. Leaves green, tender leaves purple red, heavy leathery, broad elliptic, arch shaped, midribs and lateral veins impressed, margins finely and shallowly serrate. Plant upright, tall, branches and leaves dense and growth vigorous. Blooms from late-winter to the spring of following year.

HB-22-1. 新颖 Xinying (Novelty)

杂交组合： 越南抱茎茶 × 红山茶品种'克瑞墨大牡丹'（ C. amplexicaulis × C. japonica cultivar 'Kramer's Supreme'）。

杂交苗编号： BK-No.7+8。

性状： 花朵深粉红色，中部花瓣淡粉红色，偶有白斑，牡丹型，中型至大型花，花径 9.0～12.0cm，外部大花瓣 12～13 枚，呈 2 轮排列，阔倒卵形，外翻，中部小花瓣直立，波浪状，与雄蕊混生，雄蕊簇生，开花稠密。叶片绿色，厚革质，阔椭圆形，长 15.0～17.0cm，宽 8.0～10.0cm，中脉和侧脉凹陷，叶齿细密。植株立性，高大，枝叶稠密，生长旺盛。冬末至翌年春季开花。

Abstract: Flowers deep pink, central petals light pink occasionally with white blotches, peony form, medium to large size, 9.0-12.0cm across with 12-13 large petals arranged in 2 rows, petals broad obovate, turned outward, central petals erected, wavy and clustered with stamens, bloom dense. Leaves green, thick leathery, broad elliptic, midribs and lateral veins sunken, margins finely and densely serrate. Plant upright, tall, branches and leaves dense and growth vigorous. Blooms from late-winter to the spring of following year.

HB-23-1. 新作品 Xinzuopin (New Masterwork)

杂交组合：越南抱茎茶 × 威廉姆斯杂交种'埃尔西朱瑞'（*C. amplexicaulis* × *C. x williamsii* hybrid 'Elsie Jury'）。

杂交苗编号：BAEX-No.4。

性状：花朵柔和的粉红色，渐向花瓣边缘呈模糊的粉白色，玫瑰重瓣型，中型花，花径9.0～10.0cm，花朵厚5.0cm，花瓣18～21枚，呈3轮排列，花瓣阔倒卵形，瓣面略皱，花芯处少量小花瓣与雄蕊混生，开花稠密。叶片绿色，厚革质，椭圆形，长11.5～13.0cm，宽6.0～7.0cm，中脉和侧脉凹陷，边缘叶齿细密。植株立性，高大，生长旺盛。冬初至翌年春季开花。

Abstract: Flowers soft pink, fading to faint pink white at the petal edges, rose-double form, medium size, 9.0-10.0cm across and 5.0cm deep with 18-21 petals arranged in 3 rows, petals broad obovate, petal surfaces slightly crinkled, a few small petals mixed with stamens at the center, bloom dense. Leaves green, heavy leathery, elliptic, midribs and lateral veins sunken, margins finely and densely serrate. Plant upright, tall and growth vigorous. Blooms from early-winter to the spring of following year.

HB-24-1. 古殿
Gudian (Ancient Temple)

杂交组合： 越南抱茎茶 × 红山茶品种'帝国之辉'（*C. amplexicaulis* × *C. japonica* cultivar 'Imperial Splendour'）。

杂交苗编号： BDGZH-No.1。

性状： 花朵深红色至黑红色，花芯处小花瓣淡红色，半重瓣型至牡丹型，大型花，花径 10.5～12.5cm，外部大花瓣20～25枚，呈4～5轮排列，阔倒卵形，厚质，外翻或波浪状，先端近圆，刚开放时，花芯呈球状，雄蕊簇生或基部略连生。叶片绿色，厚革质，阔椭圆形，长13.5～15.5cm，宽6.0～8.0cm，中脉和侧脉凹陷，叶齿细密。植株开张，枝叶稠密，生长旺盛。春季开花。

Abstract: Flowers deep red to dark red, central small petals light red, semi-double to peony form, large size, 10.5-12.5cm across with 20-25 exterior large petals arranged in 4-5 rows, petals broad obovate, thick texture, turned outward or wavy, apices nearly round, when just opening, a ball appears at the center, stamens clustered or slightly united at the base. Leaves green, thick leathery, broad elliptic, midribs and lateral veins sunken, margins finely and densely serrate. Plant spread, branches and leaves dense and growth vigorous. Blooms in spring.

HB-25-1. 酣悦
Hanyue (Drunken Delight)

杂交组合：越南抱茎茶 × 红山茶品种'红露珍'（*C. amplexicaulis* × *C. japonica* cultivar 'Hongluzhen'）。

杂交苗编号：BHLZ-No.2+3。

性状：花朵淡紫粉红色，偶具隐约的白斑块，玫瑰重瓣型，中型至大型花，花径9.5～12.5cm，花朵厚6.5cm，外部大花瓣20～24枚，呈4轮排列，阔倒卵形，花芯处小花瓣直立，皱褶，与少量雄蕊混生。叶片绿色，厚革质，阔椭圆形，长14.0～15.5cm，宽7.0～9.0cm，中脉和侧脉凹陷，叶齿细密。植株立性，高大，枝叶稠密，生长旺盛。冬末至翌年春季开花。

Abstract: Flowers light purple pink, occasionally with faint white markings, rose-double form, medium to large size, 9.5-12.5cm across and 6.5cm deep with 20-24 exterior large petals arranged in 4 rows, petals broad obovate, central petals erected, crinkled, mixed with a few stamens. Leaves green, thick leathery, broad elliptic, midribs and lateral veins sunken, margins finely and densely serrate. Plant upright, tall, branches and leaves dense and growth vigorous. Blooms from late-winter to the spring of following year.

HB-26-1. 风韵嫂
Fengyunsao (Charmed Sister-in Law)

杂交组合： 越南抱茎茶 × 红山茶品种'宽彩带'（*C. amplexicaulis* × *C. japonica* cultivar 'Margaret Davis'）。

杂交苗编号： BKCD-No.3。

性状： 花朵淡紫粉红色，牡丹型，中型至大型花，花径 8.0～11.5cm，花朵厚 5.8cm，大花瓣 20～24 枚，呈 4 轮排列，花瓣长倒卵形，外翻，瓣面皱褶，中部小花瓣扭曲，与雄蕊混生。叶片绿色，厚革质，长椭圆形，长 12.5～13.5cm，宽 5.5～6.0cm，中脉和侧脉凹陷，边缘叶齿密。植株紧凑，立性，高大，生长旺盛。冬初至翌年春季开花。

> **Abstract:** Flowers light purple pink, peony form, medium to large size, 8.0-11.5cm across and 5.8cm deep with 20-24 large petals arranged in 4 rows, petals long obovate, turned outward, petal surfaces crinkled, central small petals mixed with stamens. Leaves green, heavy leathery, long elliptic, midribs and lateral veins sunken, margins densely serrate. Plant compact, upright, tall and growth vigorous. Blooms from early-winter to the spring of following year.

HB-26-2. 紫红苑
Zihongyuan（Purple Red Dwelling）

杂交组合：越南抱茎茶 × 红山茶品种'宽彩带'（*C. amplexicaulis* × *C. japonica* cultivar 'Margaret Davis'）。

杂交苗编号：BKCD-No.9。

性状：刚开放时花朵紫红色，花芯球状，随着花朵的开放，渐向花芯变为粉红色，玫瑰重瓣型至完全重瓣型，中型至大型花，花径9.5～11.0cm，花朵厚4.5cm，大花瓣40～48枚，呈6～7轮排列，花瓣倒卵形，边缘内扣，花芯处小花瓣25枚以上。叶片绿色，厚革质，长椭圆形，长13.5～15.0cm，宽6.5～7.0cm，中脉和侧脉凹陷，边缘叶齿细浅。植株紧凑，生长旺盛。冬初至翌年春季开花。

Abstract: When just opening, flowers are purple red and ball-shaped at the center, as the flowers opening, petal color is fading to pink at the center, rose-double to formal double form, medium to large size, 9.5-11.0cm across and 4.5cm deep with 40-48 large petals arranged in 6-7 rows, petal edges turned inward, more than 25 small petals at the center. Leaves green, heavy leathery, long elliptic, midribs and lateral veins sunken, margins finely and shallowly serrate. Plant compact and growth vigorous. Blooms from early-winter to the spring of following year.

HB-26-3. 和睦
Hemu (Harmony)

杂交组合：越南抱茎茶 × 红山茶品种'宽彩带'（*C. amplexicaulis* × *C. japonica* cultivar 'Margaret Davis'）。

杂交苗编号：BKCD-No.12。

性状：花朵淡红色，略泛紫色调，具清晰的窄白边，玫瑰重瓣型至牡丹型，中型至大型花，花径 9.0～11.5cm，花朵厚6.5cm，大花瓣18～20枚，呈3～4轮排列，花瓣阔倒卵形，边缘内扣，中部小花瓣20枚以上，与雄蕊混生。叶片绿色，厚革质，长椭圆形，长12.5～15.5cm，宽6.0～7.5cm，中脉和侧脉凹陷，边缘叶齿细密。植株紧凑，立性，高大，生长旺盛。冬初至翌年春季开花。

Abstract: Flowers light red, slightly with purple hue and with narrow and distinct white borders, rose-double to peony form, medium to large size, 9.0-11.5cm across and 6.5cm deep with 18-20 large petals arranged in 3-4 rows, petal edges rolled inward, central small petals more than 20, mixed with stamens. Leaves green, heavy leathery, long elliptic, midribs and lateral veins sunken, margins finely and densely serrate. Plant compact, upright, tall and growth vigorous. Blooms from early-winter to the spring of following year.

HB-26-4. 粉玉露
Fenyulu (Pink Jade Dew)

杂交组合： 越南抱茎茶 × 红山茶品种'宽彩带'（*C. amplexicaulis* × *C. japonica* cultivar 'Margaret Davis'）。

杂交苗编号： BKCD-No.14+6+7。

性状： 花朵柔和的粉红色，略显模糊的紫色调，玫瑰重瓣型，中型至大型花，花径8.5～12.0cm，花朵厚5.2cm，大花瓣25～30枚，呈4～5轮排列，花瓣阔倒卵形，外翻，先端微凹，中部小花瓣与雄蕊混生。叶片绿色，厚革质，长椭圆形，长10.5～13.5cm，宽4.5～6.0cm，中脉和侧脉凹陷，边缘叶齿深尖。植株紧凑，立性，高大，生长旺盛。冬初至翌年春季开花。

Abstract: Flowers soft pink slightly with faint purple hue, rose-double form, medium to large size, 8.5-12.0cm across and 5.2cm deep with 25-30 large petals arranged in 4-5 rows, petals broad obovate, turned outward, apices emarginate, central small petals mixed with stamens. Leaves green, heavy leathery, long elliptic, midribs and lateral veins sunken, margins deeply and sharply serrate. Plant compact, upright, tall and growth vigorous. Blooms from early-winter to the spring of following year.

HB-27-1. 红诗文
Hongshiwen（Red Poem）

杂交组合： 越南抱茎茶 × 红山茶品种'全美国'（*C. amplexicaulis* × *C. japonica* cultivar 'All American'）。

杂交苗编号： BQM-No.3。

性状： 花朵红色至深红色，半重瓣型至玫瑰重瓣型，中型至大型花，花径 9.5～10.5cm，花朵厚 4.0～5.0cm，花瓣 25～30 枚，呈 5 轮排列，厚质，阔倒卵形，开花非常稠密。叶片中等绿色，嫩叶淡红色，阔椭圆形，长 14.0～15.5cm，宽 7.0～8.0cm，中脉明显，侧脉凹，叶缘齿细。植株紧凑，立性，生长旺盛。冬初至翌年春季开花。

Abstract: Flowers red to deep red, semi-double to rose-double form, medium to large size, 9.5-10.5cm across and 4.0-5.0cm deep with 25-30 thick petals arranged in 5 rows, petals broad obovate, bloom very dense. Leaves normal green, tender leaves light red, broad elliptic, midribs obvious and lateral veins sunken, margins finely serrate. Plant compact, upright and growth vigorous. Blooms from early-winter to the spring of following year.

HB-27-2. 春喜
Chunxi (Spring Delight)

杂交组合： 越南抱茎茶 × 红山茶品种'全美国'（*C. amplexicaulis* × *C. japonica* cultivar 'All American'）。

杂交苗编号： BQM-No.4。

性状： 花朵红色，中部花瓣淡红色，半重瓣型至玫瑰重瓣型，中型至大型花，花径9.5～11.5cm，花朵厚4.0～4.5cm，花瓣20～23枚，呈4～5轮排列，厚质，阔倒卵形，开花稠密。叶片中等绿色，嫩叶略泛红，椭圆形，长13.5～14.4cm，宽5.0～5.5cm，叶脉凹，叶缘齿细。植株紧凑，立性，生长旺盛。冬初至翌年春季开花。

Abstract: Flowers red, central petals light red, semi-double to rose-double form, medium to large size, 9.5-11.5cm across and 4.0-4.5cm deep with 20-23 thick petals arranged in 4-5 rows, bloom dense. Leaves normal green, tender leaves slightly red, elliptic, veins sunken, margins finely serrate. Plant compact, upright and growth vigorous. Blooms from early-winter to the spring of following year.

HB-27-3. 揽月阁
Lanyuege（Pavilion of Enjoying the Moon）

杂交组合： 越南抱茎茶 × 红山茶品种'全美国'（*C. amplexicaulis* × *C. japonica* cultivar 'All American'）。

杂交苗编号： BQM-No.5。

性状： 花朵深红色至黑红色，泛蜡光，半重瓣型，中型花，花径8.0～8.5cm，花朵厚6.5～7.5cm，花瓣15～18枚，基部连生，层叠呈塔状，花瓣阔倒卵形，雄蕊基部连生，开花极稠密。叶片中等绿色，嫩叶淡红褐色，阔椭圆形，长11.0～12.0cm，宽5.5～6.0cm，叶缘齿细。植株紧凑，立性，生长旺盛。冬初至翌年春季开花。

Abstract: Flowers deep red to dark red with waxy luster, semi-double form, medium size, 8.0-8.5cm across and 6.5-7.5cm deep with 15-18 petals arranged in tower shape, stamens united at the base, bloom very dense. Leaves normal green, tender leaves light brown red, broad elliptic, margins finely serrate. Plant compact, upright and growth vigorous. Blooms from early-winter to the spring of following year.

HB-27-4. 梦境
Mengjing (Dream Scene)

杂交组合：越南抱茎茶 × 红山茶品种'全美国'（*C. amplexicaulis* × *C. japonica* cultivar 'All American'）。

杂交苗编号：BQM-No.7。

性状：花朵深粉红色至浅红色，半重瓣型，中型花，花径9.0～9.5cm，花朵厚4.5～5.0cm，花瓣厚质，17～19枚，呈3～4轮排列，阔倒卵形，雄蕊基部连生，开花稠密。叶片中等绿色，嫩叶略泛红色，阔椭圆形，长12.0～14.0cm，宽6.5～7.0cm，叶脉凹，叶缘齿细。植株开张，枝叶稠密，生长旺盛。冬初至翌年春季开花。

Abstract: Flowers deep pink to light red, semi-double form, medium size, 9.0-9.5cm across and 4.5-5.0cm deep with 17-19 thick petals arranged in 3-4 rows, petals broad obovate, stamens united at the base, bloom dense. Leaves normal green, tender leaves slightly red, broad elliptic, veins sunken, margins finely serrate. Plant spread, branches and leaves dense and growth vigorous. Blooms from early-winter to the spring of following year.

HB-27-5. 春燕归
Chunyangui (Spring Swallows Returning)

杂交组合： 越南抱茎茶 × 红山茶品种'全美国'（*C. amplexicaulis* × *C. japonica* cultivar 'All American'）。

杂交苗编号： BQM-No.8。

性状： 花朵红色，单瓣型，微型至小型花，花径5.0～6.5cm，花瓣5枚左右，阔倒卵形，雄蕊基部连生，开花非常稠密。叶片中等绿色，嫩叶淡红色，长12.0～15.5cm，宽5.5～6.0cm，中脉凹，叶缘齿细。植株开张，枝叶繁茂，生长旺盛。冬初至翌年春季开花。

Abstract: Flowers red, single form, miniature to small size, 5.0-6.5cm across with about 5 petals, petals broad obovate, bloom very dense. Leaves normal green, tender leaves light red, margins finely serrate. Plant spread, branches and leaves dense and growth vigorous. Blooms from early-winter to the spring of following year.

HB-28-1. 新灵感 Xinlinggan (New Inspiration)

杂交组合：越南抱茎茶 × 红山茶品种'大菲丽丝'（*C. amplexicaulis* × *C. japonica* cultivar 'Francis Eugene Phillips'）。

杂交苗编号：BDFL-No.6+7。

性状：花朵粉红色至淡红色，单瓣型，小型花，花径6.0～7.0cm，花瓣6～7枚，排列呈喇叭状，雄蕊基部连生呈柱状，花丝粉红色，开花稠密。叶片淡绿色，厚革质，长椭圆形，长10.0～12.0cm，宽4.0～4.5cm，叶脉明显，叶齿尖，偶现橘红色老叶片，这是大菲丽丝品种遗传的结果（见第一部书第156页）。植株立性，生长旺盛。秋末至翌年春季开花。

Abstract: Flowers pink to light red, single form, small size, 6.0-7.0cm across with 6-7 petals arranged as trumpet-shape, stamens united into a column at the base, bloom dense. Leaves pale green, heavy leathery, long elliptic, veins obvious, margins sharply serrate, occasionally a few old leaves with orange red color, which inherited from 'Francis Eugene Phillips' (See p.156, our first book). Plant upright and growth vigorous. Blooms from late-autumn to the spring of following year.

HB-28-2. 长寿村
Changshoucun (Longevity Village)

杂交组合：越南抱茎茶 × 红山茶品种'大菲丽丝'（*C. amplexicaulis* × *C. japonica* cultivar 'Francis Eugene Phillips'）。

杂交苗编号：BDFL-No.8。

性状：花朵深粉红色至淡红色，单瓣型，中型花，花径7.5～8.5cm，花瓣7～8枚，排列呈喇叭状，阔倒卵形，瓣面非常皱褶，雄蕊基部连生呈短柱状。叶片淡绿色，厚革质，长椭圆形，长10.0～13.0cm，宽4.0～5.0cm，叶脉明显，叶齿尖深，偶有橘红色老叶片，这是'大菲丽丝'品种遗传的结果（见第一部书第156页）。植株立性，生长旺盛。秋末至翌年春季开花。

Abstract: Flowers deep pink to light red, single form, medium size, 7.5-8.5cm across with 7-8 petals arranged as a trumpet-shape, petals broad obovate, petal surfaces heavy crinkled, stamens united into a short column at the base. Leaves light green, heavy leathery, long elliptic, veins obvious, margins sharply and deeply serrate, occasionally a few old leaves with orange red, which inherited from 'Francis Eugene Phillips'(See p.156, our first book). Plant upright and growth vigorous. Blooms from late-autumn to the spring of following year.

HB-29-1. 新面貌
Xinmianmao (New Look)

杂交组合：越南抱茎茶 × 红山茶品种'花展时节'（*C. amplexicaulis* × *C. japonica* cultivar 'Show Time'）。

杂交苗编号：BHZSJ-No.2。

性状：花朵粉红色，渐向边缘变为淡粉红色，偶有白色斑块，半重瓣型至玫瑰重瓣型，中型花，花径 7.5～9.0cm，花朵厚 4.0cm，大花瓣 25 枚左右，呈 4 轮排列，花芯处有几枚直立的小花瓣，雄蕊基部连生，开花稠密。叶片绿色，厚革质，长椭圆形，长 14.5～14.8cm，宽 5.5～6.0cm，中脉和侧脉凹陷，边缘叶齿细浅。植株紧凑，立性，高大，生长旺盛。秋末至翌年春季开花。

Abstract: Flowers pink, fading to light-pink at the petal edges, occasionally with white markings, semi-double to rose-double form, medium size, 7.5-9.0cm across and 4.0cm deep with about 25 large petals arranged in 4 rows, a few small erected petals visible at the center, stamens united at the base, bloom dense. Leaves green, heavy leathery, long elliptic, midribs and lateral veins sunken, margins finely and shallowly serrate. Plant compact, upright, tall and growth vigorous. Blooms from late-autumn to the spring of following year.

HB-29-2. 春之笑
Chunzhixiao (Spring Laughter)

杂交组合： 越南抱茎茶 × 红山茶品种'花展时节'（ C. amplexicaulis × C. japonica cultivar 'Show Time'）。

杂交苗编号： BHZSJ-No.4。

性状： 花朵粉红色，略泛橘红色调，半重瓣型至牡丹型，中型花，花径9.5～10.0cm，花朵厚4.8cm，外部大花瓣18枚以上，呈5轮排列，中部小花瓣13～15枚，雄蕊基部连生。叶片绿色，嫩叶绿色，厚革质，长椭圆形，长16.5～17.0cm，宽6.5～7.0cm，主脉和侧脉凹陷，边缘叶齿细。植株立性，高大，生长旺盛。秋末至翌年春季开花。

Abstract: Flowers pink slightly with orange red hue, semi-double to peony form, medium size, 9.5-10.0cm across and 4.8cm deep with more than 18 exterior large petals arranged in 5 rows, 13-15 small petals at the center, stamens united at the base. Leaves green, tender leaves green, heavy leathery, long elliptic, midribs and lateral veins sunken, margins finely serrate. Plant upright, tall and growth vigorous. Blooms from late-autumn to the spring of following year.

HB-29-3. 粉萌妹
Fenmengmei (Pink Maidens)

杂交组合： 越南抱茎茶 × 红山茶品种'花展时节'（ *C. amplexicaulis* × *C. japonica* cultivar 'Show Time'）。

杂交苗编号： BHZSJ-No.6+7。

性状： 花朵淡粉红色至粉红色，半重瓣型，中型花，花径7.5～8.7cm，花朵厚3.7～4.2cm，大花瓣15枚左右，呈2轮排列，花瓣阔倒卵形，雄蕊基部连生，开花稠密。叶片绿色，厚革质，长椭圆形，长10.5～11.8cm，宽4.5～5.0cm，中脉和侧脉凹陷，边缘叶齿细浅。植株紧凑，立性，生长旺盛。秋末至翌年春季开花。

Abstract: Flowers light pink to pink, semi-double form, medium size, 7.5-8.7cm across and 3.7-4.2cm deep with about 15 large petals arranged in 2 rows, stamens united at the base, bloom dense. Leaves green, heavy leathery, long elliptic, midribs and lateral veins sunken, margins finely and shallowly serrate. Plant compact, upright and growth vigorous. Blooms from late-autumn to the spring of following year.

HB-30-1. 美娇
Meijiao (Beautiful Charm)

杂交组合：越南抱茎茶 × 红山茶品种'金博士'（*C. amplexicaulis* × *C. japonica* cultivar 'Dr King'）。

杂交苗编号：BJBS-No.1。

性状：花朵淡红色，半重瓣型至牡丹型，中型至大型花，花径 9.5～11.5cm，花朵厚 5.0～5.5cm，厚质花瓣 28～32 枚，呈 4～5 轮排列，阔倒卵形，开花稠密。叶片中等绿色，嫩叶淡红色，长椭圆形，长 14.5～15.8cm，宽 7.2～8.5cm，中脉明显，侧脉凹，叶缘齿细。植株紧凑，立性，生长旺盛。冬初至翌年春季开花。

Abstract: Flowers light red, semi-double to peony form, medium to large size, 9.5-11.5cm across and 5.0-5.5cm deep with 28-32 thick petals arranged in 4-5 rows, bloom dense. Leaves normal green, tender leaves light red, long elliptic, midribs obvious and lateral veins sunken, margins finely serrate. Plant compact, upright and growth vigorous. Blooms from early-winter to the spring of following year.

HB-30-2. 醉春风 Zuichunfeng (Drunk Spring Breeze)

杂交组合：越南抱茎茶 × 红山茶品种'金博士'（*C. amplexicaulis* × *C. japonica* cultivar 'Dr King'）。

杂交苗编号：BJBS-No.3+13。

性状：花朵桃红色，单瓣型，喇叭状，小型花，花径6.5～7.5cm，花瓣6～7枚，边缘略皱褶，雄蕊基部连生，开花稠密。叶片中等绿色，光亮，厚革质，长椭圆形，长15.5～17.0cm，宽6.0～7.0cm，中脉凹，叶缘齿细尖，先端略下弯。植株立性，枝条粗壮，生长旺盛。冬初至翌年春季开花。

Abstract: Flowers peach red, single form, small size, 6.5-7.5cm with 6-7 petals arranged in a trumpet shape, petals crinkled at the edges, stamens united at the base, bloom dense. Leaves normal green, shiny, long elliptic, midribs sunken, margins finely and sharply serrate and apices down ward. Plant upright, branches sturdy and growth vigorous. Blooms from early-winter to the spring of following year.

HB-30-3. 越松子
Yuesongzi (Surpass Pine Cone)

杂交组合： 越南抱茎茶 × 红山茶品种'金博士'(*C. amplexicaulis* × *C. japonica* cultivar 'Dr King')。

杂交苗编号： BJBS-No.4。

性状： 花朵深粉红色，半重瓣型，中型花，花径7.5～8.0cm，花朵厚6.0cm，花瓣25～30枚，呈5～6轮螺旋状排列，阔倒卵形，外翻，雄蕊基部连生，花朵整朵掉落，开花稠密。叶片绿色，厚革质，阔椭圆形，长12.0～13.0cm，宽5.0～6.0cm，中脉凹，叶缘齿尖而稠密。植株立性，枝条稠密，生长旺盛。冬初至翌年春季开花。

Abstract: Flowers deep pink, semi-double form, medium size, 7.5-8.0cm across and 6.0cm deep with 25-30 petals arranged spirally in 5-6 rows, petals broad obovate, rolled outward, stamens united at the base, flowers fall whole, bloom dense. Leaves green, heavy leathery, broad elliptic, midribs sunken, margins sharply and densely serrate. Plant upright, branches dense and growth vigorous. Blooms from early-winter to the spring of following year.

HB-30-4. 春丽 Chunli（The Beauty of Spring）

杂交组合： 越南抱茎茶 × 红山茶品种'金博士'（*C. amplexicaulis* × *C. japonica* cultivar 'Dr King'）。

杂交苗编号： BJBS-No.6。

性状： 花朵深桃红色，中部花瓣具白色边缘，半重瓣型至玫瑰重瓣型，有时呈牡丹型，中型花，花径 8.5～9.5cm，花朵厚 5.2cm，花瓣 25～30 枚，呈 3～5 轮螺旋状排列，基部略连生，阔倒卵形，边缘外翻，开花稠密。叶片中等绿色，厚革质，阔椭圆形，长 13.5～14.0cm，宽 5.5～6.0cm，中脉凹，叶缘齿尖细。植株立性，枝条稠密，生长旺盛。冬初至翌年春季开花。

Abstract: Flowers deep peach red, central petals with white borders, semi-double to rose-double form, sometimes peony form, medium size, 8.5-9.5cm across and 5.2cm deep with 25-30 petals arranged spirally in 3-5 rows, petal edges turned outward, bloom dense. Leaves normal green, broad elliptic, midribs sunken, margins sharply and finely serrate. Plant upright, branches dense and growth vigorous. Blooms from early-winter to the spring of following year.

HB-30-5. 越秀 Yuexiu (Surpass Beauty)

杂交组合：越南抱茎茶 × 红山茶品种'金博士'（*C. amplexicaulis* × *C. japonica* cultivar 'Dr King'）。

杂交苗编号：BJBS-No.12。

性状：花朵粉红色至深粉红色，半重瓣型，中型花，花径8.5～10.0cm，花朵厚5.0cm，花瓣18～21枚，呈3～4轮有序排列，瓣面皱褶，雄蕊基部连生，花朵整朵掉落，开花稠密。叶片绿色，厚革质，阔椭圆形，长12.5～13.5cm，宽5.5～6.5cm，中脉凹，叶缘齿尖而稠密。植株立性，枝条稠密，生长旺盛。冬初至翌年春季开花。

Abstract: Flowers pink to deep pink, semi-double form, medium size, 8.5-10.0cm across and 5.0cm deep with 18-21 petals arranged orderly in 3-4 rows, petal surfaces wrinkled, stamens united at the base, flowers fall whole, bloom dense. Leaves green, heavy leathery, broad elliptic, midribs sunken, margins sharply and densely serrate. Plant upright, branches dense and growth vigorous. Blooms from early-winter to the spring of following year.

HB-31-1. 紫胭脂
Ziyanzhi (Purple Rouge)

杂交组合：越南抱茎茶 × 红山茶杂交品种'金黑绒'（*C. amplexicaulis* × *C. japonica* hybrid 'Jinheirong'）。

杂交苗编号：BJHR-No.10+11+2。

性状：花朵黑紫红色，单瓣型，中型花，花径 8.0～9.5cm，花瓣 5～6 枚，阔倒卵形，边缘略内卷，瓣面可见深红色脉纹，雄蕊基部连生，呈柱状，花朵整朵掉落。叶片深绿色，厚革质，长椭圆形，长 12.0～13.0cm，宽 4.0～5.0cm，主脉黄绿色，侧脉不明显，叶齿钝。植株立性，高大，生长旺盛。冬末始花，春季盛花。

Abstract: Flowers dark purple red, single form, medium size, 8.0-9.5cm across with 5-6 petals, petals broad obovate with deep red veins on the surfaces, stamens united into a column at the base, flowers fall whole. Leaves deep green, heavy leathery, long elliptic, midribs light yellow and lateral veins unconspicuous, margins obtusely serrate. Plant upright, tall and growth vigorous. Starts to bloom from late-winter and fully blooms in spring.

HB-31-2. 少女裙
Shaonüqun (Maiden Skirt)

杂交组合： 越南抱茎茶 × 红山茶杂交品种'金黑绒'（*C. amplexicaulis* × *C. japonica* hybrid 'Jinheirong'）。

杂交苗编号： BJHR-No.13。

性状： 花朵红色，随着花朵开放，边缘出现隐约的淡红色，花芯处偶有白斑，半重瓣型，大型至巨型花，花径11.0～13.2cm，花朵厚6.5cm，大花瓣16枚以上，呈3～4轮排列，花瓣外翻，肉质，呈沟槽，边缘略波浪状，花朵整朵掉落。嫩叶淡红色，成熟叶绿色，厚革质，长椭圆形，长14.0～16.0cm，宽5.0～5.5cm，主脉黄色，侧脉不明显，叶齿钝，叶柄长1.0cm。植株立性，生长旺盛。冬末始花，春季盛花。

Abstract: Flowers red, as flowers opening, faint white edges are visible, occasionally some white markings on central petals, semi-double form, large to very large size, 11.0-13.2cm across and 6.5cm deep with more than 16 large petals arranged in 3-4 rows, petals rolled outward, groove-shaped, edges slightly wavy, flowers fall whole. Tender leaves light red, mature leaves green, heavy leathery, long elliptic, midribs yellow and lateral veins unconspicuous, margins obtusely serrate. Plant upright and growth vigorous. Starts to bloom from late-winter and fully blooms in spring.

HB-32-1. 大跨越
Dakuayue (Great Stride Leap)

杂交组合： 越南抱茎茶 × 红山茶品种'金盘荔枝'的机遇苗（C. amplexicaulis × a chance seedling of C. japonica cultivar 'Jinpan Lizhi'）。

杂交苗编号： BJPS-No.12+16。

性状： 花朵深桃红色，渐向边缘变为粉红色，半重瓣型，中型至大型花，花径8.5～10.5cm，花朵厚5.5cm，大花瓣13～15枚，呈2轮排列，花瓣阔倒卵形，花瓣边缘略皱，雄蕊基部略连生。叶片中等绿色，厚革质，阔椭圆形，长13.0～14.0cm，宽6.5～7.0cm，中脉和侧脉明显，叶缘齿尖密。植株立性，枝条开张，生长旺盛。春季开花。

Abstract: Flowers deep peach red, fading to pink at the edges, semi-double form, medium to large size, 8.5-10.5cm across and 5.5cm deep with 13-15 large petals arranged in 2 rows, petals broad obovate, edges slightly crinkled, stamens slightly united at the base. Leaves normal green, thick leathery, broad elliptic, midribs and lateral veins visible, margins sharply and densely serrate. Plant upright, branches spread and growth vigorous. Blooms in spring.

HB-32-2. 小螺旋
Xiaoluoxuan (Small Spiral)

杂交组合：越南抱茎茶 × 红山茶品种'金盘荔枝'的机遇苗（*C. amplexicaulis* × a chance seedling of *C. japonica* cultivar 'Jinpan Lizhi'）。

杂交苗编号：BJPS-No.25。

性状：花朵桃红色，单瓣型，小型至中型花，花径6.5～8.5cm，大花瓣5～6枚，呈螺旋状排列，花瓣阔倒卵形，外翻，先端近圆，雄蕊基部略连生，呈筒状。叶片中等绿色，厚革质，阔椭圆形，长11.0～12.5cm，宽5.0～6.5cm，中脉和侧脉明显，叶缘齿尖密。植株立性，枝条开张，生长旺盛。春季开花。

Abstract: Flowers peach red, single form, small to medium size, 6.5-8.5cm across with 5-6 large petals arranged spirally, petals broad obovate, rolled outward, stamens slightly united into a column at the base. Leaves normal green, thick leathery, broad elliptic, midribs and lateral veins visible, margins sharply and densely serrate. Plant upright, branches spread and growth vigorous. Blooms in spring.

HB-32-3. 大视野
Dashiye (Broad Field of Vision)

杂交组合：越南抱茎茶×红山茶品种'金盘荔枝'的机遇苗（*C. amplexicaulis* × a chance seedling of *C. japonica* cultivar 'Jinpan Lizhi'）。

杂交苗编号：BJPS-No.27。

性状：花朵深红色，渐渐变为深粉红色，偶有白斑，半重瓣型至牡丹型，大型至巨型花，花径11.5～13.3cm，花朵厚6.0cm，大花瓣15～18枚，呈3轮松散排列，中部小花瓣扭曲。嫩叶略泛红，成熟叶中等绿色，厚革质，阔椭圆形，长14.5～15.5cm，宽6.0～6.5cm，中脉和侧脉明显，叶缘略波浪状，叶齿尖。植株立性，枝条开张，生长旺盛。春季开花。

> **Abstract:** Flowers deep red, gradually changed into deep pink, occasionally with white markings, semi-double to peony form, large to very large size, 11.5-13.3cm across and 6.0cm deep with 15-18 large petals arranged loosely in 3 rows, central small petals twisted. Leaves normal green, thick leathery, broad elliptic, midribs and lateral veins obvious, margins slightly wavy and sharply serrate. Plant upright, branches spread and growth vigorous. Blooms in spring.

HB-33-1. 惬意粉
Qieyifen (Pleasant Pink)

杂交组合：越南抱茎茶 × 红山茶品种'红衣大皇冠'(*C. amplexicaulis* × *C. japonica* cultivar 'Betty Sheffield Blush Supreme')。

杂交苗编号：BHYDHG-No.3。

性状：花朵粉红色至深粉红色，花芯处小花瓣具白斑，半重瓣型至玫瑰重瓣型，中型花，花径 8.5～10.0cm，花朵厚 6.0cm，大花瓣 25～28 枚，呈 5 轮有序排列，花瓣阔倒卵形，雄蕊基部略连生。叶片绿色，厚革质，阔椭圆形，长 12.0～13.0cm，宽 5.5～6.0cm，中脉和侧脉明显，叶缘齿尖密。植株立性，高大，生长旺盛。冬末至翌年春季开花。

Abstract: Flowers pink to deep pink with white markings on the central small petals, semi-double to rose-double form, medium size, 8.5-10.0cm across and 6.0cm deep with 25-28 large petals arranged orderly in 5 rows, petals broad obovate, stamens slightly united at the base. Leaves green, thick leathery, broad elliptic, midribs and lateral veins visible, margins sharply and densely serrate. Plant upright, tall and growth vigorous. Blooms from late-winter to the spring of following year.

HB-34-1. 小酒杯 Xiaojiubei (Small Wine Cup)

杂交组合： 越南抱茎茶 × 非云南山茶杂交种'龙火珠'（*C. amplexicaulis* × Non-*reticulata* hybrid 'Dragon Fireball'）。

杂交苗编号： BLHZ-No.1。

性状： 花朵深粉红色至桃红色，偶具白斑，单瓣型，微型至小型花，花径 5.5～7.0cm，花瓣 6～7 枚，呈螺旋状排列，花瓣阔倒卵形，雄蕊基部连生，开花稠密。叶片绿色，厚革质，椭圆形，长 10.5～11.5cm，宽 4.0～5.0cm，中脉和侧脉凸起，基部楔形，边缘叶齿细尖。植株立性，生长旺盛。冬末至翌年春季开花。

Abstract: Flowers deep pink to peach red, occasionally with white markings, single form, miniature to small size, 5.5-7.0cm across with 6-7 petals arranged spirally, petals broad obovate, bloom dense. Leaves green, heavy leathery, elliptic, midribs and lateral veins raised, cuneate at the base, margins finely and sharply serrate. Plant upright and growth vigorous. Blooms from late-winter to the spring of following year.

HB-34-2. 凤凰亭
Fenghuangting (Phoenix Pavilion)

杂交组合：越南抱茎茶 × 非云南山茶杂交种'龙火珠'（*C. amplexicaulis* × Non-*reticulata* hybrid 'Dragon Fireball'）。

杂交苗编号：BLHZ-No.7。

性状：花朵柔和的桃红色，偶有隐约的白斑，雄蕊瓣具小白斑，随着花朵开放，花色逐渐变为粉红色，托桂型至牡丹型，小型至中型花，花径6.5～8.0cm，花朵厚4.5cm，大花瓣11～13枚，呈2轮排列，雄蕊瓣与雄蕊混生。叶片绿色，厚革质，椭圆形，长12.5～13.5cm，宽4.5～5.0cm，中脉和侧脉凸起，基部楔形，边缘叶齿细尖。植株紧凑，立性，生长旺盛。秋末至翌年春末开花。

Abstract: Flowers soft peach red, occasionally with faint white markings, petaloids with small white markings, as flowers opening, flower color gradually changes into pink, anemone to peony form, small to medium size, 6.5-8.0cm across and 4.5cm deep with 11-13 large petals arranged in 2 rows, petaloids mixed with stamens. Leaves green, heavy leathery, elliptic, midribs and lateral veins raised, cuneate at the base, margins finely and sharply serrate. Plant compact, upright and growth vigorous. Blooms from late-autumn to the late-spring of following year.

HB-35-1. 春红 Chunhong (Spring Red)

杂交组合：越南抱茎茶 × 非云南山茶杂交种'香四射'（*C. amplexicaulis* × Non-*reticulata* hybrid 'Xiangsishe'）。

杂交苗编号：BXSS-No.2。

性状：花朵红色，略泛紫色调，具微香，半重瓣型至牡丹型，中型至大型花，花径 8.0～11.5cm，花朵厚 5.5～6.8cm，花瓣 22～24 枚，呈 5 轮排列，花瓣阔倒卵形，先端近圆形，雄蕊基部略连生，花丝近白色，花药黄色，开花稠密。叶片绿色，厚革质，阔椭圆形，长 13.0～14.0cm，宽 7.0～8.0cm，中脉凸起，侧脉凹陷，叶齿细尖。植株立性，高大，枝条稠密，生长旺盛。春季开花。

Abstract: Flowers red slightly with purple hue and with faint fragrance, semi-double to peony form, medium to large size, 8.0-11.5cm across and 5.5-6.8cm deep with 22-24 petals arranged in 5 rows, petals broad obovate, apices nearly round, stamens slightly united at the base, filaments nearly white, anthers yellow, bloom dense. Leaves green, heavy leathery, broad elliptic, midribs raised and lateral veins sunken, margins finely and sharply serrate. Plant upright, tall, branches dense and growth vigorous. Blooms in spring.

HB-36-1. 阿婆六水星
Apoliu Shuixing（Apoliu's Mercury）

杂交组合：越南抱茎茶 × 威廉姆斯杂交种'超星'（*C. amplexicaulis* × *C. x williamsii* hybrid 'Super Star'）。

杂交苗编号：BCX-No.1。

性状：花朵白色，渐向花瓣基部呈淡黄色，半重瓣型，中型花，花径8.0～9.5cm，花朵厚5.0cm，花瓣厚质，18～23枚，呈3～4轮松散、星状排列，花瓣狭长倒卵形，边缘内卷，呈沟槽状，雄蕊基部近离生。叶片中等绿色，薄革质，阔椭圆形，长12.0～13.5cm，宽5.5～6.5cm，中脉明显，叶缘齿钝。植株开张，枝条细软，略匍匐，生长旺盛。春初至春末开花。

注：本品种系2019年由广东省从化区阿婆六茶花谷培育成功，并分别在中国花卉协会茶花分会和国际山茶协会登录。

Abstract: Flowers white, fading to light yellow at the petal bases, semi-double form, medium size, 8.0-9.5cm across and 5.0cm deep with 18-23 thick petals arranged loosely in 3-4 rows as a star shape, petals narrow-long obovate, edges turned inward, stamens nearly separated at the base. Leaves normal green, thin leathery, broad elliptic, midribs obvious, margins obtusely serrate. Plant spread, branches thin and soft, slightly creeping and growth vigorous. Blooms from early-spring to late-spring.

Mark: The hybrid was bred by Apoliu Camellia Valley, Conghua District, Guangzhou City, Guangdong Province, China and registered in Chinese Camellia Branch Society, Chinese Flowers Association and in ICS in 2019 respectively.

HB-37-1. 胖大姐
Pangdajie (Fat Eldest Sister)

杂交组合：越南抱茎茶 × 崇左金花茶（*C. amplexicaulis* × *C. chuangtsoensis*）。

杂交苗编号：BCHZ-No.2。

性状：花朵粉红色至深粉红色，边缘略显朦胧的白边，花瓣基部外侧紫红色，半重瓣型，中型花，花径 8.0～9.0cm，花朵厚 7.0cm，花瓣 13～15 枚，呈 3 轮排列，花瓣阔圆形，厚肉质，先端近圆，雄蕊多数，基部连生，呈 1.5cm 的短管状，花朵整朵掉落。嫩叶紫红色，成熟叶绿色，厚革质，长椭圆形，长 14.0～17.0cm，宽 5.5～6.0cm，主脉和侧脉凹，先端下弯，基部没有抱茎的耳叶，叶齿尖浅，叶柄长 0.8cm。植株立性，枝条开张，生长旺盛。冬末始花，春季盛花。

Abstract: Flowers pink to deep pink with hazy white edges, the base of petal back surfaces purple-red, semi-double form, medium size, 8.0-9.0cm across and 7.0cm deep with 13-15 petals arranged in 3 rows, petals broad round, thick fleshy, stamens united into a short column, flowers fall whole. Tender leaves purple-red, mature leaves green, heavy leathery, long elliptic, midribs and lateral veins sunken, no lobes at the base, margins shallowly and sharply serrate, petioles 0.8cm in length. Plant upright, branches spread and growth vigorous. Starts to bloom from late-winter, fully blooms in spring.

HB-38-1. 初探
Chutan (First Exploration)

杂交组合： 越南抱茎茶 × 尖萼红山茶（*C. amplexicaulis* × *C. edithae*）。

杂交苗编号： BJE-No.4。

性状： 花朵淡粉红色，略泛紫色调，单瓣型，小型至中型花，花径6.5～8.5cm，花瓣5～6枚，阔倒卵形，瓣面略皱，外翻，雄蕊基部连生呈柱状。叶片灰绿色，革质，嫩叶和嫩枝被柔毛，长椭圆形，长10.5～12.0cm，宽5.5～6.5cm，叶脉明显，基部心形，边缘叶齿细尖，枝条被柔毛。植株立性，生长中等。冬初至翌年春季开花。

Abstract: Flowers light pink with some purple hue, single form, small to medium size, 6.5-8.5cm across with 5-6 petals, petals broad obovate, petal surfaces slightly crinkled, turned outward, stamens united into a column at the base. Leaves greyish green, tender leaves and branches villous, leathery, long elliptic, veins visible, the base heart shaped, margins finely and sharply serrate. Plant upright and growth normal. Blooms from early-winter to the spring of following year.

HB-39-1. 新思路
Xinsilu (New Thoughts)

杂交组合： 越南抱茎茶 × 南山茶（*C. amplexicaulis* × *C. semiserrata*）。

杂交苗编号： BNSC-No.17+3。

性状： 花朵桃红色至红色，单瓣型至半重瓣型，中型至大型花，花径9.0～11.0cm，花瓣11～13枚，厚肉质，阔倒卵形，松散排列，先端圆，边缘内扣，雄蕊200枚以上，中部以下连生，呈红色，雌蕊长4.0cm左右，花朵整朵掉落，开花稠密，不能自然坐果。叶片浓绿色，嫩叶紫红色，硬革质，长椭圆形，长17.0～18.0cm，宽3.0～4.0cm，叶面主脉和侧脉深凹陷，叶背面则凸起，边缘叶齿尖。植株立性，枝条半开张，生长极旺盛。冬末至春季开花。

> **Abstract:** Flowers peach red to red, single to semi-double form, medium to large size, 9.0-11.0cm across with 11-13 fleshy petals, apices round, stamens over 200 at the center, flowers fall whole, bloom dense, no natural fruit. Leaves dark green, tender leaves purple red, hard leathery, long elliptic, midribs and lateral veins deeply sunken, margins sharply serrate. Plant upright, branches semi-spread and growth very vigorous. Blooms from late-winter to the spring of following year.

HB-40-1. 新发现 Xinfaxian (New Discovery)

杂交组合：越南抱茎茶 × 三江瘤果茶（*C. amplexicaulis* × *C. pyxidiacea*）。

杂交苗编号：BSJLGC-No.2。

性状：花朵淡粉红色，而后变为粉白色，渐向上部呈白色，单瓣型，中型至大型花，花径8.5～12.0cm，花瓣8枚，阔倒卵形，瓣面略皱褶，边缘内卷，雄蕊基部连生呈柱状，花丝和花药黄色。叶片绿色，嫩叶紫红色，厚革质，长椭圆形，长28.0～35.0cm，宽9.0～10.0cm，叶面主脉和侧脉凹陷，叶背面则凸起，边缘叶齿深尖，基部近圆，不抱茎，叶柄长1.0cm。植株立性，侧枝披斜，叶片近羽状排列，生长极旺盛。冬末至春季开花。

Abstract: Flowers light pink, then gradually changing into white-pink, fading to white at the top part, single form, medium to large size, 8.5-12.0cm across with 8 broad obovate petals, petal surfaces slightly wrinkled, edges rolled inward, filaments and anthers yellow. Leaves green, tender leaves purple red, heavy leathery, long elliptic, midribs and lateral veins deeply sunken, margins deeply and sharply serrate, the base nearly round and no clasped stem, petioles 1.0cm long. Plant upright, branches hang down and growth very vigorous. Blooms from late-winter to the spring of following year.

HB-41-1. 夜市
Yeshi (Night Market)

杂交组合： 云南山茶品种'珍妮珀赛尔'×越南抱茎茶（*C. reticulata* cultivar 'Jean Pursel' × *C. amplexicaulis*）。

杂交苗编号： ZNPSEB-No.1。

性状： 花朵红色至黑红色，花芯处小花瓣粉白色，具白斑，玫瑰重瓣型，大型花，花径11.0～12.5cm，花朵厚5.5cm，大花瓣50多枚，呈8～10轮有序排列，花瓣近圆形，边缘内卷，花芯处小花瓣约30枚。叶片绿色，嫩叶紫红色，革质，长椭圆形，长10.0～11.0cm，宽4.5～5.5cm，叶齿细浅，叶柄长0.4cm。植株紧凑，矮性，生长中等。冬末始花，春季盛花。

Abstract: Flowers red to dark red, central small petals pink-white with white markings, rose-double form, large size, 11.0-12.5cm across and 5.5cm deep with more than 50 large petals arranged orderly in 8-10 rows, petals nearly round, edges rolled inward, about 30 small petals at the center. Leaves green, tender leaves purple-red, leathery, long elliptic, margins finely and shallowly serrate, petioles 0.4cm in length. Plant compact, dwarf, growth normal. Starts to bloom from late-winter and fully blooms in spring.

常规茶花杂交新品种育种概况
The Breeding Outline on the Normal Camellia New Hybrids

为了与我们的四季茶花新品种区别开来，我们把常规茶花品种定义为世界上现有的不含杜鹃红山茶血统的全部茶花品种。常规茶花品种数量成千上万，至今依然是世界上栽培的主流品种。它们冬、春开花，花色、花型丰富多彩，深受人们的青睐。

> In order to distinguish from our new camellia hybrids that bloom year-round, we define the normal camellia cultivars here as the world's all existing camellia cultivars which are without the ancestry of *C. azalea*. There are thousands and thousands of normal camellia cultivars, which have been cultivated as the main cultivars in the world until now. The normal camellia cultivars bloom in winter and spring and flower colors and flower forms are rich and colorful, which are deeply accepted by people.

常规茶花品种杂交育种示意图
A sketch of normal camellia crosses breeding

第一部书中，曾介绍过20个常规茶花品种间的杂交新品种，它们涉及8个杂交组合（见第一部书第491～518页）。

本书将介绍12个常规茶花杂交组合及从中获得的19个杂交新品种，其中有4个杂交组合是与第一部书重复的。

如果加上第一部书列出的杂交组合和获得的杂交新品种，到目前为止，常规茶花品种间的杂交组合总数已达 16 个，获得的杂交新品种总数为 39 个。

这些杂交新品种的性状，取决于其杂交亲本的性状。杂交新品种的开花期与常规茶花品种的相同，均为冬、春季节开花。这些杂交新品种，有的花色奇特，有的开花稠密，都是盆栽莳养和环境美化的好品种。

应该指出的是，凡在我们第一部书中介绍过的杂交组合，本书的组合编号将与第一部书相同，其新品种的编号将延续第一部书的序列编号。

Twenty new hybrids between normal camellia cultivars were introduced in our first book and the new hybrids were involved in eight cross-combinations (See p.491-518, our first book).

Twelve cross-combinations of normal camellia cultivars and 19 new hybrids obtained from them will be introduced in this book. Of the cross-combinations, there are four combinations appeared in our first book.

If the cross-combinations and the new hybrids listed in our first book are added, we have carried 16 cross-combinations and obtained 39 new hybrids in total now.

The characteristics of these new hybrids depend on the characteristics of their cross parents. The blooming period of the new hybrids is as same as normal camellias' which bloom from winter to the spring of following year. These new hybrids, some of them have unique flower colors and some of them have dense flowers, are all good varieties for potted cultivation and landscaping.

It should be pointed that the numbers of the cross-combinations introduced in our first book will be as same as those of our first book and the new hybrids numbers in this book will follow the serial numbers of our first book.

紫绮
(Ziqi) HC-02-6

黑皱锦
(Heizhoujin) HC-09-1

勇士
(Yongshi) HC-17-1

紫靓丽
(Ziliangli) HC-11-1

常规茶花杂交新品种详解
The Detailed Descriptions for the Normal Camellia New Hybrids

HC-02-4. 迷人红 Mirenhong（Charming Red）

杂交组合：红山茶品种'媚丽'×红山茶品种'阿兰'（*C. japonica* cultivar 'Tama Beauty'× *C. japonica* cultivar 'Mark Alan'）。

杂交苗编号：MA-No.1A。

性状：花朵深红色，泛紫色调，偶有少量白色斑点，完全重瓣型，中型花，花径8.5～10.0cm，花朵厚4.5cm，花瓣40～50枚，呈6轮覆瓦状紧实排列，花瓣倒卵形，外翻。叶片中等绿色，革质，椭圆形，长8.0～9.5cm，宽4.0～5.0cm，先端尖，边缘齿细密。植株开张，灌丛状，生长旺盛。冬季至翌年春季开花。

Abstract: Flowers dark red with purple hue and occasionally with a few white spots, formal double form, medium size, 8.5-10.0cm across and 4.5cm deep with 40-50 petals arranged imbricately and tightly in 6 rows, petals obovate, turned outward. Leaves normal green, leathery, elliptic, margins densely and finely serrate. Plant spread, bushy and growth vigorous. Blooms from winter to the spring of following year.

HC-02-5. 复色花容
Fuse Huarong (Flower Face Variegated)

杂交组合：红山茶品种'媚丽' × 红山茶品种'阿兰'（*C. Japonica* cultivar 'Tama Beauty' × *C. Japonica* cultivar 'Mark Alan'）。

杂交苗编号：MA-No.2MU。

性状：本品种系'花容'品种（见第一部书第495～496页）的复色突变。花朵紫红色，具白色斑块，半重瓣型，中型花，花径9.5～10.0cm，花朵厚5.0cm，大花瓣25枚，呈2～3轮排列，花瓣阔倒卵形，波浪状或扭曲，雄蕊散射状。叶片绿色，偶有黄斑，革质，椭圆形，长9.2cm，宽4.5cm，边缘齿细密。植株开张，生长旺盛。冬季至翌年春季开花。

Abstract: It is a variegated mutation from 'Huarong' (See p.495-496, our first book). Flowers purple red, blotched white, semi-double form, medium size, 9.5-10.0cm across and 5.0cm deep with 25 large petals that arranged in 2-3 rows, petals broad obovate, wavy or twisted, stamens radial. Leaves green, occasionally with yellow markings, leathery, elliptic, margins densely and finely serrate. Plant spread and growth vigorous. Blooms from winter to the spring of following year.

HC-02-6. 紫绮
Ziqi (Purple Silk Cheongsam)

杂交组合：红山茶品种'媚丽'×红山茶品种'阿兰'（ *C. japonica* cultivar 'Tama Beauty' × *C. japonica* cultivar 'Mark Alan'）。

杂交苗编号：MA-No.1B。

性状：花朵多顶生，紫红色，花瓣边缘具较窄的白色镶边，完全重瓣型，中型花，花径8.0～9.5cm，花朵厚4.5cm，花瓣40枚以上，呈6～8轮覆瓦状排列，花瓣阔倒卵形，外翻，花芯常呈球状。叶片中等绿色，革质，椭圆形，长8.0～9.0cm，宽4.5～5.0cm，先端尖，边缘齿细密。植株开张，灌丛状，生长旺盛。冬季至翌年春季开花。

Abstract: Solitary terminal flowers, purple red with narrow white borders at the petal edges, formal-double form, medium size, 8.0-9.5cm across and 4.5cm deep with more than 40 petals arranged imbricately in 6-8 rows, petals broad obovate, turned outward, usually a small ball visible at the center. Leaves normal green, leathery, elliptic, margins finely and densely serrate. Plant spread, bushy and growth vigorous. Blooms from winter to the spring of following year.

HC-02-7. 秋纹
Qiuwen (Miss Qiuwen)

杂交组合： 红山茶品种'媚丽'×红山茶品种'阿兰'（*C. japonica* cultivar 'Tama Beauty'× *C. japonica* cultivar 'Mark Alan'）。

杂交苗编号： MA-No.12。

性状： 花朵黑红色，泛紫色调，花瓣边缘具白色窄边，单瓣型，小型至中型花，花径7.0～8.0cm，花瓣5枚，阔倒卵形，外翻，略皱褶，雄蕊基部连生呈圆柱状，开花稠密。叶片绿色，革质，椭圆形，长7.8～8.5cm，宽3.5～4.0cm，边缘齿细密。植株开张，生长旺盛。冬季至翌年春季开花。

Abstract: Flowers dark red with purple hue and narrow white borders, single form, small to medium size, 7.0-8.0cm across with 5 petals, petals broad obovate, rolled outward, slightly wrinkled, stamens united into a column, bloom dense. Leaves green, leathery, elliptic, margins densely and finely serrate. Plant spread and growth vigorous. Blooms from winter to the spring of following year.

HC-04-2. 金黑绒
Jinheirong (Golden Black Charpie)

杂交组合： 红山茶品种'白斑康乃馨'×红山茶品种'黑摩萨'(*C. japonica* cultivar 'Ville de Nantes' × *C. japonica* cultivar 'Heimosa')。

杂交苗编号： NTHM-No.3 。

性状： 花朵黑红色，有绒质感，半重瓣型，大型花，花径 10.0～11.5cm，花朵厚 6.0cm，花瓣 13～14 枚，呈 3 轮排列，花瓣近圆形，瓣面具黑红色脉纹，雄蕊基部略连生，花丝红色。叶片浓绿色，革质，椭圆形，长 8.5～9.5 cm，宽 4.0～5.0 cm，中脉明显，先端下弯，边缘叶齿稀钝。植株紧凑，枝叶稠密，生长旺盛。冬初至翌年春季开花。

Abstract: Flowers dark-red with terry feeling, semi-double form, large size, 10.0-11.5cm across and 6.0cm deep with 13-14 petals arranged in 3 rows, petals nearly round, stamens slightly united at the base, filaments red. Leaves deep green, leathery, elliptic, midrib visible, margins obtusely and sparsely serrate. Plant compact, branches dense and growth vigorous. Blooms from early-winter to the spring of following year.

HC-05-2. 红舞台 Hongwutai（Red Stage）

杂交组合：红山茶品种'凯夫人'×云南山茶品种'山茶之都'（ C. japonica cultivar 'Lady Kay' × C. reticulata cultivar 'Massee Lane'）。

杂交苗编号：KAIS–No.2。

性状：花朵艳红色，半重瓣型，大型至巨型花，花径10.5～13.5cm，花朵厚5.0cm，花瓣12～14枚，呈2轮排列，花瓣阔倒卵形，大波浪状，雄蕊基部略连生，开花稠密。叶片绿色，厚革质，阔椭圆形，长7.5～8.5cm，宽5.0～5.5cm，叶齿尖密。植株紧凑，枝叶稠密，生长旺盛。冬末至翌年春季开花。

> **Abstract:** Flowers bright red, semi-double form, large to very large size, 10.5-13.5cm across and 5.0cm deep with 12-14 petals arranged in 2 rows, petals broad obovate and big wavy, stamens slightly united at the base, bloom dense. Leaves green, heavy leathery, broad elliptical, margins sharply and densely serrate. Plant compact, branches dense and growth vigorous. Blooms from late-winter to the spring of following year.

HC-09-1. 墨皱锦
Mozhoujin (Black & Crinkled Brocade)

杂交组合：红山茶品种'金盘荔枝'×红山茶品种'黑魔法'(*C. japonica* cultivar 'Jinpan Lizhi' × *C. japonica* cultivar 'Black Magic')。

杂交苗编号：JPHM-No.2。

性状：花朵黑红色，具蜡光，单瓣型，中型至大型花，花径9.0～12.5cm，花瓣6枚，肉质，阔倒卵形，边缘皱褶或波浪状，雄蕊基部略连生，花丝红色。叶片绿色，光亮，厚革质，阔椭圆形，长7.5～8.0cm，宽4.5～5.0cm，主脉和侧脉凸起，叶齿尖密。植株开张，枝叶稠密，生长旺盛。冬末至翌年春季开花。

Abstract: Flowers dark red with waxy luster, single form, medium to large size, 9.0-12.5cm across with 6 petals, petals fleshy, broad obovate and edges crinkled or wavy, stamens slightly united at the base, filaments red. Leaves green, shiny, heavy leathery, broad elliptical, midrib and lateral veins raised, margins sharply and densely serrate. Plant spread, branches dense and growth vigorous. Blooms from late-winter to the spring of following year.

HC-10-5. 玉润
Yurun（Jade Moisten）

杂交组合： 红山茶品种'媚丽'× 云南山茶品种'山茶之都'（*C. japonica* cultivar 'Tama Beauty'× *C. reticulata* cultivar 'Massee Lane'）。

杂交苗编号： MLS-No.2。

性状： 花朵白色，花瓣基部略显粉色，半重瓣型至牡丹型，大型至巨型花，花径11.0～13.5cm，花朵厚6.0cm，大花瓣18～21枚，呈3～4轮排列，花瓣倒卵形，外翻，中部13～15枚雄蕊瓣扭曲并与雄蕊混生，雄蕊簇生，花丝白色。叶片绿色，光亮，革质，椭圆形，长8.5～9.5cm，宽4.0～4.5cm，叶齿密。植株开张，矮性，生长旺盛。冬末至翌年春季开花。

Abstract: Flowers white with slight pink at the bases of petals, semi-double to peony form, large to very large size, 11.0-13.5cm across and 6.0cm deep with 18-21 large petals arranged in 3-4 rows, petals obovate, turned outward, 13-15 central petaloids twisted and mixed with stamens, stamens clustered, filaments white. Leaves green, shiny, leathery, elliptic, margins densely serrate. Plant spread, dwarf and growth vigorous. Blooms from late-winter to the spring of following year.

HC-10-6. 冰艺 Bingyi (Ice Art)

杂交组合： 红山茶品种'媚丽'×云南山茶品种'山茶之都'（ C. japonica cultivar 'Tama Beauty' × C. reticulata cultivar 'Massee Lane'）。

杂交苗编号： MLS-No.3。

性状： 花朵白色，托桂型至牡丹型，中型至大型花，花径8.5～11.0cm，花朵厚5.5cm，外部大花瓣6枚，呈1轮排列，阔倒卵形，波浪状，中部无数小花瓣簇拥成球，有少量黄色雄蕊外露，花丝白色。叶片绿色，光亮，革质，椭圆形，长8.0～9.0cm，宽4.5～5.0cm，叶齿明显。植株开张，枝条稠密，生长旺盛。冬末至翌年春季开花。

Abstract: Flowers white, anemone to peony form, medium to large size, 8.5-11.0cm across and 5.5cm deep with 6 exterior large petals arranged in 1 row, petals broad obovate, wavy, a lot of small petals surround into a ball at the center, some yellow stamens visible, filaments white. Leaves green, shiny, leathery, elliptic, margins obviously serrate. Plant spread, branches dense and growth vigorous. Blooms from late-winter to the spring of following year.

HC-10-7. 唯美 Weimei (Only Beauty)

杂交组合： 红山茶品种'媚丽'× 云南山茶品种'山茶之都'（*C. japonica* cultivar 'Tama Beauty' × *C. reticulata* cultivar 'Massee Lane'）。

杂交苗编号： MLS-Mu, No.6B + 7B。

性状： 该品种是'蜂露'品种（见第一部书第504～505页）的一个突变。花朵浅紫红色，花瓣边缘具宽窄不等的白边，半重瓣型至玫瑰重瓣型，中型花，花径8.5～9.5cm，花朵厚5.0cm，花瓣40～55枚，呈7～8轮有序排列，花瓣倒卵形，内扣，先端近圆形，中部雄蕊瓣扭曲，与雄蕊混生。叶片绿色，光亮，革质，椭圆形，长11.0～12.5cm，宽4.5～5.5cm，偶有白斑，叶齿密。植株开张，矮性，生长旺盛。冬末至翌年春季开花。

> **Abstract:** It is a mutation from 'Fenglu' (See p.504-505, our first book). Flowers light purple red with white borders of varying widths, semi-double to rose-double form, medium size, 8.5-9.5cm across and 5.0cm deep with 40-55 petals arranged orderly in 7-8 rows, petals obovate, rolled inward, apices nearly round, central petaloids twisted and mixed with stamens. Leaves green, shiny, leathery, elliptic, margins densely serrate. Plant spread, dwarf and growth vigorous. Blooms from late-winter to the spring of following year.

HC-11-1. 紫靓丽
Ziliangli (Purple Beauty)

杂交组合：红山茶品种'大红金心' × 非云南山茶杂交种'甜凯特'（*C. japonica* cultivar 'Dahong Jinxin' × Non-*reticulata* hybrid 'Sweet Emily Kate'）。

杂交苗编号：DHTKT-No.3。

性状：花朵紫红色，偶有少量白斑，半重瓣型，中型花，花径8.5～9.5cm，大花瓣18～21枚，呈3轮排列，阔倒卵形，外翻，波浪状，雄蕊簇生，通常5簇，基部略连生，雌蕊近退化。叶片绿色，革质，阔椭圆形，长7.5～8.5cm，宽4.5～5.0cm，叶脉清晰，先端尖，叶齿稀。植株开张，矮灌状，枝条稠密，生长旺盛。冬末至春季开花。

Abstract: Flowers purple red, occasionally with a few white markings, semi-double form, medium size, 8.5-9.5cm across with 18-21 large petals arranged in 3 rows, petals broad obovate, rolled outward and wavy, stamens clustered, usually 5 clusters which united at the base, pistil nearly degraded. Leaves green, leathery, broad obovate, veins visible, apices pointed, margins sparsely serrate. Plant spead, dwarf and growth vigorous. Blooms from late-winter to the spring of following year.

HC-12-1. 美秋
Meiqiu (Beautiful Autumn)

杂交组合： 红山茶品种'锯叶椿'×红山茶品种'克瑞墨大牡丹'（*C. japonica* cultivar 'Nokogiriba-tsubaki' × *C. japonica* cultivar 'Kramer's Supreme'）。

杂交苗编号： JYK-No.12。

性状： 花朵淡桃红色，略泛橘红色调，具芳香，半重瓣型至牡丹型，中型花，花径8.0～9.0cm，花朵厚4.5cm，外部大花瓣12～15枚，呈2轮排列，阔倒卵形，瓣面边缘略皱褶，先端微凹，中部有少量雄蕊瓣，雄蕊簇状。叶片中等绿色，革质，椭圆形，长6.5～8.0cm，宽4.0～4.5cm，边缘叶齿明显。植株紧凑，立性，枝叶稠密，生长旺盛。秋末始花，冬季盛花。

Abstract: Flowers lightly peach red, slightly with orange hue, with fragrance, semi-double to peony form, medium size, 8.0-9.0cm across and 4.5cm deep with 12-15 exterior large petals arranged in 2 rows, petals broad obovate, edges slightly crinkled, a few petaloids at the center, stamens clustered. Leaves normal green, leathery, elliptic, margins obviously serrate. Plant compact, upright, branches dense and growth vigorous. Starts to bloom from late-autumn and fully blooms in winter.

HC-12-2. 野秋
Yeqiu (Wild Autumn)

杂交组合： 红山茶品种'锯叶椿'× 红山茶品种'克瑞墨大牡丹'（*C. japonica* cultivar 'Nokogiriba-tsubaki' × *C. japonica* cultivar 'Kramer's Supreme'）。

杂交苗编号： JYK-No.14。

性状： 花朵淡桃红色至橘红色，略带芳香，瓣面具深红色脉纹，玫瑰重瓣型至完全重瓣型，中型至大型花，花径8.5～11.5cm，花朵厚5.0cm，花瓣30枚，呈6轮以上有序排列，阔倒卵形，边缘内卷，先端圆。叶片中等绿色，叶背面灰绿色，革质，椭圆形，长7.5～8.5cm，宽4.5～5.0cm，边缘具叶齿。植株紧凑，立性，枝叶稠密，生长旺盛。秋末始花，冬季盛花。

Abstract: Flowers lightly peach red to orange red with deep red veins on the petal surfaces, slight fragrance, rose-double to formal double form, medium to large size, 8.5-11.5cm across and 5.0cm deep with 30 petals arranged orderly in over 6 rows, petals broad obovate, edges rolled inward, apices round. Leaves normal green, leathery, elliptic, margins obviously serrate. Plant compact, upright, branches dense and growth vigorous. Starts to bloom from late-autumn and fully blooms in winter.

HC-13-1. 阿婆六之晨
Apoliu Zhichen (Apoliu's Morning)

杂交组合：红山茶品种'羽衣'×红山茶品种'佛会'（ *C. japonica* cultivar 'Hagoromo' × *C. japonica* cultivar 'Ryuge'）。

杂交苗编号：YYFH-No.1。

性状：花朵极淡的肉粉红色，渐向花瓣边缘粉色变粉白色，玫瑰重瓣型，中型花，花径8.5～9.5cm，花朵厚4.8cm，花瓣30～36枚，呈4～5轮松散、有序排列，花瓣长倒卵形，先端凹，雄蕊少量，花丝白色，花药淡黄色。叶片中等绿色，光亮，革质，阔椭圆形，长6.5～8.0cm，宽3.8～4.5cm，中脉明显，叶齿钝。植株紧凑，枝叶繁茂，生长旺盛。春初至春末开花。

注：本品种2019年由广东省从化区阿婆六茶花谷培育成功，并分别在中国花卉协会茶花分会和国际山茶协会登录。

Abstract: Flowers very lightly meat-pink, fading to pink-white at petal edges, rose-double form, medium size, 8.5-9.5cm across and 4.8cm deep with 30-36 petals arranged loosely and orderly in 4-5 rows, petals long obovate, stamens a few, filaments white. Leaves normal green, leathery, broad elliptic, margins obtusely serrate. Plant compact, branches dense and growth vigorous. Blooms from early-spring to late-spring.

Mark: The hybrid was bred by Apoliu Camellia Valley, Conghua District, Guangzhou City, Guangdong Province, China and registered in Chinese Camellia Branch Society, Chinese Flowers Association and in ICS in 2019 respectively.

HC-14-1. 荣归
Ronggui (Gloriously Return to Homeland)

杂交组合： 云南山茶品种'上海女士'×红山茶品种'香神'（*C. reticulata* cultivar 'Shanghai Lady' × *C. japonica* cultivar 'Scentsation'）。

杂交苗编号： SHNSXS-No.1。

性状： 花朵桃红色至淡红色，泛淡紫色调，渐向花芯呈粉白色，半重瓣型，巨型花，花径15.0～16.5cm，花朵厚4.5cm，大花瓣18～20枚，呈3轮排列，花瓣近卵圆形，厚质，边缘皱褶或波浪状，雄蕊簇生，花丝乳白色，花药黄色。叶片绿色，革质，长椭圆形，长12.0～13.0cm，宽4.0～4.5cm，叶面光滑，中脉凹陷，侧脉不明显，边缘叶齿细尖。植株立性，生长旺盛。冬末至春季开花。

Abstract: Flowers peach red to light red, with light purple hue, fading to white pink at the center, semi-double form, very large size, 15.0-16.5cm across and 4.5cm deep with 18-20 large petals arranged in 3 rows, petals nearly round, thick texture, edges crinkled or wavy, stamens clustered, filaments milk-white, anthers yellow. Leaves green, leathery, long elliptic, surfaces smooth, midribs sunken, lateral veins unconspicuous, margins finely and sharply serrate. Plant upright and growth vigorous. Blooms from late-winter to the spring of following year.

HC-15-1. 阿婆六女神
Apoliu Nüshen（Apoliu's Goddess）

杂交组合：云南山茶品种'梦女'×云南山茶品种'雪娇'（*C. reticulata* cultivar 'Dream Girl'× *C. reticulata* cultivar 'Xuejiao'）。

杂交苗编号：MNXJ-No.1。

性状：花朵深粉红色至淡红色，略泛淡紫色调，半重瓣型至牡丹型，中型至大型花，花径9.0～11.5cm，花朵厚5.5cm，外部大花瓣15～18枚，呈3轮松散排列，花瓣阔倒卵形，外翻，波浪状，瓣面皱褶，中部小花瓣直立、扭曲，与雄蕊混生，花丝和花药淡黄色，开花稠密。叶片深绿色，厚革质，叶面略粗糙，长椭圆形，长7.5～9.0cm，宽3.0～3.5cm，叶齿细尖。植株开张，枝叶稠密，生长旺盛。冬末至翌年春季开花。

注：本品种2019年由广东省从化区阿婆六茶花谷培育成功，并分别在中国花卉协会茶花分会和国际山茶协会登录。

Abstract: Flowers deep pink to light red, slightly with pale purple hue, semi-double to peony form, medium to large size, 9.0-11.5cm across and 5.5cm deep with 15-18 exterior large petals loosely arranged in 3 rows, petals broad obovate, rolled outward, wavy, petal surfaces wrinkled, central small petals erected, twisted and mixed with stamens, bloom dense. Leaves deep green, heavy leathery, slightly rough on surfaces, long elliptic, margins finely and sharply serrate. Plant spread, branches dense and growth vigorous. Blooms from late-winter to the spring of following year.

Mark: The hybrid was bred by Apoliu Camellia Valley, Conghua District, Guangzhou City, Guangdong Province, China and registered in Chinese Camellia Branch Society, Chinese Flowers Association and in ICS in 2019 respectively.

HC-16-1. 留芳亭
Liufangting (Gloriette with Fragrance)

杂交组合： 尖萼红山茶 × 红山茶品种'香屋'（*C. edithae* × *C. japonica* cultivar 'Scentsation'）。

杂交苗编号： JEXW-No.1。

性状： 花朵淡红色，泛淡紫色调，略具芳香，单瓣型，小型至中型花，花径6.5～8.5cm，花瓣5枚，阔倒卵形，边缘皱褶，雄蕊圆筒状，基部与花瓣连生，子房被柔毛，开花稠密。嫩叶被毛，成熟叶绿色，革质，长倒卵形，长10.3～10.5cm，宽3.8～4.4cm，叶脉凹陷，基部心形，先端尖，叶齿细尖，叶柄被柔毛。植株紧凑，立性，枝条稠密，生长旺盛。冬末至春季开花。

Abstract: Flowers light red, with light purple hue and slight fragrant, single form, small to medium size, 6.5-8.5cm across with 5 petals, petals broad obovate, edges crinkled, stamens columnar, united with petals at the base, ovary pubescent, bloom dense. Tender leaves pubescent, mature leaves green, leathery, long obovate, veins sunken, base heart-shaped, petioles pubescent, apices pointed, margins finely and sharply serrate. Plant compact, upright, branches dense and growth vigorous. Blooms from late-winter to the spring of following year.

HC-17-1. 勇士 Yongshi（Warrior）

杂交组合： 浙江红山茶 × 红山茶品种'黑魔法'（*C. chekiangoleosa* × *C. japonica* cultivar 'Black Magic'）。

杂交苗编号： ZHMF-No.2 + 4。

性状： 花朵黑红色，具蜡光，偶有白色条纹，半重瓣型，中型至大型花，花径9.5～12.5cm，花朵厚5.5cm，大花瓣13～15枚，呈2轮松散排列，花瓣阔倒卵形，外翻，瓣面略皱褶，雄蕊基部略连生，开花稠密。叶片绿色，厚革质，长椭圆形，长10.0～11.5cm，宽4.5～5.5cm，两侧内翻，呈沟槽状，略扭曲，边缘叶齿深尖。植株开张，生长旺盛。冬末始花，春季盛花。

> **Abstract:** Flowers dark red with waxy luster, occasionally with white stripes, semi-double form, medium to large size, 9.5-12.5cm across and 5.5cm deep with 13-15 large petals arranged loosely in 2 rows, petals broad obovate, turned outward, petal surfaces slightly crinkled, stamens slightly united at the base, bloom dense. Leaves green, heavy leathery, long elliptic, both sides turned inward into a groove, slightly twisted, margins deeply and sharply serrate. Plant spread and growth vigorous. Starts to bloom in late-winter and fully blooms in spring.

HC-17-2. 娘子军
Niangzijun (Women Army)

杂交组合： 浙江红山茶 × 红山茶品种'黑魔法'（*C. chekiangoleosa* × *C. japonica* cultivar 'Black Magic'）。

杂交苗编号： ZHMF-No.10。

性状： 花朵黑红色，具蜡光，渐向基部呈红色，偶有白色条纹，花瓣边缘近黑色，半重瓣型，大型至巨型花，花径12.5～13.5cm，花朵厚5.5cm，大花瓣14～15枚，呈2轮松散排列，花瓣阔倒卵形，厚质，上部边缘内卷，雄蕊基部略连生，散射状，开花稠密。叶片绿色，厚革质，椭圆形，长9.0～9.5cm，宽4.0～5.0cm，中脉凹，边缘叶齿细。植株开张，生长旺盛。冬末始花，春季盛花。

Abstract: Flowers dark red with waxy luster, fading to red at the petal bases, occasionally with white stripes, edges nearly black, semi-double form, large to very large size, 12.5-13.5cm across and 5.5cm deep with 14-15 large petals arranged loosely in 2 rows, petals broad obovate, thick texture, upper parts rolled inward, stamens slightly united at the base, scattered, bloom dense. Leaves green, heavy leathery, elliptic, midribs sunken, margins finely serrate. Plant spread and growth vigorous. Starts to bloom in late-winter and fully blooms in spring.

茶花机遇苗新品种选育概况
The Outline on the New Camellia Varieties Selected from Chance Seedlings

机遇苗新品种实际上就是不知道父本的自然杂交新品种。正如我们在第一部书中所述的，从茶花实生苗中选择具有较好性状的机遇苗，是茶花育种的一条捷径。在第一部书中，我们曾介绍过31个从机遇苗选育出的优秀茶花品种。这些新品种是从24个母本品种的实生苗中获得的（见第一部书第519～556页）。

> Chance new varieties are actually natural new hybrids that do not know their male-parents. As what we wrote in our first book, to select better chance seedlings with better characteristics from camellia seedlings was a shortcut for camellia breeding. Thirty one excellent camellia varieties selected from seedlings had been introduced in our first book. These new varieties were obtained from the seedlings of 24 female cultivars (See p.519-556, our first book).

机育苗新品种 New varieties obtained from seedlings

开花后，从实生苗中选育新品种
Selected new varieties from the seedlings after they bloom

机遇苗品种选育示意图
A sketch of new camellia varieties selected from chance seedlings

本书将介绍45个机遇苗新品种，它们是从25个茶花品种的实生苗中选育出来的。其中，母本与第一部书相同的有8个，不同的有17个。如果加上第一部书列出的机遇苗新品种，到目前为止，所获得的机遇苗新品种已达76个，涉及的母本有41个。

因为所获得的机遇苗新品种都属于常规茶花品种，所以它们的花期也与常规茶花品种相同，即冬末至翌年春季开花。

最后应该提及的是，凡在我们第一部书中已有的机遇苗品种的母本，本续集所获得的机遇苗新品种的编号将延续第一部书的序列编号。比如，HD-01，HD-02，HD-10，HD-11，HD-13，HD-14，HD-20，HD-23 这 8 个母本，它们的机遇苗品种的编号都是接着第一部书的编号排序的。

This book will introduce 45 chance new varieties that selected from the seedlings of 25 camellia cultivars. Of them, there are eight female cultivars that are as same as what in our first book and 17 female cultivars are different from what in our first book. If the new chance varieties of our first book are all added, 76 new chance varieties, which involve in 41 female cultivars would have obtained in total until now.

The new varieties selected from chance seedlings all belong to normal camellias, so their blooming period is as same as normal camellias that is from late-winter to the spring of following year.

Finally, it should be mentioned that in all female cultivars that had introduced in our first book, the numbers of the new chance varieties in this book will follow the serial numbers of our first book, such as the eight female cultivars: HD-01, HD-02, HD-10, HD-11, HD-13, HD-14, HD-20 and HD-23, they are all numbered in succession to our first book.

小惊喜
(Xiaojingxi) HD-10-2

超丽
(Chaoli) HD-20-6

攀登
(Pandeng) HD-23-4

六角亭
(Liujiaoting) HD-29-1

茶花机遇苗新品种详解

The Detailed Descriptions for the New Camellia Varieties Selected from Chance Seedlings

HD-01-2. 婚纱
Hunsha (Wedding Dress)

性状：第一部书中本组曾介绍过1个品种（见第一部书第521～522页），其编号接续第一部书。本品种选自红山茶品种'大菲丽丝'的机遇苗 DFLS-No.1。花朵洁白色，偶有少量红条纹，半重瓣型至牡丹型，中型至大型花，花径8.5～11.0cm，花朵厚6.0cm。大花瓣约30枚，呈5～6轮排列，外翻，波浪状，花瓣边缘深度皱褶，中部少量小花瓣，雄蕊簇生。叶片绿色，革质，长椭圆形，略扭曲，边缘叶齿深而尖。植株立性，枝条细软，生长中等。冬初至翌年春季开花。

Abstract: A chance seedling DFLS-No.1 selected from *C. japonica* cultivar 'Francis Eugene Phillis'. Flowers pure white, semi-double to peony form, medium to large size, 8.5-11.0cm across and 6.0cm deep with about 30 large petals arranged in 5-6 rows, large petals turned outward, wavy, petal edges heavily crinkled, a few small petals at the center, stamens clustered. Leaves green, slightly twisted, margins deeply and sharply serrate. Plant upright, branches soft and growth normal. Blooms from early-winter to the spring of following year.

HD-02-2. 金源
Jinyuan (Gold Fountainhead)

性状：第一部书中本组曾介绍过1个品种（见第一部书第522～523页），其编号接续第一部书。本品种选自红山茶品种'佛蕾德'的机遇苗FLD-No.1。花朵黑红色，泛蜡光，半重瓣型，小型至中型花，花径7.0～9.0cm，花朵厚6.0cm，外部大花瓣15枚以上，呈2轮松散排列，外翻或波浪状，花芯处偶有具白斑的雄蕊瓣，雄蕊散射状。叶片绿色，光亮，革质，长椭圆形，边缘齿浅尖。植株开张，枝条稠密，生长旺盛。冬末至翌年春季开花。

Abstract: A chance seedling FLD-No.1 selected from *C. japonica* cultivar 'Fred Sander'. Flowers dark red with waxy luster, semi-double form, small to medium size, 7.0-9.0cm across and 6.0cm deep with more than 15 exterior large petals arranged loosely in 2 rows, the petals turned outward or wavy, occasionally with white markings on the petaloids at the center, stamens scattered. Leaves green, shiny, long elliptic, margins shallowly and sharply serrate. Plant spread, branches dense and growth vigorous. Blooms from late-winter to the spring of following year.

HD-10-2. 小惊喜 Xiaojingxi (Little Surprise)

性状： 第一部书中本组曾介绍过1个品种（见第一部书第531页），其编号接续第一部书。本品种选自红山茶品种'媚丽'的机遇苗ML-No.21。花朵深红色至黑红色，有蜡质感，边缘具宽白边，完全重瓣型，微型至小型花，花径5.5～6.5cm，花朵厚4.5cm，花瓣50枚以上，呈5轮以上排列，外翻，开花稠密。叶片绿色，革质，椭圆形，边缘叶齿细密。植株开张，生长旺盛。冬初至翌年春季开花。

Abstract: A chance seedling ML-No. 21 selected from *C. japonica* cultivar 'Tama Beauty'. Flowers deep red to dark red with broad white borders at petal edges, formal double form, miniature to small size, 5.5-6.5cm across and 4.5cm deep with more than 50 petals arranged in over 5 rows, petals turned outward, bloom dense. Leaves green, leathery, elliptic, margins densely and finely serrate. Plant spread and growth vigorous. Blooms from early-winter to the spring of following year.

HD-10-3. 小玉镯
Xiaoyuzhuo (Small Jade Bracelet)

性状： 其编号接续上一个品种。本品种选自红山茶品种'媚丽'的机遇苗ML-No.28。花朵深红色，边缘具窄白边，单瓣型，微型至小型花，花径5.5～6.0cm，花瓣5～6枚，扇形，瓣面略波浪状，雄蕊基部连生，开花稠密。叶片绿色，革质，长椭圆形，先端尖，边缘叶齿细密。植株开张，枝条略下垂，生长旺盛。冬初至翌年春季开花。

Abstract: A chance seedling ML-No. 28 selected from *C. japonica* cultivar 'Tama Beauty'. Flowers deep red with narrow white borders at petal edges, single form, miniature to small size, 5.5-6.0cm across with 5-6 fan-shaped petals, petal surfaces slightly wavy, stamens united at the base, bloom dense. Leaves green, leathery, long elliptic, tips pointed, margins densely and finely serrate. Plant spread, branches slightly hanging down and growth vigorous. Blooms from early-winter to the spring of following year.

HD-10-4. 粉玉镯 Fenyuzhuo（Pink Jade Bracelet）

性状： 其编号接续上一个品种。本品种选自红山茶品种'媚丽'的机遇苗 ML-No.37。花朵粉红色，边缘具模糊的窄白边，半重瓣型，中型花，花径 8.5～9.0cm，花朵厚 4.0cm，花瓣 18 枚，呈 2～3 轮排列，阔倒卵形，略波浪状，雄蕊基部连生，开花稠密。叶片绿色，革质，椭圆形，边缘叶齿尖密。植株开张，枝条软，生长旺盛。冬末至翌年春季开花。

Abstract: A chance seedling ML-No.37 selected from *C. japonica* cultivar 'Tama Beauty'. Flowers pink with faint white borders at petal edges, semi-double form, medium size, 8.5-9.0cm across and 4.0cm deep with 18 petals arranged in 2-3 rows, petals broad obovate, slightly wavy, stamens united at the base, bloom dense. Leaves green, leathery, long elliptic, margins sharply and densely serrate. Plant spread, branches soft and growth vigorous. Blooms from late-winter to the spring of following year.

HD-10-5. 大玉镯
Dayuzhuo (Large Jade Bracelet)

性状：其编号接续上一个品种。本品种选自红山茶品种'媚丽'的机遇苗 ML-No.45。花朵深红色，边缘具明显的白边，半重瓣型，中型至大型花，花径 9.0～12.0cm，花朵厚 4.8cm，花瓣 12～13 枚，呈 2 轮排列，阔倒卵形，略波浪状，雄蕊基部连生，开花稠密。叶片绿色，革质，长椭圆形，略扭曲，边缘叶齿尖而密。植株开张，枝条硬挺，生长旺盛。冬末至翌年春季开花。

Abstract: A chance seedling ML-No. 45 selected from *C. japonica* cultivar 'Tama Beauty'. Flowers deep red with obviously white borders at petal edges, semi-double form, medium to large size, 9.0-12.0cm across and 4.8cm deep with 12-13 petals arranged in 2 rows, petal broad obovate, slightly wavy, stamens united at the base, bloom dense. Leaves green, leathery, long elliptic, slightly twisted, margins sharply and densely serrate. Plant spread, branches hard and growth vigorous. Blooms from late-winter to the spring of following year.

HD-11-2. 洪福
Hongfu（Vast Happiness）

性状： 第一部书中本组曾介绍过1个品种（见第一部书第532～533页），其编号接续第一部书。本品种选自红山茶品种'南京玛瑙'的机遇苗NJMN-No.1。花朵红色，泛紫色调，具少量白斑块，半重瓣型，中型花，花径8.5～9.0cm，花朵厚约4.5cm。花瓣15～17枚，呈2～3轮排列，瓣面略皱褶，雄蕊基部略连生，开花稠密。叶片绿色，革质，椭圆形，边缘叶齿钝。植株立性，生长旺盛。冬初至翌年春季开花。

Abstract: A chance seedling NJMN-No.1 selected from *C. japonica* cultivar 'Nanjing Manao'. Flowers red with purple hue and a few white markings, semi-double form, medium size, 8.5-9.0cm across and about 4.5cm deep with 15-17 petals arranged in 2-3 rows, petal surfaces slightly crinkled, stamens slightly united at the base, bloom dense. Leaves green, leathery, elliptic, margins obtusely serrate. Plant upright and growth vigorous. Blooms from early-winter to the spring of following year.

HD-13-2. 雪山 Xueshan (Snow Mountain)

性状： 第一部书中本组曾介绍过1个品种（见第一部书第534～535页），其编号接续第一部书。本品种选自红山茶品种'月光湾'的机遇苗YGW-No.22。花朵洁白色，半重瓣型，中型至大型花，花径9.5～12.0cm，花朵厚4.5cm，大花瓣23～25枚，呈3～5轮排列，外翻，雄蕊簇生。叶片绿色，光亮，革质，椭圆形，边缘齿密。植株立性，枝条稠密，生长中等。冬末至翌年春季开花。

Abstract: A chance seedling YGW-No.22 selected from *C. japonica* cultivar 'Moonlight Bay'. Flowers pure white, semi-double form, medium to large size, 9.5-12.0cm across and 4.5cm deep with 23-25 large petals arranged in 3-5 rows, petals turned outward, stamens clustered. Leaves green, shiny, leathery, elliptic, margins densely serrate. Plant upright, branches dense and growth normal. Blooms from late-winter to the spring of following year.

HD-13-3. 北极冰 Beijibing (Arctic Ice)

性状： 其编号接续上一个品种。本品种选自红山茶品种'月光湾'的机遇苗 YGW-No.30。花朵白色，花瓣基部略显淡黄色，半重瓣型至牡丹型，中型花，花径 8.5～9.5cm，花朵厚 5.0cm，外部大花瓣 16～18 枚，呈 2～3 轮排列，花瓣近圆形，中部小花瓣 10～15 枚，半直立，扭曲，略皱褶，与雄蕊混生，雄蕊簇生。叶片绿色，光亮，革质，椭圆形，边缘叶齿钝。植株紧凑，生长中等。春初至春末开花。

Abstract: A chance seedling YGW-No.30 selected from *C. japonica* cultivar 'Moonlight Bay'. Flowers white, changing to light yellow at the base of the petals, semi-double to peony form, medium size, 8.5-9.5cm across and 5.0cm deep with 16-18 exterior large petals arranged in 2-3 rows, petals nearly round, 10-15 small petals semi-erected, twisted and crinkled at the center and mixed with stamens. Leaves green, shiny, leathery, elliptic, margins obtusely serrate. Plant compact and growth normal. Blooms from early-spring to late-spring.

HD-14-3. 陶醉 Taozui (Intoxication)

性状： 第一部书中本组曾介绍过2个品种（见第一部书第535～536页和第540～541页），其编号接续第一部书。本品种选自红山茶品种'白斑康乃馨'的机遇实生苗NT-No.3。花朵红色至黑红色，泛淡紫色调，完全重瓣型，中型至大型花，花径9.5～11.5cm，花朵厚约5.5cm，花瓣60枚以上，呈6轮以上排列，刚开放时，花瓣排列紧实，全开放后，花瓣排列松散，花瓣外翻，近圆形，长5.5cm，宽5.0cm。叶片绿色，厚革质，椭圆形，长7.5～9.0cm，宽4.0～4.5cm，边缘叶齿钝。植株开张，枝条稠密，生长旺盛。冬初至翌年春季开花。

Abstract: A chance seedling NT-No.3 selected from *C. japonica* cultivar 'Ville de Nantes'. Flowers red to dark red with light purple hue, formal double form, medium to large size, 9.5-11.5cm across and about 5.5cm deep with more than 60 petals arranged in over 6 rows, when flowers just opening, the petals arrange tightly and after they fully opened, the petals arrange loosely and turn outward, petals nearly round. Leaves green, heavy leathery, elliptic, margins obtusely serrate. Plant spread, branches dense and growth vigorous. Blooms from early-winter to the spring of following year.

HD-20-3. 初恋
Chulian (First Love)

性状： 第一部书中本组曾介绍过 2 个品种（见第一部书第 542 ～ 544 页），其编号接续第一部书。本品种选自红山茶品种'羽衣'的机遇苗 YY-No.4。花朵白色，泛极淡的羞红色，渐向花芯呈白色，完全重瓣型，中型花，花径 8.5 ～ 9.5cm，花朵厚 4.5cm，花瓣 80 枚以上，呈 10 轮以上覆瓦状排列，花瓣阔倒卵形，开花稠密。叶片绿色，光亮，硬革质，椭圆形，略波浪状，边缘叶齿尖。植株紧凑，枝条坚挺，生长旺盛。冬末至翌年春季开花。

Abstract: A chance seedling YY-No.4 selected from *C. japonica* cultivar 'Hagoromo'. Flowers very lightly blush red, fading to white at the center, formal double form, medium size, 8.5-9.5cm across and 4.5cm deep with more than 80 petals arranged imbricately in over 10 rows, petals broad obovate, bloom dense. Leaves green, shiny, hard leathery, elliptic, margins sharply serrate. Plant compact, branches strong and growth vigorous. Blooms from late-winter to the spring of following year.

HD-20-4. 白娘子 Bainiangzi (White Snake Lady)

性状： 其编号接续上一个品种。本品种选自红山茶品种'羽衣'的机遇苗 YY-No.7。花朵洁白色，花芯部分泛淡黄色，半重瓣型，中型至大型花，花径 9.0～11.5cm，花朵厚约 4.8cm，花瓣约 18 枚，呈 3 轮松散排列，花瓣阔倒卵形，雄蕊基部连生。叶片绿色，革质，椭圆形，边缘叶齿细尖。植株紧凑，灌丛状，生长旺盛。冬末至翌年春季开花。

> **Abstract:** A chance seedling YY-No.7 selected from *C. japonica* cultivar 'Hagoromo'. Flowers pure white with light yellow at the center, semi-double form, medium to large size, 9.0-11.5cm across and about 4.8cm deep with about 18 petals arranged loosely in 3 rows, petal broad obovate, stamens united at the base. Leaves green, leathery, elliptic, margins finely and sharply serrate. Plant compact, bushy and growth vigorous. Blooms from late-winter to the spring of following year.

HD-20-5. 和顺
Heshun (Harmony & Smooth)

性状：其编号接续上一个品种。本品种选自红山茶品种'羽衣'的机遇苗 YY-No.10 + 14。花朵白色，具红色条纹，半重瓣型，中型花，花径 7.5～8.5cm，花朵厚 4.5cm，花瓣 12～18 枚，呈 2～3 轮排列，花瓣阔倒卵形，雄蕊基部连生，开花稠密。叶片绿色，光亮，革质，椭圆形，边缘叶齿钝。植株开张，灌丛状，生长旺盛。冬末至翌年春季开花。

Abstract: A chance seedling YY-No.10+14 selected from *C. japonica* cultivar 'Hagoromo'. Flowers white with red stripes, semi-double form, medium size, 7.5-8.5cm across and 4.5cm deep with 12-18 petals arranged in 2-3 rows, petals broad obovate, stamens united at the base, bloom dense. Leaves green, shiny, leathery, elliptic, margins obtusely serrate. Plant spread, bushy and growth vigorous. Blooms from late-winter to the spring of following year.

HD-20-6. 超丽
Chaoli (Surpassed Beauty)

性状： 其编号接续上一个品种。本品种选自红山茶品种'羽衣'的机遇苗 YY-No.11。花朵柔和的淡紫粉红色，渐向花瓣基部变为淡粉红色，有时会出现白色带红条纹的花朵，完全重瓣型，中型至大型花，花径 9.5～12.5cm，花朵厚 5.0cm，花瓣 90 枚以上，呈 13～15 轮覆瓦状排列，中部花瓣边缘内卷。叶片绿色，光亮，革质，阔椭圆形，边缘叶齿浅。植株立性，枝条稠密，生长旺盛。冬末至翌年春季开花。

Abstract: A chance seedling YY-No.11 selected from *C. japonica* cultivar 'Hagoromo'. Flowers soft and light purple pink, fading to light pink at the petal bases, sometimes white flowers with red stripes, formal double form, medium to large size, 9.5-12.5cm across and 5.0cm deep with over 90 petals arranged imbricately in 13-15 rows, the edges of central petals rolled inward. Leaves green, shiny, leathery, broad elliptic, margins shallowly serrate. Plant upright, branches dense and growth vigorous. Blooms from late-winter to the spring of following year.

HD-20-7. 千秋旺 Qianqiuwang (Forever Prosperous)

性状： 其编号接续上一个品种。本品种选自红山茶品种'羽衣'的机遇苗YY-No.12＋15。花朵艳红色，中部花瓣淡红色，偶有少量白条纹，半重瓣型至牡丹型，中型至大型花，花径9.0～12.5cm，花朵厚约5.3cm，大花瓣18～21枚，呈3轮松散排列，大波浪状，边缘皱褶，雄蕊基部连生。叶片绿色，光亮，厚革质，椭圆形，边缘叶齿细。植株紧凑，立性，生长旺盛。冬末至翌年春季开花。

Abstract: A chance seedling YY-No.12+15 selected from *C. japonica* cultivar 'Hagoromo'. Flowers bright red, central petals light red, occasionally with a few white stripes, semi-double to peony form, medium to large size, 9.0-12.5cm across and about 5.3cm deep with 18-21 large petals arranged loosely in 3 rows, large petals big wavy, edges crinkled, stamens united at the base. Leaves green, shiny, heavy leathery, elliptic, margins finely serrate. Plant compact, upright and growth vigorous. Blooms from late-winter to the spring of following year.

HD-20-8. 紫绸
Zichou (Purple Silk)

性状： 其编号接续上一个品种。本品种选自红山茶品种'羽衣'的机遇苗 YY-No.13。花朵红色，泛紫色调，半重瓣型，中型至大型花，花径9.0～12.5cm，花朵厚5.0cm，花瓣12～14枚，呈2轮排列，花瓣阔倒卵形，波浪状，雄蕊基部略连生。叶片绿色，光亮，革质，椭圆形，边缘叶齿细。植株开张，灌丛状，生长旺盛。冬末至翌年春季开花。

Abstract: A chance seedling YY-No.13 selected from *C. japonica* cultivar 'Hagoromo'. Flowers red with purple hue, semi-double form, medium to large size, 9.0-12.5cm across and 5.0cm deep with 12-14 petals arranged in 2 rows, petals broad obovate, wavy, stamens united at the base. Leaves green, shiny, leathery, elliptic, margins finely serrate. Plant spread, bushy and growth vigorous. Blooms from late-winter to the spring of following year.

HD-20-9. 羞女 Xiunü (Shy Girl)

性状： 其编号接续上一个品种。本品种选自红山茶品种'羽衣'的机遇苗YY-No.18。花朵柔和的羞粉红色，半重瓣型，中型至大型花，花径8.5～12.0cm，花朵厚5.0cm，花瓣13～16枚，呈2～3轮排列，瓣面略皱褶，雄蕊基部连生，开花稠密。叶片绿色，光亮，革质，阔椭圆形，边缘叶齿细。植株开张，灌丛状，生长旺盛。冬末至翌年春季开花。

Abstract: A chance seedling YY-No.18 selected from *C. japonica* cultivar 'Hagoromo'. Flowers softly blush pink, semi-double form, medium to large size, 8.5-12.0cm across and 5.0cm deep with 13-16 petals arranged in 2-3 rows, petal surfaces slightly crinkled, stamens united at the base, bloom dense. Leaves green, shiny, leathery, broad elliptic, margins finely serrate. Plant spread, bushy and growth vigorous. Blooms from late-winter to the spring of following year.

HD-20-10. 笑脸
Xiaolian (Smile Face)

性状： 其编号接续上一个品种。本品种选自红山茶品种'羽衣'的机遇苗 YY-No.21。花朵白色，具许多红条纹和红色斑点，半重瓣型，中型花，花径 8.0～9.5cm，花朵厚 4.0cm，花瓣 18～24 枚，呈 3～4 轮排列，中部花瓣边缘内卷，先端尖。叶片绿色，光亮，革质，椭圆形，边缘叶齿细。植株紧凑，生长旺盛。冬末至翌年春季开花。

Abstract: A chance seedling YY-No.21 selected from *C. japonica* cultivar 'Hagoromo'. Flowers white with many red stripes and red spots, semi-double form, medium size, 8.0-9.5cm across and 4.0cm deep with 18-24 petals arranged in 3-4 rows, central petals rolled inward at the edges. Leaves green, shiny, leathery, elliptic, margins finely serrate. Plant compact and growth vigorous. Blooms from late-winter to the spring of following year.

HD-23-4. 攀登 Pandeng (Climbing)

性状： 第一部书中本组曾介绍过 3 个品种（见第一部书第 549～552 页），其编号接续第一部书。本品种选自红山茶品种'石达琳'的机遇苗 SDL-No.3。花朵深粉红色，泛淡紫色调，偶有少量小白斑，玫瑰重瓣型至完全重瓣型，小型花，花径 6.5～7.5cm，花朵厚 4.0cm，花瓣 50 枚以上，呈 10 轮以上螺旋状排列，外部花瓣外翻，开花稠密。叶片绿色，革质，椭圆形，边缘叶齿细密。植株紧凑，立性，生长旺盛。冬季至翌年春季开花。

Abstract: A chance seedling SDL-No.3 selected from *C. japonica* cutivar 'Shidalin'. Flowers deep pink with lightly purple hue, occasionally a few small white markings visible, rose-double to formal-double form, small size, 6.5-7.5cm across and 4.0cm deep with more than 50 petals arranged helically in over 10 rows, exterior petals turned outward, bloom dense. Leaves green, leathery, elliptic, margins densely and finely serrate. Plant compact, upright and growth vigorous. Blooms from winter to the spring of following year.

HD-23-5. 恬静
Tianjing（Riposato）

性状： 其编号接续上一个品种。本品种选自红山茶品种'石达琳'的机遇实生苗SDL-No.2＋4。花朵柔和的粉红色，单瓣型，中型花，花径7.5～8.5cm，花瓣5～6枚，长4.2cm，宽3.0cm，瓣面略皱，雄蕊基部连生呈柱状，开花稠密。叶片绿色，革质，长椭圆形，略扭曲，边缘叶齿细密。植株紧凑，生长旺盛。冬季至翌年春季开花。

Abstract: A chance seedling SDL-No.2+4 selected from *C. japonica* cutivar 'Shidalin'. Flowers soft pink, single form, medium size, 7.5-8.5cm across with 5-6 petals, petal surfaces slightly crinkled, stamens united into a column at the base, bloom dense. Leaves green, leathery, long elliptic, slightly twisted, margins densely and finely serrate. Plant compact and growth vigorous. Blooms from winter to the spring of following year.

HD-25-1. 大玉 Dayu (Large Jade)

性状： 选自非云南山茶杂交种'劳克'的机遇苗LK-No.3。花朵洁白色，外部花瓣略显极淡的粉红色，半重瓣型，中型至大型花，花径8.5～11.0cm，花朵厚4.5cm，花瓣约20枚，呈3轮排列，雄蕊基部连生，呈柱状，开花稠密。叶片绿色，革质，长椭圆形，边缘叶齿细密，先端尖。植株开张，生长旺盛。冬季至翌年春季开花。

Abstract: A chance seedling LK-No.3 selected from Non-*reticulata* hybrid 'Loki Schmidt'. Flowers pure white with very light pink on the exterior petals, semi-double form, medium to large size, 8.5-11.0cm across and 4.5cm deep with about 20 petals arranged in 3 rows, stamens united at the base into a column, bloom dense. Leaves green, leathery, long elliptic, margins densely and finely serrate, apices pointed. Plant spread and growth vigorous. Blooms from winter to the spring of following year.

HD-25-2. 小玉
Xiaoyu (Small Jade)

性状：选自非云南山茶杂交种'劳克'的机遇苗 LK-No.4。花朵洁白色，半重瓣型，小型至中型花，花径6.5～8.0cm，花朵厚4.0cm，花瓣18枚左右，呈2～3轮排列，雄蕊簇状，花丝奶白色，花药黄色，开花稠密。叶片绿色，革质，长椭圆形，略扭曲，边缘叶齿细密。植株开张，生长旺盛。冬季至翌年春季开花。

Abstract: A chance seedling LK-No.4 selected from Non-*reticulata* hybrid 'Loki Schmidt'. Flowers pure white, semi-double form, small to medium size, 6.5-8.0cm across and 4.0cm deep with about 18 petals arranged in 2-3 rows, stamens clustered, bloom dense. Leaves green, leathery, long elliptic, slightly twisted, margins densely and finely serrate. Plant spread and growth vigorous. Blooms from winter to the spring of following year.

HD-26-1. 小礼物
Xiaoliwu (Small Gift)

性状： 选自非云南山茶杂交种'达格玛'的机遇苗 DGM-No.1。花朵浅红色，渐向花瓣边缘呈粉白色，瓣面可见深红色脉纹，单瓣型，微型至小型花，花径 5.0～7.0cm，花瓣 5～6 枚，阔倒卵形，边缘略皱褶，雄蕊基部连生呈粗柱状，开花稠密。叶片绿色，光亮，革质，椭圆形，边缘齿尖。植株立性，枝条稠密，生长旺盛。冬末至翌年春季开花。

Abstract: A chance seedling DGM-No.1 selected from Non-*reticulata* hybrid 'Dagmar Bergholf'. Flowers light red, fading to white pink at the edges of petals, deep red veins visible on the petal surfaces, single form, miniature to small size, 5.0-7.0cm across with 5-6 petals, petals broad obovate, edges slightly crinkled, stamens united into a rough column, bloom dense. Leaves green, shiny, leathery, elliptic, margins sharply serrate. Plant upright, branches dense and growth vigorous. Blooms from late-winter to the spring of following year.

HD-27-1. 粉星
Fenxing (Pink Stars)

性状： 选自非云南山茶杂交种'莫塔教授'的机遇苗 MTJS-No.1。花朵粉红色，略泛橘红色调，单瓣型，微型花，花径 5.5～6.0cm，花瓣 5～6 枚，其中 2 枚花瓣较大而导致花型不规整，花瓣倒卵形，薄质，边缘波浪状，雄蕊基部连生呈柱状，开花稠密。叶片绿色，薄革质，椭圆形，边缘叶齿细密。植株开张，灌丛状，枝条细软，略下垂，生长旺盛。冬末至翌年春季开花。

Abstract: A chance seedling MTJS-No.1 selected from Non-*reticulata* hybrid 'Professor Gianmario Motta'. Flowers pink with some orange hue, single form, miniature size, 5.5-6.0cm across with 5-6 petals, two of the petals are larger, petals obovate, thin texture, edges wavy, stamens united into a column, bloom dense. Leaves green, thin leathery, elliptic, margins finely and densely serrate. Plant spread, bushy, branches soft and hanging down and growth vigorous. Blooms from late-winter to the spring of following year.

HD-27-2. 小紫童 Xiaozitong（Little Purple Child）

性状：选自非云南山茶杂交种'莫塔教授'的机遇苗 MTJS-No.2。花朵紫红色，单瓣型，微型至小型花，花径 5.5～6.5cm，花瓣 5～6 枚，瓣面略皱，可见深红色脉纹，雄蕊连生，呈柱状，开花稠密。叶片绿色，光亮，革质，小椭圆形，边缘叶齿浅而尖。植株开张，枝条稠密，生长旺盛。冬末至翌年春季开花。

Abstract: A chance seedling MTJS-No.2 selected from Non-*reticulata* hybrid 'Professor Gianmario Motta'. Flowers purple red, single form, miniature to small size, 5.5-6.5cm across with 5-6 petals, slightly crinkled on the petal surfaces, stamens united into a column, bloom dense. Leaves green, shiny, leathery, small elliptic, margins shallowly and sharply serrate. Plant spread, branches dense and growth vigorous. Blooms from late-winter to the spring of following year.

HD-27-3. 红耳坠
Hongerzhui（Red Eardrop）

性状： 选自非云南山茶杂交种'莫塔教授'的机遇苗 MTJS-No.5。花朵艳红色，单重瓣型，微型花，花径 5.0～6.0cm，花瓣 5～6 枚，倒卵形，波浪状，雄蕊基部连生，花丝淡红色，柱头高出雄蕊 0.5～1.0cm，开花稠密。叶片绿色，革质，小椭圆形，边缘叶齿细。植株立性，紧凑，生长旺盛。冬末至翌年春季开花。

Abstract: A chance seedling of MTJS-No.5 selected from Non-*reticulata* hybrid 'Professor Gianmario Motta'. Flowers bright red, single form, miniature size, 5.0-6.0cm across with 5-6 petals, petals obovate, wavy, stamens united at the base, filaments pale red, pistils 0.5-1.0cm higher than stamens, bloom dense. Leaves green, leathery, small elliptic, margins finely serrate. Plant upright, compact and growth vigorous. Blooms from late-winter to the spring of following year.

HD-27-4. 小粉童 Xiaofentong（Little Pink Child）

性状： 选自非云南山茶杂交种'莫塔教授'的机遇苗 MTJS-No.6。花朵淡粉红色，单瓣型，微型花，花径 5.5～6.0cm，花瓣 5～6 枚，阔倒卵形，瓣面略皱，边缘波浪状，雄蕊连生，呈柱状，开花稠密。叶片绿色，光亮，革质，椭圆形，边缘叶齿浅而尖。植株立性，枝条稠密，生长旺盛。冬末至翌年春季开花。

Abstract: A chance seedling MTJS-No.6 selected from Non-*reticulata* hybrid 'Professor Gianmario Motta'. Flowers light pink, single form, miniature size, 5.5-6.0cm across with 5-6 petals, petals broad obovate, slightly crinkled on the petal surfaces, edges wavy, stamens united into a column, bloom dense. Leaves green, shiny, leathery, elliptic, margins shallowly and sharply serrate. Plant upright, branches dense and growth vigorous. Blooms from late-winter to the spring of following year.

HD-28-1. 日出 Richu (Sunup)

性状：选自非云南山茶杂交种'香太阳'的机遇苗 XTY-No.1。花朵淡橘红色，泛紫色调，瓣面可见深色细脉纹，单瓣型，中型花，花径8.5～9.5cm，花瓣6～7枚，厚质，边缘波浪状，雄蕊基部连生，呈柱状，开花稠密。叶片绿色，革质，椭圆形，基部楔形，边缘齿密。植株立性，枝条稠密，生长旺盛。冬末至翌年春季开花。

Abstract: A chance seedling XTY-No.1 selected from Non-*reticulata* hybrid 'Scented Sun'. Flowers lightly orange red with purple hue, deep color veins visible on the petal surfaces, single form, medium size, 8.5-9.5cm across with 6-7 petals, petals thick texture, edges wavy, stamens united into a column at the base, bloom dense. Leaves green, leathery, elliptic, cuneate at the base, margins densely serrate. Plant upright, branches dense and growth vigorous. Blooms from late-winter to the spring of following year.

HD-29-1. 六角亭
Liujiaoting (Hexagonal Pavillon)

性状： 选自红山茶品种'超级南天武士'的机遇苗 CJNT-No.3。花朵下部花瓣黑红色，具蜡光，上部花瓣深粉红色，完全重瓣型，小型至中型花，花径 6.5～8.0cm，花朵厚约 5.0cm，下部花瓣 35～40 枚，呈 5～6 轮紧实排列，先端圆，上部花瓣 40 枚以上，呈 7 轮六角状排列，边缘内卷。叶片绿色，光亮，革质，椭圆形，先端尖，边缘叶齿密。植株紧凑，生长旺盛。春初至春末开花。

Abstract: A chance seedling CJNT-No.3 selected from *C. japonica* cultivar 'Dixie Knight Supreme'. Flowers dark red with waxy luster at the lower petals, deep pink at the upper petals, formal double form, small to medium size, 6.5-8.0cm across and 5.0cm deep, lower part 35-40 petals arranged tightly in 5-6 rows, apices round, upper part more than 40 petals arranged hexagonally in 7 rows, the edges of the petals rolled inward. Leaves green, shiny, leathery, elliptic, tips pointed, margins densely serrate. Plant compact and growth vigorous. Blooms from early-spring to late-spring.

HD-30-1. 紫花巷
Zihuaxiang (Purple Flower Alley)

性状： 选自红山茶品种'卜伴'的机遇苗 BB-No.1。花朵黑紫红色，泛蜡光，偶有少量白条纹，半重瓣型，小型至中型花，花径 6.5～8.5cm，花朵厚 3.5cm，花瓣 16～18 枚，呈 2～3 轮松散排列，花瓣窄倒卵形，长 5.0cm，宽 2.0cm，外翻，雄蕊少量，近离生，花丝淡红色，花药黄色，开花稠密。叶片绿色，光亮，革质，椭圆形，长 7.0～8.0cm，宽 3.0～3.5cm，略波浪状，边缘齿密。植株紧凑，枝条稠密，生长旺盛。冬末至翌年春季开花。

Abstract: A chance seedling BB-No.1 selected from *C. japonica* cultivar 'Bokuhan'. Flowers dark purple red with waxy luster, with a few white stripes, semi-double form, small to medium size, 6.5-8.5cm across and 3.5cm deep with 16-18 petals arranged loosely in 2-3 rows, petals narrow obovate, turned outward, stamens a few, nearly free at the base, filaments pale red, anthers yellow, bloom dense. Leaves green, shiny, leathery, elliptic, slightly wavy, margins densely serrate. Plant compact, branches dense and growth vigorous. Blooms from late-winter to the spring of following year.

HD-30-2. 七仙女 Qixiannü (The Seventh Fairy)

性状： 选自红山茶品种'卜伴'的机遇实生苗 BB-No.4。花朵嫩粉红色，略泛淡紫色调，渐向花芯呈淡粉红色，中部雄蕊瓣具白条纹，半重瓣型至牡丹型，微型至小型花，花径 5.5～7.0cm，花朵厚 3.0cm，大花瓣 27～30 枚，呈 5 轮松散排列，花瓣倒卵形，长 3.0cm，宽 2.0cm，先端近圆形，中部小花瓣波浪状，与雄蕊混生，花丝白色，花药黄色。叶片绿色，光亮，革质，小椭圆形，长 7.0～8.0cm，宽 3.0～3.5cm，边缘齿密。植株紧凑，枝条稠密，生长旺盛。冬末至翌年春季开花。

Abstract: A chance seedling BB-No.4 selected from *C. japonica* cultivar 'Bokuhan'. Flowers soft pink with light purple hue, fading to light pink at the center, central petaloids with white stripes, semi-double to peony form, miniature to small size, 5.5-7.0cm across and 3.0cm deep with 27-30 large petals arranged loosely in 5 rows, petals obovate, apices nearly round, central small petals wavy, mixed with stamens, filaments white, anthers yellow. Leaves green, shiny, leathery, small elliptic, margins densely serrate. Plant compact, branches dense and growth vigorous. Blooms from late-winter to the spring of following year.

HD-30-3. 绣凤 Xiufeng (The Phoenix with Embroider)

性状： 选自红山茶品种'卜伴'的机遇实生苗 BB-No.6。花朵粉红色，泛紫色调，半重瓣型至牡丹型，中型花，花径 8.0～8.5cm，花朵厚 3.5cm，花瓣约 16 枚，呈 2 轮排列，倒卵形，长 4.5cm，宽 3.5cm，外翻，雄蕊簇生，花丝白色，花药黄色。叶片绿色，光亮，革质，小椭圆形，长 6.0～6.5cm，宽 3.0～3.5cm，边缘齿浅密。植株紧凑，枝条稠密，生长旺盛。冬末至翌年春季开花。

Abstract: A chance seedling BB-No.6 selected from *C. japonica* cultivar 'Bokuhan'. Flowers pink with purple hue, semi-double to peony form, medium size, 8.0-8.5cm across and 3.5cm deep with about 16 petals arranged in 2 rows, petals obovate, turned outward, stamens clustered, filaments white, anthers yellow. Leaves green, shiny, leathery, small elliptic, margins shallowly and densely serrate. Plant compact, branches dense and growth vigorous. Blooms from late-winter to the spring of following year.

HD-31-1. 云朵飘
Yunduopiao（Cloudlets Drifting）

性状：选自红山茶品种'冰霜王后'的机遇苗 BSWH-No.1。花朵白色，花瓣基部偶有少量红条纹，完全重瓣型，中型花，花径 7.5～8.5cm，花朵厚 4.5cm，花瓣 80 枚以上，呈 9 轮以上排列，刚开放时，花芯珠球状，外轮花瓣排列紧实，外翻。叶片绿色，光亮，革质，阔椭圆形，边缘齿密。植株紧凑，枝条稠密，生长旺盛。冬末至翌年春季开花。

Abstract: A chance seedling BSWH-No.1 selected from *C. japonica* cultivar 'Frost Queen'. Flowers white, occasionally with a few red stripes at petal bases, formal double form, medium size, 7.5-8.5cm across and 4.5cm deep with over 80 petals arranged in over 9 rows, when just opening, a petal ball always appears at the center, exterior petals arranged tightly, turned outward. Leaves green, shiny, leathery, broad elliptic, margins densely serrate. Plant compact, branches dense and growth vigorous. Blooms from late-winter to the spring of following year.

HD-32-1. 花灯笼
Huadenglong (Colorful Lantern)

性状： 选自红山茶品种'波特'的机遇苗 BOT-No.1。花朵淡橘粉红色，花瓣和雄蕊瓣具白色斑块，托桂型至牡丹型，中型至大型花，花径 9.5～11.0cm，花朵厚 5.5cm，外部大花瓣 8～9 枚，呈 1～2 轮排列，外翻，波浪状，花芯处有无数雄蕊瓣松散地集聚成球，开花稠密。叶片绿色，光亮，稠密，革质，小椭圆形，边缘齿浅。植株紧凑，枝条稠密，生长中等。冬末始花，春初盛花。

Abstract: A chance seedling BOT-No.1 selected from *C. japonica* cultivar 'Peter Pan'. Flowers light orange pink, blotched white at both petals and petaloids, anemone to peony form, medium to large size, 9.5-11.0cm across and 5.5cm deep with 8-9 exterior large petals arranged in 1-2 rows, the petals turn outward and are wavy, countless petaloids gather loosely into a ball at the center, bloom dense. Leaves green, shiny, dense, leathery, small elliptic, margins shallowly serrate. Plant compact, branches dense and growth normal. Starts to bloom in late-winter and fully blooms in the spring.

HD-33-1. 黑亮哥 Heiliangge（Black-bright Brother）

性状： 选自红山茶品种'小松子'的机遇苗SZ-No.1。花朵黑红色，泛蜡光，半重瓣型，中型花，花径7.5～9.5cm，花朵厚4.5cm，花瓣15～18枚，呈3轮排列，花瓣厚质，倒卵形，瓣面略皱和波浪状，雄蕊基部连生呈柱状。叶片绿色，光亮，革质，阔椭圆形，边缘齿浅。植株立性，枝条稠密，生长旺盛。冬末至翌年春季开花。

Abstract: A chance seedling SZ-No.1 selected from *C. japonica* cultivar 'Xiaosongzi'. Flowers dark red with waxy luster, semi-double form, medium size, 7.5-9.5cm across and 4.5cm deep with 15-18 petals arranged in 3 rows, the petals thick texture, obovate, petal surfaces slightly crinkled and wavy, stamens united into a column. Leaves green, shiny, leathery, broad elliptic, margins shallowly serrate. Plant upright, branches dense and growth vigorous. Blooms from late-winter to the spring of following year.

HD-34-1. 黑红塔
Heihongta (Black-red Tower)

性状： 选自红山茶品种'达婷'的机遇苗 DT-No.1。花朵黑红色，具蜡质感，瓣面可见黑色脉纹，半重瓣型，中型至大型花，花径 8.0～11.0cm，花朵厚约 7.0cm，花瓣 16～18 枚，呈 3～4 轮松散排列，花瓣厚质，近圆形，雄蕊基部连生。叶片绿色，厚革质，椭圆形，略波浪状，边缘叶齿密。植株紧凑，立性，生长旺盛。冬末至翌年春季开花。

Abstract: A chance seedling DT-No.1 selected from *C. japonica* cultivar 'Mary Agnes Patin'. Flowers dark red with waxy feelings, dark veins visible on the petal surfaces, semi-double form, medium to large size, 8.0-11.0cm across and about 7.0cm deep with 16-18 petals arranged loosely in 3-4 rows, petals thick texture, nearly round, stamens united at the base. Leaves green, heavy leathery, elliptic, slightly wavy, margins densely serrate. Plant compact, upright and growth vigorous. Blooms from late-winter to the spring of following year.

HD-35-1. 团聚
Tuanju (Reunion)

性状： 选自红山茶品种'佛会'的机遇苗 FH-No.1。花朵淡粉红色，泛橘红色调，牡丹型，微型至小型花，花径 5.0～7.0cm，花朵厚 5.0cm，外轮无花瓣，雄蕊全部瓣化，簇拥成团，雄蕊瓣 180 枚以上，小旗瓣皱褶或卷曲，偶有白色隐斑，雄蕊散射状。叶片中等绿色，光亮，革质，长椭圆形，边缘齿细浅。植株立性，枝条稠密，生长旺盛。冬末至翌年春季开花。

> **Abstract:** A chance seedling FH-No.1 selected from *C. japonica* cultivar 'Ryuge'. Flowers light pink with orange hue, peony form, miniature to small size, 5.5-7.0cm across and 5.0cm deep, no exterior petals, all the stamens are petaloids that unit together into a ball, petaloids over 180, petaloids crinkled or curly occasionally with faint white markings, stamens scattered. Leaves green, leathery, long elliptic, margins finely and shallowly serrate. Plant upright, branches dense and growth vigorous. Blooms from late-winter to the spring of following year.

HD-36-1. 春满园 Chunmanyuan (Spring Fully in Gardens)

性状：选自红山茶品种'哈里斯顿'的机遇苗HLSD-No.1。花朵深粉红色，泛紫色调，略具芳香，半重瓣型至牡丹型，中型至大型花，花径8.5～11.5cm，花朵厚约5.5cm。外部大花瓣18～25枚，呈3轮排列，外翻，阔倒卵形，边缘略皱褶，中部小花瓣半直立，与雄蕊混生。叶片绿色，革质，椭圆形，边缘叶齿钝。植株开张，生长旺盛。冬初至翌年春季开花。

Abstract: A chance seedling HLSD-No.1 selected from *C. japonica* cultivar 'Hallston'. Flowers deep pink with purple hue and slight fragrance, semi-double to peony form, medium to large size, 8.5-11.5cm across and about 5.5cm deep with 18-25 exterior large petals arranged in 3 rows, exterior petals turned outward, broad obovate, slightly crinkled at the edges, central petals erected and mixed with stamens. Leaves green, leathery, elliptic, margins obtusely serrate. Plant spread and growth vigorous. Blooms from early-winter to the spring of following year.

HD-36-2. 铁娘子
Tieniangzi (Iron Lady)

性状： 选自红山茶品种'哈里斯顿'的机遇苗 HLSD-No.2。萼片大，厚革质，被白柔毛，似盔甲。花朵深粉红色至红色，略具芳香，瓣面可见深色脉纹，半重瓣型，中型至大型花，花径 9.5～12.5cm，花朵厚约5.5cm，外部大花瓣15～18枚，呈2～3轮排列，外翻，阔倒卵形，长5.5cm，宽5.2cm，厚质，边缘皱褶，雄蕊基部连生。叶片绿色，厚革质，长椭圆形，龟背状，边缘叶齿浅。植株开张，灌丛状，生长旺盛。冬末至翌年春季开花。

Abstract: A chance seedling HLSD-No.2 selected from *C. japonica* cultivar 'Hallstone'. Sepals large, thick and puberulent which like armours. Flowers deep pink to red, slightly fragrant, deep veins visible on the petal surfaces, semi-double form, medium to large size, 9.5-12.5cm across and about 5.5cm deep with 15-18 exterior large petals arranged in 2-3 rows, petals turned outward, broad obovate, thick texture, slightly crinkled at the edges, stamens united at the base. Leaves green, heavy leathery, long elliptic, turtleback shaped, margins shallowly serrate. Plant spread, bushy and growth vigorous. Blooms from late-winter to the spring of following year.

HD-37-1. 尽朝晖
Jinzhaohui（Morning Sunlight over the Sky）

性状： 选自红山茶品种'肯肯'的机遇苗 KK-No.1。花朵淡橘红色，具无数纵向的白色和深红色条纹，花芯处白斑较多，牡丹型，中型至大型花，花径9.5～11.5cm，花朵厚5.5cm，外部大花瓣12～15枚，呈2～3轮排列，花瓣近圆形，外翻，先端近圆，中部小花瓣与雄蕊混生。叶片绿色，光亮，革质，椭圆形，边缘叶齿细浅。植株立性，生长旺盛。冬末至翌年春季开花。

Abstract: A chance seedling KK-No.1 selected from *C. japonica* cultivar 'Can Can'. Flowers light orange pink with a lot of longitudinal white and deep red stripes, blotched white at the center, peony form, medium to large size, 9.5-11.5cm across and 5.5cm deep with 12-15 exterior large petals arranged in 2-3 rows, petals turned outward, apices nearly round, central small petals mixed with stamens. Leaves green, shiny, leathery, elliptic, margins finely and shallowly serrate. Plant upright and growth vigorous. Blooms from late-winter to the spring of following year.

HD-37-2. 如梦
Rumeng（Like a Dream）

性状： 选自红山茶品种'肯肯'的机遇苗KK-No.2。花朵淡粉红色，泛紫色调，渐向花朵中部变为粉白色，偶有隐约白斑，牡丹型，中型至大型花，花径9.5～12.5cm，花朵厚5.0cm，外部大花瓣15～18枚，呈3轮紧实排列，中部小花瓣皱褶，与雄蕊混生。叶片绿色，光亮，革质，阔椭圆形，边缘叶齿细。植株立性，生长旺盛。冬末至翌年春季开花。

Abstract: A chance seedling KK-No.2 selected from *C. japonica* cultivar 'Can Can'. Flowers light pink with purple hue, fading to white pink at the center, occasionally blotched faint white, peony form, medium to large size, 9.5-12.5cm across and 5.0cm deep with 15-18 exterior large petals arranged tightly in 3 rows, central small petals crinkled and mixed with stamens. Leaves green, shiny, leathery, broad elliptic, margins finely serrate. Plant upright and growth vigorous. Blooms from late-winter to the spring of following year.

HD-38-1. 黑皱 Heizhou (Dark Crinkles)

性状： 选自红山茶品种'毛缘大黑红'的机遇苗 MYDHH-No.1。花朵黑红色，泛紫色调，瓣面可见黑色脉纹，半重瓣型，中型花，花径8.0～9.0cm，花朵厚约4.5cm，花瓣18～20枚，呈2～3轮紧实排列，瓣面皱褶，雄蕊簇生。叶片绿色，革质，椭圆形，边缘叶齿细密。植株紧凑，矮灌状，生长旺盛。冬初至翌年春季开花。

Abstract: A chance seedling MYDHH-No.1 selected from *C. japonica* cultivar 'Clark Hubbs'. Flowers dark red with purple hue, dark veins visible on the petal surfaces, semi-double form, medium size, 8.0-9.0cm across and about 4.5cm deep with 18-20 petals arranged tightly in 2-3 rows, petal surfaces crinkled, stamens clustered. Leaves green, leathery, elliptic, margins finely and densely serrate. Plant compact, bushy and growth vigorous. Blooms from early-winter to the spring of following year.

HD-39-1. 香纽扣
Xiangniukou (Scented Button)

性状：选自红山茶品种'香屋'的机遇苗 XW-No.3。花朵红色，花瓣中部偶有一条纵向的白条纹，略带芳香，完全重瓣型，微型花，花径 4.5～5.5cm，花朵厚 3.5～4.5cm，花瓣 50 枚以上，呈 8 轮以上排列。叶片绿色，光亮，革质，阔椭圆形，边缘齿尖。植株立性，枝条稠密，生长旺盛。冬末至翌年春季开花。

Abstract: A chance seedling XW-No.3 selected from *C. japonica* cultivar 'Scented Treasure'. Flowers red with a longitudinal white stripe at the mid-part of a petal, slightly fragrant, formal double form, miniature size, 4.5-5.5cm across and 3.5-4.5cm deep with over 50 petals arranged in over 8 rows. Leaves green, shiny, leathery, broad elliptic, margins sharply serrate. Plant upright, branches dense and growth vigorous. Blooms from late-winter to the spring of following year.

HD-40-1. 大气魄
Daqipo (Great Courage)

性状： 选自红山茶品种'伯恩赛德博士'的机遇苗 BESD-No.1。花朵红色，半重瓣型至牡丹型，巨型花，花径 13.0～14.0cm，花朵厚 6.0cm，大花瓣 16～17 枚，呈 2 轮排列，花瓣近圆形，边缘皱褶或略波浪状，中部偶有数枚雄蕊瓣，雄蕊基部略连生。叶片中等绿色，光亮，叶脉明显，厚革质，阔椭圆形，叶缘锯齿钝。植株立性，枝条较软，生长旺盛。冬末至翌年春季开花。

Abstract: A chance seedling BESD-No.1 selected from *C. japonica* cultivar 'Dr Burnside'. Flowers red, semi-double to peony form, very large size, 13.0-14.0cm and 6.0cm deep with 16-17 large petals arranged in 2 rows, petals nearly round, edges crinkled or slightly wavy, occasionally some petaloids at the center, stamens slightly united at the base. Leaves normal green, shiny, veins visible, heavy leathery, broad elliptic, margins obtusely serrate. Plant upright, branches soft and growth vigorous. Blooms from late-winter to the spring of following year.

HD-41-1. 玉珠 Yuzhu (Jade Bead)

性状： 选自攸县油茶杂交种'玉梅'的机遇苗 YM-No.1。花朵白色，渐向花芯呈粉红色，具芳香，半重瓣型至托桂型，小型至中型花，花径 6.5～8.5cm，花朵厚 4.5cm，大花瓣 5～7 枚，阔倒卵形，边缘略波浪状，带红色隐斑，中部雄蕊瓣通常簇拥呈珠状，花瓣逐片掉落，开花极稠密。叶片绿色，革质，小椭圆形，叶齿细尖，先端略下弯。植株立性，枝叶稠密，生长旺盛。秋初至冬末开花。

Abstract: A chance seedling YM-No.1 selected from *C. yuhsienensis* hybrid 'Yume'. Flowers white, changing to pink at the center, fragrant, semi-double to anemone form, small to medium size, 6.5-8.5cm across and 4.5cm deep with 5-7 large petals, large petals broad obovate, wavy at the edges, petaloids usually congregated into a bead at the center, petals fall one by one, bloom very dense. Leaves green, leathery, small elliptic, margins finely and sharply serrate. Plant upright, branches dense and growth vigorous. Blooms from early-autumn to late-winter.

中国主要茶花苗圃选登
The Introductions of Selected Major Camellia Nurseries in China

肇庆棕榈谷花园有限公司简介
A Brief Introduction of Zhaoqing Palm Valley Garden Co., Ltd.

肇庆棕榈谷花园有限公司占地3 000余亩，拥有完善的生产和研发配套设施，建设有500亩多花黄花风铃木园、500亩多花紫花风铃木园、1 500亩四季茶花园，推动农业生态特色旅游发展，助力乡村振兴。

2006年，公司在基地建设四季茶花育种中心，并组建茶花育种团队，致力于山茶属资源收集与评价、多季茶花育种研究及新品种产业化推广。现已收集国内外山茶属栽培品种及物种600多个，培育出各类茶花新品种700多个，出版世界首部四季茶花专著《四季茶花杂交新品种彩色图集》，填补了世界上茶花不能夏季开花的空白。

作为四季茶花育种中心、四季茶花新品种的发源地，公司正在积极推动四季茶花在庭院、公共绿地的应用，欢迎您前来洽谈业务。

地址：广东省肇庆市高要区回龙镇棕榈谷花园四季茶花育种基地

联系人：刘信凯　电话：+ 8613929891021　邮箱：lxk1000@163.com

Abstract: The company covers an area of more than 200 hectares, has a perfect production and research facilities, the construction of 33.3 hectares of *Tabebuia chrysantha* with yellow flowers, 33.3 hectares of *Tabebuia chrysantha* with purple flowers and 100 hectares of camellia new hybrids that bloom year-round, promote the development of agricultural ecological characteristics tourism. It has collected more than 600 camellia cultivars from the world, bred more than 700 camellia new hybrids and published the world's first monograph *Illustrations of the New Camellia Hybrids that Bloom Year-round*. The company is actively promoting the application of the ever-blooming camellia hybrids in courtyard and public green space. You are welcome to negotiate business with it.

Contact: Mr. Liu Xinkai　Tel: +8613929891021　E-mail: lxk1000@163.com

肇庆棕榈谷花园有限公司办公室门前景观
A landscape at the front of the office of Zhaoqing Palm Vellay Garden Co., Ltd.

四季茶花育种基地一角
A corner of the new camellia hybrid breeding base

四季茶花品种大树
Big trees of the ever-blooming camellia hybrids

四季茶花产品展示
The display of the ever-blooming camellia

广东阿婆六茶花谷
Apoliu Camellia Valley, Guangdong Province

阿婆六茶花谷位于广州从化区良口镇阿婆六自然村，总面积约 400 公顷，茶花种植区面积约 20 公顷，现收集茶花品种已达 600 多个。阿婆六茶花谷分为 12 个园区，包括云茶园、红山茶园、茗茶园、香花茶园、茶梅园、精品茶花园、山茶物种园、四季茶花园、金花茶园、国际山茶友谊园、农耕文化园、杜鹃园。

阿婆六茶花谷一直致力于山茶品种和物种的收集、保存、栽培、育种研究与推广应用，并于 2019 年成功举办了"中国从化国际山茶协会古山茶保育大会"。

阿婆六茶花谷于 2020 年被国际山茶协会认证为"国际杰出茶花园"，是广州首家被认证为"国际杰出茶花园"的公园。创办人侯文卿先生被授予国际山茶协会主席勋章奖。阿婆六茶花谷欢迎您前来参观！

地址：广东省广州市从化区良口镇阿婆六村

联系人：侯文卿　电话：+ 8613702435167　邮箱：737831871@qq.com

Abstract: The Valley is located in Apoliu village, Liangkou Town, Conghua District, Guangzhou City, with area about 400 hectares. The camellia planting area covers an area of about 20 hectares. Now the collection of camellia varieties has reached to more than 600 kinds, and the park is divided into 12 gardens. It includes *C. reticulata* garden, *C. japonica* garden, Tea garden, Fragrant camellia garden, *C. sasanqua* garden, Boutique camellia garden, Camellia species Garden, Ever-blooming camellia Garden, Yellow camellia garden, International camellia friendship garden, Farming culture garden and Rhododendron garden.

The Conference of the History Camellia Conservation, ICS was successfully held at the Valley in 2019,

The garden was honored as an International Camellia Garden of Excellence by ICS in 2020 and the manager, Mr. Hou Wenqing was awarded the ICS President's Medal. You are welcome to visit the Valley.

Contact: Mr. Hou Wenqing　Tel: +8613702435167　E-mail: 737831871@qq.com

阿婆六茶花谷大门
The gate of Apoliu Camellia Valley

2020 年阿婆六茶花谷"国际杰出茶花园"授牌仪式
An awarding ceremony of International Camellia Garden of Excellence by ICS for the garden in 2020

佛山市林业科学研究所（佛山植物园）
Forestry Research Institute of Foshan City (Foshan Botanical Garden)

佛山市林业科学研究所（佛山植物园）位于广东省佛山市南海区狮山镇，为公益性质的科研事业单位，主要从事山茶科植物的引种栽培、开发利用，城市林业优良树种的筛选应用，城市生态修复技术研究，林业调查规划，林业科技推广，林业科普教育以及植物园建设管理等方面的工作。

研究所自20世纪80年代初开始进行山茶科植物品种和原生种的收集，开展与山茶相关的引种、栽培、育种、病虫害防治等科学研究以及推广应用。园内目前收集了来自世界各地的名贵茶花品种和山茶原生种共1 525种，保存了广东省最大的金花茶种群，建立了张氏红山茶异地保存库，被国家林业局授予"国家山茶种质资源库"，获得了国际杰出茶花园、全国绿化模范单位、国家林业和草原局山茶花产业国家创新联盟常务理事单位等称号，是集植物保育、科研科普、观光休闲等功能于一体的地方性植物园。

地址：广东省佛山市南海区狮山镇兴业西路17号

联系人：田雪琴　电话：+ 8675786666243　邮箱：185190299@qq.com

Abstract: It is a public welfare scientific research institution, mainly engaged in the introduction, cultivation, development and utilization of *Theaceae* plants, selection and application of excellent tree species in urban forestry, research on urban ecological restoration technology, forestry investigation and planning, forestry science and technology extension, forestry science popularization education and botanical garden construction and management.

At present, the park has collected a total of 1, 525 kinds of rare camellia varieties and camellia species from all over the world, preserving the largest yellow camellia population in Guangdong Province, and establishing the remote preservation bank of *C. changii*, which has been awarded the National Camellia Germplasm Bank by the State Forestry Administration. It has won the titles of International Camellia Garden of Excellence, National Model Greening Unit, Standing Director Unit of Camellia Industry National Innovation Alliance of the State National Forestry and Grassland Administration, etc. It is a local botanical garden integrating plant conservation, scientific research and popular science, sightseeing and leisure.

Contact: Miss. Tian Xueqin　Tel: +8675786666243　E-mail: 185190299@qq.com

佛山市林业科学研究所（佛山植物园）门口
The gate of Forestry Research Institute of Foshan City (Foshan Botanical Garden)

佛山植物园内茶花景观
A landscape in Foshan Botanical Garden

浙江彩园居生态农业发展有限公司
Zhejiang Caiyuanju Ecological Agriculture Development Co., Ltd.

浙江彩园居生态农业发展有限公司，成立于2016年，总部位于浙江省湖州市德清县莫干山国际旅游渡假区内，在山东、安徽和上海设有分支机构，种植总面积1 000余亩。公司在新品种、容器成品苗、本土树种全冠移植、山茶花等高新品种培育与园林应用等多个领域，均达到国内领先水平。公司立足江浙沪皖，面向全球，以匠心精神致力于园林园艺植物材料的标准化及新优品种的推广与应用。欢迎洽谈业务！

地址：浙江省湖州市德清县莫干山镇何村村

联系人：符秀玉　电话：+8618688959796　邮箱：caiyuanju2016@163.com

Abstract: It founded in 2016, and its headquarter locates in Moganshan International Tourism Resort, Deqing County, Huzhou City, Zhejiang Province, with branches in Shandong, Anhui and Shanghai, with a total planting area of more than 66.7 hectares. The company has reached the domestic leading level in many fields such as new varieties, container finished seedlings, whole crown transplantation of local tree species, cultivation of super new varieties such as camellia cultivations and garden application. Based in Jiangsu, Zhejiang, Shanghai and Anhui provinces, the company is committed to the standardization of garden and horticultural plant materials and the promotion and application of new superior varieties with the spirit of ingenuity. Welcome to contact us with business!

Contact: Mr. Fu Xiuyu　Tel: +8618688959796　E-mail: caiyuanju2016@163.com

浙江彩园居创始人符秀玉
Mr. Fu Xiuyu, the originator of Zhejiang Caiyuanju Ecological Agriculture Development Co., Ltd.

浙江彩园居创始人和茶花育种专家高继银研究员
The originator and professor Gao Jiyin together

上海星源农业实验场上海茶花园
Shanghai Camellia Garden, Shanghai Xingyuan Agricultural Experimental Farm

上海茶花园是上海地区唯一的一个以收集和生产茶花新品种为主的观光园，地处淀山湖边，占地 100 多亩，生产设施齐全。拥有国内外茶花最新品种 900 多个，自主培育并获得国家林业和草原局知识产权保护的茶花品种有 6 个。近年来，与兄弟单位合作，先后引进多季茶花新品种 20 多个，年生产盆栽茶花 5 万盆以上。承担和完成上海市农业委员会下达的科研项目多个。茶花园主要面向上海园林绿化和家庭栽培的高端市场，生产精品盆栽和地栽茶花，同时提供科普、休闲和观光服务。欢迎参观和洽谈业务。

地址：上海市青浦区商榻锦商公路 379 号

联系人：周和达　电话：+ 8618930295611　邮箱：1030142407@qq.com

Abstract: It is the only sightseeing garden in Shanghai that focuses on the collection and production of new camellia varieties. There are more than 900 latest camellia varieties at home and abroad, and 6 camellia varieties are bred by the garden and they have obtained the patent rights from National Forestry and Grassland Administration. In recent years, more than 20 camellia new hybrids with ever-blooming trait have been introduced. The garden's annual production of potted camellia plants can reach over 50,000 pots. It undertakes and completes many research projects issued by Shanghai Agriculture Committee. The garden is aimed at the high-end market of Shanghai's landscaping and home cultivation, producing fine potted and ground camellia plants, as well as providing science popularization, leisure and sightseeing. You are welcome to visit and negotiate business with us.

Contact: Mr. Zhou Heda　Tel: +8618930295611　E-mail: 1030142407@qq.com

上海茶花园大门
The gate of Shanghai Camellia Garden

上海茶花园实验区
An experiment area of camellias in the garden

德庆县莫村镇大农茶花种植场
Large Farm Camellia Plantation of Mo Village, Deqing County

大农茶花种植场总面积达 1 000 亩以上，是目前世界上生产四季茶花新品种的最大的苗圃。该场拥有 30 亩现代化大棚，还经营油茶、金花茶、黄花梨等苗木。每年繁殖培育各种规格的苗木 80 万株。四季茶花等各类苗木规格齐全，可供电商平台和出口之用。

公司每年为周边农户提供农业技能培训，并解决周边农户就业 100 余人，成为当地乡村振兴的典范，是广东省"一村一品、一镇一业"依托单位。公司将致力于打造集标准化种植示范、规模化生产、观光旅游于一体的综合性农业基地。

地址：广东省肇庆市德庆县莫村镇扶赖村委会良义村

联系人：张铭　电话：+ 8618666776161　邮箱：61725308@qq.com

Abstract: The plantation covers more than 66.7 hectares, mainly produces ever-blooming camellia new hybrids, *C. oleifera*, yellow camellias and *Dalbergia odorifere*. It is the largest camellia nursery where produces the ever-blooming camellia new hybrids in the world. It can produce 800,000 plants of camellias each year. The plantation will be committed to build a comprehensive agricultural base integrating standardized planting demonstration, large-scale production and sightseeing tourism.

Contact: Mr. Zhang Ming　Tel: +8618666776161　E-mail: 61725308@qq.com

大农茶花种植场基地全貌
Overview of Large Farm Camellia Plantation

大农茶花种植场基地一角
A corner of Large Farm Camellia Plantation

盆栽的四季茶花
The potted ever-blooming camellia hybrids

邛崃市龚花园
Mr. Gong's Garden of Qionglai City

龚花园由龚兴国于1966年创立，是成都市较早从事花卉苗木生产、销售的大户，龚氏第二代五兄妹秉承父业，现共有花卉苗木生产基地520亩，以茶花、罗汉松、腊梅、金弹子及各类盆景苗木为主，其中茶花苗圃面积300余亩，品种多达800个左右，并创建品牌"龚氏茶花园"。

现任邛崃市花木盆景协会会长的龚国文（龚兴国之子），为川派盆景大师、四川省民间手工艺大师、茶花专家，是我国第一批中国茶花协会及国际山茶协会会员，其作品多次参加国家、省、市盆景展览，获得了金、银、铜等奖。

龚国文于20世纪90年代开始在成都郫都区春天花卉市场、三圣花乡、邛崃花木盆景市场等地经营门市店8个，年销售额达到1 000万元，他牵头成立了邛崃市国祥花卉苗木种植专业合作社，任理事长，合作社现有会员152户，种植各类苗木花卉2.3万亩，带动园林盆景从业人员500人以上，每年解决300余人的就业问题，带动会员销售各类花木盆景两亿元以上。

地址：四川省成都市邛崃市文南路170号路的尧花店

联系人：龚国文　电话：+ 8613699033618　邮箱：510034570@qq.com

Abstract: The second generation of Gong's five siblings inherited their father's business, and now have a total production base with 34.7 hectares for flowers and seedlings, mainly camellias, Podocarpus macrophyllus, Chimonanthus praecox, Diospyros cathayensis and various kinds of bonsai seedlings. Among them, camellia nursery covers an area of more than 20 hectares, with about 800 varieties. Gong's Camellia Garden has also been established.

Mr. Gong Guowen is the president of Qionglai Flowers and Plants Bonsai Association, is also a camellia expert. He has received many awards. He ran eight bonsai outlets in Chengdu area since the 1990's and now annual sales reach 10 million yuan. The Qionglai Guoxiang Flower Seedling Planting Professional Cooperation he organized has 152 households, grows 1,533 hectares of various flowers, solves the employment problem of 300 people every year and brings the income of over 200 million yuan.

Contact: Mr. Gong Guowen　Tel: +8613699033618　E-mail: 510034570@qq.com

龚花园大门
The gate of Mr. Gong's Garden

龚花园内的盆栽茶花
The potted camellias of Mr. Gong's Garden

龚花园内盛开的茶花
The fully blooming camellias of Mr. Gong's Garden

福建省龙岩市秀峰茶花有限公司
——中国最大的茶花品种原生苗繁育基地之一
Xiufeng Camellia Co., Ltd., Longyan City, Fujiang Province
One of the Largest Camellia Original Plants Production Base in China

秀峰茶花有限公司于1979年创立,一直致力于国内外名优茶花品种的收集、培育和推广。目前已收集国内外名优山茶花品种1 000多个,杜鹃品种300多个,其中华东山茶花品种100多个,有60个华东山茶花品种系福建省内或全国独有的珍稀品种,是目前福建省内茶花品种最多、保存最完整的茶花良种场。

公司现获得植物新品种专利权10个,是国内最大的茶花品种原生苗繁育基地之一,每年可繁育国内外名优茶花品种原生小苗300多个品种,40万株。另外,苗圃还有名优杜鹃苗木20万株,10～30年生茶花大树20 000株。

地址:福建省漳平市永福镇西山村十里花街秀峰茶花场

联系人:张陈环　电话:+8613616913611　邮箱:1369581343@qq.com

Abstract: It founded in 1979 and has been committed to the collection, cultivation and promotion of famous camellia varieties at home and abroad. It has collected over 1,000 famous camellia varieties, and over 300 azalea varieties, of them, including more than 100 *C. japonica* varieties originated from East China, and 60 camellia varieties are unique and rare varieties in Fujian province or in the whole country. It is the camellia nursery with the most quantities of camellia varieties and the most complete preservation in Fujian province.

It has obtained 10 patent rights. There are more than 300 camellia varieties that produced 400,000 rooted cutting plants each year. Also, there are 200,000 super azalea plants and 20,000 big camellia trees for sale.

Contact: Mr. Zhang Chenhuan　Tel: +8613616913611　E-mail: 1369581343@qq.com

秀峰茶花有限公司大面积的茶花苗木
A large area of camellia plants cultured by Xiufeng Camellia Co., Ltd.

秀峰茶花有限公司大量的盆栽茶花
A lot of potted camellia plants at the Xiufeng Camellia Co., Ltd.

苗圃名称	简介	联系方式
广西岑溪市成园花木场 (Chengyuan Flower Nursery, Cenxi City, Guangxi)	该场占地50余亩，致力于四季茶花容器苗、古桩茶花种植和销售，产品品质优良，深受客户青睐。欢迎各地经销商联系合作。 地址：广西壮族自治区梧州市岑溪市大隆镇大峡村 The nursery covers an area of more than 3.3 hectares, focus on four-season camellia container seedlings, the planting and sales of the ancient camellias with old trunks, their product quality is good, favored by customers. Welcome dealers around the contact and cooperation. Address: Daxia Village, Dalong Town, Cenxi City, Guangxi	联系人（Contacts）：周学成 (Zhou Xuecheng) 电话（Tel.）：+8613647744681 微信（WeChat）:
广西岑溪市九千亿茶花有限公司 (Jiuqianyi Camellia Co., Ltd., Cenxi City, Guangxi)	该公司是一家专注四季茶花种植的农业公司，占地120亩，年产优质四季茶花盆栽苗10多万盆。公司始终坚持品质第一，产品深受客户青睐。欢迎全国各地经销商前来采购。 地址：广西壮族自治区梧州市岑溪市糯垌镇塘坡村 The company covers an area of eight hectares, with an annual output of more than 100,000 high-quality four season camellia potted plants. The company always takes quality as the first, and the products are favored by customers. Dealers all over China are welcome to purchase. Address: Tangpo Village, Nuodong Town, Cengxi City, Guangxi	联系人（Contacts）：钟彬（Zhong Bin） 电话（Tel.）：+8618275897314 微信（WeChat）:

参考文献
Bibliography

一、图书文献 Books and Monographs

［1］程金水.园林植物遗传育种学［M］.2版.北京：中国林业出版社，2010.

［2］高继银，杜跃强，CLIFFORD R P，等.山茶属植物主要原种彩色图集［M］.杭州：浙江科学技术出版社，2005.

［3］高继银，刘信凯，赵强民.四季茶花杂交新品种彩色图集［M］.杭州：浙江科学技术出版社，2016.

［4］管开云，李纪元，王仲朗，等.中国茶花图鉴［M］.杭州：浙江科学技术出版社，2014.

［5］闵天禄.世界山茶属的研究［M］.昆明：云南科学技术出版社，2000.

［6］沈荫椿.山茶［M］.北京：中国林业出版社，2009.

［7］张乐初，游慕贤，陈德松，等.中国茶花文化［M］.上海：上海文化出版社，2003.

［8］张宏达.中国植物志：第四十九卷第三分册［M］.北京：科学出版社，1998.

［9］张宏达.山茶属植物的系统研究［M］.广州：中山大学学报编辑部，1981.

［10］横山三郎，桐野秋豐.新装版日本の椿花［M］.京都：株式会社日本京都淡交社，2005.

［11］日本ツベキ協会，日本ツベキサ.ツベキンヵ名鑑［M］.东京：株式会社誠文堂新光社，1998.

［12］ARTHUR A G. Camellia nomenclature［M］. California: The Southern California Camellia Society Inc., 1999.

［13］OREL G, ANTHONY S C. A long-term study 1999-2017 and taxomic review of *Camellia amplexicaulis*（Pitard）Cohen Stuart sensu Itato［M］. Sydney: Theaceae Exploration Associates, 2017.

［14］DAVID L F, MILTON H B. The camellia: its history, culture, genetics and look into its future development［M］. South Carolina: The R. L. Bryan Company, 1978.

［15］NEVILLE H. The international camellia register: second supplement 1990-2010［M］. Dorset: The Minster Press, 2011.

［16］SEALY J R. A revision of the Genus *Camellia*［M］. London: The Royal Horticultural Society, 1958.

［17］THOMAS J S. The international camellia register: volume one［M］. Sydney: Fine Arts Press Pty Limited Press, 1993.

［18］THOMAS J S. The international camellia register: volume two［M］. Sydney: Fine Arts Press Pty Limited Press, 1993.

［19］THOMAS J S. The international camellia register: supplement to volume one and two［M］. Sydney: Fine Arts Press Pty Limited Press, 1997.

［20］WILLIAM L A. Beyond camellia belt: breeding, propagating and growing cold-hardy camellias

[M]. Illinots: Ball Publishing, 2007.

二、中文期刊文献 Journals and magazines in Chinese

[1] 刘信凯，黄万坚，钟乃盛，等．杜鹃红山茶种间杂交亲和性及5个杂交种真实性的研究[C]．中国温州首届海峡两岸山茶花资源保护与开发利用学术研讨会暨2011年中国茶花育种年会论文集，2011：128-138.

[2] 卫兆芬．中国山茶属一新种[J]．植物研究，1986，6（4）：141-143.

[3] 叶创兴．山茶属一新种[J]．广东省植物学会会刊，1985（3）．

[4] 叶创兴．山茶属一新种[J]．中山大学学报，1986（3）：25.

[5] 叶创兴．山茶属三新种[J]．中山大学学报，1987（1）：17-20.

[6] 黄连冬，梁盛业，叶创兴．四季花金花茶：金花茶一新种[J]．园林植物研究与应用，2014（1）：69-70.

三、外文期刊文献 Journals and magazines in foreign language

[1] OREL G, MARCHANT D. *Camellia amplexicaulis*（Pitard）Coh. St.（Theaceae）molecular and morphological comparison of selected samples from Viet Nam, China and the USA[J]. International Camellia Journal, 2010: 89-94.

[2] GAO J Y, HUANG W J, ZHONG N S. *Camellia changii*, the ever-blooming camellia[J]. International Camellia Journal, 2008: 89-96.

[3] GAO J Y. *Camellia azalea*: a unique species[J]. Amer. Camell. Yearbook, 2000: 11-12.

[4] GAO J Y. Talking about *Camellia amplexicaulis*[J]. Amer. Camell. Yearbook, 2002: 22-24.

[5] GAO J Y, LIU X K, ZHONG N S. Research on inter-species-crosses of *Camellia azalea*, and the first 5 summer-blooming hybrids[J]. International Camellia Journal, 2011: 47-53.

[6] GAO J Y, LIU X K, ZHONG N S. The first camellias to bloom from summer to December. New hybrids from China, using *C. azalea* as a parent[J]. International Camellia Journal, 2011: 54-59.

[7] GAO J Y, LIU X K, ZHONG N S. 20 summer-blooming hybrids, a Chinese breakthrough in camellia breeding[J]. The Camellia Journal, 2012: 9-12.

[8] GAO J Y, LIU X K, ZHONG N S. A new generation of camellia hybrids[J]. International Camellia Journal, 2014: 49-51.

[9] GAO J Y, LIU X K, ZHONG N S. *Camellia chuangtsoensis*, another re-blooming species discovered after *Camellia azalea* in China[J]. International Camellia Journal, 2010: 84-89.

[10] LIU X K, HUANG W J, ZHONG N S. A significant achievement on camellia breeding in China[J]. International Camellia Journal, 2012: 101-108.

[11] LIU X K, ZHONG N S, HUANG W J. A preliminary study on genetic characteristics of inter-specific hybrids of *Camellia amplexicaulis*[J]. International Camellia Journal, 2014: 47-49.

茶花新品种索引
Index of the New Camellia Hybrids

1. 品种名称笔画索引（Index of the New Camellia Hybrids in Chinese Characters）

笔画 Chinese Characters	品种名称 Name in Chinese	拼音名称 Name in Pinyin	英文含义 Meaning in English	所属组合 Combination	页码 Page
1	一见钟情	Yijian Zhongqing	Love at First Sight	HA-36-11	87
	一等颜值	Yideng Yanzhi	First Class Pretty	HAR-03-5	258
2	十月红透	Shiyue Hongtou	Red Through in October	HAR-10-2	299
	十月烟火	Shiyue Yanhuo	October's Fireworks	HAR-14-21	347
	七夕礼品	Qixi Lipin	Double-Seventh Day's Gift	HAR-08-6	291
	七月开幕	Qiyue Kaimu	In July Curtain-up	HAR-10-7	304
	七仙女	Qixiannü	The Seventh Fairy	HD-30-2	550
	八月红浪	Bayue Honglang	August's Red Waves	HAR-14-23	349
	八月飘雪	Bayue Piaoxue	August's Snowing	HAR-03-8	261
	八月踏浪	Bayue Talang	Walking in Waves in August	HAR-10-6	303
	儿童乐园	Ertong Leyuan	Children's Playground	HAR-14-6	332
	九九同心	Jiujiu Tongxin	Nine Persons with One Heart	HAR-35-6	422
	九九艳阳	Jiujiu Yanyang	The Bright Sun of Sept. Ninth	HA-01-14	20
	九天瑶池	Jiutian Yaochi	Jade Pool at the Ninth Heaven	HAR-14-1	327
	九月惊艳	Jiuyue Jingyan	September's Amazement	HAR-10-5	302
3	三千墨丽	Sanqian Moli	Three Thousand Ink Works	HAR-35-10	426
	大气魄	Daqipo	Great Courage	HD-40-1	563
	大玉	Dayu	Large Jade	HD-25-1	540
	大玉镯	Dayuzhuo	Large Jade Bracelet	HD-10-5	525
	大视野	Dashiye	Broad Field of Vision	HB-32-3	486
	大唐宫灯	Datang Gongdeng	Palace Lantern in Great Tang Dynasty	HAR-32-1	411
	大粉丝团	Da Fensituan	Big Group of Fans	HA-63-2	139
	大家闺秀	Dajia Guixiu	Young Girl from Respectable Family	HAR-06-3	275
	大菲升级	Dafei Shengji	'Francis Eugene Phillis' Upgrade	HA-06-3	36
	大喜临门	Daxi Linmen	Great Happy Event is Coming	HAR-02-2	227
	大跨越	Dakuayue	Great Stride Leap	HB-32-1	484
	大福大贵	Dafu Dagui	Big Happiness & Big Millionaire	HAR-02-20	245
	万事顺景	Wanshi Shunjing	Everything Good Fortune	HA-55-1	111
	万家灯火	Wanjia Denghuo	Myriad Families' Lights	HA-10-16	46
	小玉	Xiaoyu	Small Jade	HD-25-2	541
	小玉镯	Xiaoyuzhuo	Small Jade Bracelet	HD-10-3	523
	小鸟依人	Xiaoniao Yiren	Sweet & Helpless Birds	HAR-14-10	336
	小礼物	Xiaoliwu	Small Gift	HD-26-1	542
	小粉童	Xiaofentong	Little Pink Child	HD-27-4	546
	小酒杯	Xiaojiubei	Small Wine Cup	HB-34-1	488
	小惊喜	Xiaojingxi	Little Surprise	HD-10-2	522

续表

笔画 Chinese Characters	品种名称 Name in Chinese	拼音名称 Name in Pinyin	英文含义 Meaning in English	所属组合 Combination	页码 Page
3	小紫童	Xiaozitong	Little Purple Child	HD-27-2	544
	小螺旋	Xiaoluoxuan	Small Spiral	HB-32-2	485
	千秋旺	Qianqiuwang	Forever Prosperous	HD-20-7	534
	夕阳余晖	Xiyang Yuhui	Afterglow of the Sunset	HA-10-17	47
	广场红艺	Guangchang Hongyi	Square Red Art	HA-53-2	106
	广场舞步	Guangchang Wubu	Dance Steps in Square	HA-10-10	40
	女模时装	Nümo Shizhuang	Female Models' Fashionable Dress	HA-10-11	41
	飞溅火花	Feijian Huohua	Flying Sparks	HAR-02-17	242
	乡村色彩	Xiangcun Secai	Rural Color	HAR-12-7	319
4	天伦之乐	Tianlun Zhile	Enjoying Family's Happiness	HA-10-18	48
	云朵飘	Yunduopiao	Cloudlets Drifting	HD-31-1	552
	艺术世家	Yishu Shijia	Artistic Family	HAR-20-2	380
	五谷丰登	Wugu Fengdeng	Abundant Harvest of All Crops	HA-81-5	200
	五福临门	Wufu Linmen	Five Blessings Arriving at Home	HA-68-1	160
	不夜红城	Buye Hongcheng	Ever-bright Red City	HA-80-2	194
	戈壁驼铃	Gebi Tuoling	Gobi Camel Bells	HAR-35-8	424
	少女裙	Shaonüqun	Maiden Skirt	HB-31-2	483
	日月重光	Riyue Chongguang	The Sun & The Moon Appear Together	HAR-03-7	260
	日出	Richu	Sunup	HD-28-1	547
	日照香炉	Rizhao Xianglu	Sunshine Censer	HAR-15-2	353
	中秋红喜	Zhongqiu Hongxi	Mid-Autumn Festival's Red Delight	HA-16-4	65
	水墨丹青	Shuimo Danqing	Chinese Ink Painting	HA-53-3	107
	午夜灯火	Wuye Denghuo	Midnight Lights	HAR-10-10	307
	午夜金光	Wuye Jinguang	Midnight Golden Light	HA-67-2	150
	长寿村	Changshoucun	Longevity Village	HB-28-2	473
	长辫姑娘	Changbian Guniang	Long Braid Girl	HAR-04-3	265
	今日喜儿	Jinri Xier	Today's Miss Xier	HAR-13-2	323
	月下瑶台	Yuexia Yaotai	Superb Balcony under the Moon	HAR-11-2	310
	月月玫红	Yueyue Meihong	Monthly Rose Red	HAR-14-3	329
	风华正茂	Fenghua Zhengmao	Prime of Life	HAR-12-5	317
	风和日丽	Fenghe Rili	Breezy & Sunny	HAR-24-1	394
	风韵嫂	Fengyunsao	Charmed Sister-in Law	HB-26-1	463
	凤冠霞帔	Fengguan Xiapei	Phoenix Coronet & Robes	HAR-02-12	237
	凤凰亭	Fenghuangting	Phoenix Pavilion	HB-34-2	489
	六六大顺	Liuliu Dashun	Double Sixs Making Everything Smooth	HAR-06-7	279
	六角亭	Liujiaoting	Hexagonal Pavillon	HD-29-1	548
	六脉神剑	Liumai Shenjian	Six Pulses God Sword	HAR-04-6	268
	六朝脂粉	Liuchao Zhifen	Six Dynasties' Face Powder	HAR-03-6	259
	火山丽景	Huoshan Lijing	Beautiful Scene of Volcano	HAR-38-1	443
	火红双节	Huohong Shuangjie	Fire Red Double Festivals	HA-11-4	53
	火红年代	Huohong Niandai	Fire Red Era	HAR-03-3	256
	心醉神迷	Xinzui Shenmi	Mental Intoxication	HA-01-8	14
	孔雀炫丽	Kongque Xuanli	Peacock Flaunting Its Beauty	HAR-34-1	415
	双红荔城	Shuanghong Licheng	Double Red Litchi City	HAR-02-15	240

续表

笔画 Chinese Characters	品种名称 Name in Chinese	拼音名称 Name in Pinyin	英文含义 Meaning in English	所属组合 Combination	页码 Page
4	双面佳人	Shuangmian Jiaren	Beautiful Maiden with Double Faces	HAR-02-9	234
	书香之家	Shuxiang Zhijia	Literary Family	HAR-36-1	436
5	玉珠	Yuzhu	Jade Bead	HD-41-1	564
	玉浦新秀	Yupu Xinxiu	Tama-no-ura's New Rookie	HA-81-7	202
	玉润	Yurun	Jade Moisten	HC-10-5	506
	古朴清幽	Gupu Qingyou	Ancient Beauty & Tranquility	HAR-37-2	439
	古庙红钟	Gumiao Hongzhong	Red Bell of Ancient Temple	HA-80-1	193
	古殿	Gudian	Ancient Temple	HB-24-1	461
	左家娇女	Zuojia Jiaonü	Tender & Cute Girl	HAR-35-1	417
	龙火贺喜	Longhuo Hexi	Dragon-fire Congratulations	HA-67-6	154
	龙章凤姿	Longzhang Fengzi	Dragon's Literate & Phoenix's Posture	HAR-37-4	441
	北极冰	Beijibing	Arctic Ice	HD-13-3	528
	田园风光	Tianyuan Fengguang	Rural Scenery	HAR-35-14	430
	四季风情	Siji Fengqing	Four-season Amorous Feelings	HA-21-8	68
	四季合韵	Siji Heyun	The Rhyme Conforming Four Seasons	HAR-22-3	386
	生日宴会	Shengri Yanhui	Birthday Party	HA-67-10	158
	生命阳光	Shengming Yangguang	Sunny Life	HA-01-12	18
	仙女下凡	Xiannü Xiafan	Fairy from Sky Down to the Earth	HA-05-4	32
	白娘子	Bainiangzi	White Snake Lady	HD-20-4	531
	白富美	Baifumei	White, Rich & Beautiful Girl	HB-17-2	456
	外婆童谣	Waipo Tongyao	Grandmother's Nursery Rhymes	HAR-35-2	418
	处处欢腾	Chuchu Huanteng	Exult Everywhere	HA-14-5	61
	冬夏出彩	Dongxia Chucai	Colorfull in Winter & Summer	HAR-06-8	280
	闪光风车	Shanguang Fengche	Glittering Windmill	HAR-16-1	355
	半卷红帘	Banjuan Honglian	Semi-rolled Red Curtain	HA-86-1	214
	汉森之悦	Hansen Zhiyue	Mr Waldemar Max Hansen's Delight	HAR-08-3	288
	民族服饰	Minzu Fushi	National Dress	HA-81-8	203
	加冕红毯	Jiamian Hongtan	The Red Carpet for Crowning	HA-73-1	174
6	吉日良辰	Jiri Liangchen	Auspicious Day	HA-43-4	91
	吉星高照	Jixing Gaozhao	Lucky Stars Shining Brightly	HAR-21-1	382
	吉祥如意	Jixiang Ruyi	Good Lucky as Desired	HA-67-1	149
	西子晚霞	Xizi Wanxia	West Lake's Sunset Glow	HA-69-1	165
	百媚千娇	Baimei Qianjiao	Enchanting Beauty	HAR-08-5	290
	成功之喜	Chenggong Zhixi	Exultation of Success	HAR-14-2	328
	当代俏丽	Dangdai Qiaoli	Contemporary Pretty Girl	HAR-02-13	238
	团聚	Tuanju	Reunion	HD-35-1	556
	回龙颂歌	Huilong Songge	Huilong Town's Ode	HA-22-2	71
	回龙晨曦	Huilong Chenxi	Huilong Town's Morning Light	HAR-01-1	224
	回龙新貌	Huilong Xinmao	Huilong Town's New Look	HAR-10-9	306
	回眸之丽	Huimou Zhili	Beautiful Glancing Back	HAR-02-21	246
	年年有余	Niannian Youyu	Having Surplus Every Year	HA-74-1	176
	乔之千金	Qiaozhi Qianjin	Dr Georg Ziemes' Beloved Daughter	HAR-13-4	325
	伟大复兴	Weida Fuxing	Great Renaissance	HAR-15-1	352
	似曾相见	Siceng Xiangjian	Seems to Have Met Before	HAR-14-24	350

续表

笔画 Chinese Characters	品种名称 Name in Chinese	拼音名称 Name in Pinyin	英文含义 Meaning in English	所属组合 Combination	页码 Page
6	后起之秀	Houqi Zhixiu	Up-rising Star	HA-64-1	141
	后浪奔涌	Houlang Benyong	Back Wavy Surges	HAR-14-18	344
	行为艺术	Xingwei Yishu	Performance Art	HAR-35-18	434
	旭日东升	Xuri Dongsheng	The Sun Rising from East	HA-70-2	168
	多季红冠	Duoji Hongguan	Multiseasonal Red Crown	HAR-18-10	370
	多季阿兰	Duoji Alan	Multiseasonal Mark Alan	HA-62-2	136
	多季玲珑	Duoji Linglong	Multiseasonal Exquisiteness	HA-72-1	172
	多季耐冬	Duoji Naidong	Multiseasonal Naidong	HA-60-1	131
	多季绝美	Duoji Juemei	Multiseasonal Absolute Beauty	HA-65-3	145
	多季润香	Duoji Runxiang	Multiseasonal Moist Fragrance	HA-65-1	143
	多季盛情	Duoji Shengqing	Multiseasonal Great Kindness	HAR-02-10	235
	多姿丽影	Duozi Liying	Multi-Poses & Pretty Shadow	HAR-02-26	251
	争奇斗艳	Zhengqi Douyan	Contend in Beauty & Fascination	HA-21-9	69
	冰艺	Bingyi	Ice Art	HC-10-6	507
	刘村彩虹	Liucun Caihong	Liu Village's Rainbow	HAR-10-8	305
	尽朝晖	Jinzhaohui	Morning Sunlight over the Sky	HD-37-1	559
	如画梦境	Ruhua Mengjing	Picturesque Dream	HA-81-1	196
	如梦	Rumeng	Like a Dream	HD-37-2	560
	欢庆	Huanqing	Festivity	HB-02-9	451
	红云闪光	Hongyun Shanguang	Lightening Red Clouds	HA-65-2	144
	红龙舞天	Honglong Wutian	Red Dragon Dancing in the Sky	HA-53-1	105
	红发模特	Hongfa Mote	Red Hair Models	HAR-35-7	423
	红耳坠	Hongerzhui	Red Eardrop	HD-27-3	545
	红色主题	Hongse Zhuti	Red Theme	HA-59-3	128
	红色畅想	Hongse Changxiang	Red Imagination	HA-10-13	43
	红色星空	Hongse Xingkong	Red Starry Sky	HAR-10-3	300
	红色勋章	Hongse Xunzhang	Red Medal	HA-79-3	190
	红衣仙女	Hongyi Xiannü	The Fairy Dressing Red	HAR-08-4	289
	红红火火	Honghong Huohuo	Prosperity & Jollification	HAR-04-1	263
	红林瓦寨	Honglin Wazhai	Tile Village in Red Forest	HA-71-1	170
	红诗文	Hongshiwen	Red Poem	HB-27-1	467
	红城金库	Hongcheng Jinku	Red City's Gold Vault	HAR-35-17	433
	红城夜景	Hongcheng Yejing	Red City's Night Scenes	HAR-35-11	427
	红城漫金	Hongcheng Manjin	Red City Overflowing Gold	HAR-35-13	429
	红星奖章	Hongxing Jiangzhang	Red Star Medal	HA-82-1	205
	红院满福	Hongyuan Manfu	Red Courtyard with Full Happiness	HAR-14-20	346
	红绒公主	Hongrong Gongzhu	Red Velvet Princess	HAR-35-16	432
	红铃报喜	Hongling Baoxi	Happy News from Red Bells	HAR-06-6	278
	红粉舞会	Hongfen Wuhui	Red-pink Dancing Party	HAR-06-4	276
	红浪滔天	Honglang Taotian	Surge Red Waves	HAR-09-2	296
	红盘托金	Hongpan Tuojin	Red Plate Holding up Gold	HAR-22-2	385
	红绸舞浪	Hongchou Wulang	Red Silk Dancing Waves	HA-10-14	44
	红装伊人	Hongzhuang Yiren	Beloved Lady Dressed in Red	HA-67-8	156
	红蜡雕塑	Hongla Diaosu	Sculpture with Red Wax	HAR-14-14	340

续表

笔画 Chinese Characters	品种名称 Name in Chinese	拼音名称 Name in Pinyin	英文含义 Meaning in English	所属组合 Combination	页码 Page
6	红舞台	Hongwutai	Red Stage	HC-05-2	504
	红漫金山	Hongman Jinshan	Red Covering Golden Mountains	HAR-14-11	337
	红瓣金心	Hongban Jinxin	Red Petals & Golden Heart	HAR-10-4	301
	红瓣香心	Hongban Xiangxin	Red Petals with Fragrant Heart	HA-77-2	183
	红瓣竞秀	Hongban Jingxiu	Red Petals' Beauty Competitions	HAR-33-1	413
7	赤红花海	Chihong Huahai	Crimson Flower Sea	HA-75-1	178
	赤诚之心	Chicheng Zhixin	Sincere Heart	HAR-18-11	371
	赤诚红心	Chicheng Hongxin	Sincere Red Heart	HA-03-46	29
	赤焰炉火	Chiyan Luhuo	Red Flame Stove Fire	HA-77-3	184
	花衣小旋	Huayi Xiaoxuan	Miss Xiaoxuan Dressed Colorful Clothes	HA-14-6	62
	花灯笼	Huadenglong	Colorful Lantern	HD-32-1	553
	花季对歌	Huaji Duige	Coupled Singing in Blooming Season	HA-56-3	115
	花城闹市	Huacheng Naoshi	Guangzhou's Busy Streets	HA-26-6	76
	花样年华	Huayang Nianhua	Diversity of Life	HAR-06-1	273
	花容月貌	Huarong Yuemao	Beautiful Feature as Flower & Moon	HAR-02-18	243
	杏粉花雨	Xingfen Huayu	Apricot Pink & Flower Rain	HA-58-3	122
	时尚小妹	Shishang Xiaomei	Stylish Little Sister	HAR-35-5	421
	佛植华章	Fozhi Huazhang	Foshan Botanical Garden's Brilliant Works	HA-01-18	24
	佛植盈瑞	Fozhi Yingrui	Foshan Botanical Garden's Full Auspiciousness	HA-78-1	186
	余霞成绮	Yuxia Chengqi	The Sunset Likely Beautiful Silk	HA-16-5	66
	邻家红姐	Linjia Hongjie	Neighbor's Red Sister	HAR-08-8	293
	狂欢舞曲	Kuanghuan Wuqu	Orgiastic Dance Tune	HAR-19-1	377
	彤海咏秋	Tonghai Yongqiu	Red Sea Singing Autumn	HA-12-2	55
	闲花映池	Xianhua Yingchi	Wild Flowers Reflected from Pool	HAR-02-24	249
	灿烂阳光	Canlan Yangguang	Brilliant Sunshine	HAR-02-6	231
	灿烂辉煌	Canlan Huihuang	Splendid & Glorious	HA-10-12	42
	良口红秋	Liangkou Hongqiu	Liangkou Town's Red Autumn	HA-77-1	182
	初恋	Chulian	First Love	HD-20-3	530
	初探	Chutan	First Exploration	HB-38-1	493
	层层诗意	Cengceng Shiyi	Layers of Poetic Flavor	HAR-31-1	409
	层叠美景	Cengdie Meijing	Layering Beautiful Scenery	HAR-18-2	362
	张灯结彩	Zhangdeng Jiecai	Decorated scenes with Lanterns & Streamers	HA-87-1	216
	阿婆六之晨	Apoliu Zhichen	Apoliu's Morning	HC-13-1	512
	阿婆六女神	Apoliu Nüshen	Apoliu's Goddess	HC-15-1	514
	阿婆六水星	Apoliu Shuixing	Apoliu's Mercury	HB-36-1	491
8	青春之歌	Qingchun Zhige	Youth's Song	HAR-14-19	345
	坦荡胸怀	Tandang Xionghuai	Magnanimous Mind	HAR-22-1	384
	幸福之家	Xingfu Zhijia	Happy Family	HAR-18-6	366
	幸福玛黛琳	Xingfu Madailin	Happy Madeline	HAR-02-8	233
	幸福时代	Xingfu Shidai	The Happy Era	HAR-11-3	311
	披红俏女	Pihong Qiaonü	The Pretty Draped Red	HAR-22-4	387
	英雄本色	Yingxiong Bense	Hero's True Quality	HAR-02-19	244

续表

笔画 Chinese Characters	品种名称 Name in Chinese	拼音名称 Name in Pinyin	英文含义 Meaning in English	所属组合 Combination	页码 Page
8	雨后暮景	Yuhou Mujing	Evening Scene after Rain	HAR-12-4	316
	奔放舞者	Benfang Wuzhe	Unrestrained Dancer	HAR-27-1	401
	奇妙形色	Qimiao Xingse	Fantastic Shape & Color	HAR-18-14	374
	卓越风姿	Zhuoyue Fengzi	Remarkable Posture	HAR-12-3	315
	旺财	Wangcai	Flourish Treasures	HB-02-7	449
	旺德福	Wangdefu	Wonderful	HB-09-4	453
	国色天姿	Guose Tianzi	Possess Surpassing Beauty	HA-13-6	58
	明星风范	Mingxing Fengfan	Star Manner	HAR-07-2	284
	知足常乐	Zhizu Changle	Contentment Being Happiness	HA-67-3	151
	乖巧女孩	Guaiqiao Nühai	Well-behaved & Clever Girl	HA-81-3	198
	和顺	Heshun	Harmony & Smooth	HD-20-5	532
	和谐家园	Hexie Jiayuan	Harmonious Homeland	HA-79-1	188
	和睦	Hemu	Harmony	HB-26-3	465
	季节色彩	Jijie Secai	Seasonal Colors	HA-01-13	19
	欣欣向荣	Xinxin Xiangrong	Flourishment	HA-10-9	39
	金灿红云	Jincan Hongyun	Gold Shining in Red Clouds	HAR-11-1	309
	金艳靓影	Jinyan Liangying	Jinyan's Pretty Shadow	HA-79-2	189
	金黑绒	Jinheirong	Golden Black Charpie	HC-04-2	503
	金源	Jinyuan	Gold Fountainhead	HD-02-2	521
	肥后二喜	Feihou Erxi	Higo's Second Pleasure	HA-44-2	94
	变装魔女	Bianzhuang Monü	Magic Girl in Changing Dresses	HA-01-17	23
	夜市	Yeshi	Night Market	HB-41-1	496
	炎夏红伞	Yanxia Hongsan	Red Umbrella in Hot Summer	HAR-17-2	358
	炎夏红浪	Yanxia Honglang	Red Waves in Hot Summer	HA-16-3	64
	炎夏魔红	Yanxia Mohong	Hot Summer's Magic Red	HAR-08-2	287
	波吉特爱心	Bojite Aixin	Mrs Birgit Linthe's Loving Heart	HAR-03-1	254
	波光粼粼	Boguang Linlin	Sparkling Ripples	HAR-14-9	335
	波特新生	Bote Xinsheng	Peter Pan's Rebirth	HA-57-1	117
	波特新姿	Bote Xinzi	Peter Pan's New Posture	HA-57-2	118
	宝石流霞	Baoshi Liuxia	Gem Bathed in Flowing Rosy Clouds	HAR-18-9	369
9	春之笑	Chunzhixiao	Spring Laughter	HB-29-2	475
	春红	Chunhong	Spring Red	HB-35-1	490
	春丽	Chunli	The Beauty of Spring	HB-30-4	480
	春喜	Chunxi	Spring Delight	HB-27-2	468
	春满园	Chunmanyuan	Spring Fully in Gardens	HD-36-1	557
	春燕归	Chunyangui	Spring Swallows Returning	HB-27-5	471
	玲珑黑妹	Linglong Heimei	Exquisite Black Sister	HA-66-1	147
	珊瑚田歌	Shanhu Tiange	Coral's Field Songs	HA-48-3	102
	珊瑚姑娘	Shanhu Guniang	Coral Girl	HA-48-4	103
	珊瑚新貌	Shanhu Xinmao	Coral's New Look	HA-48-2	101
	荣归	Ronggui	Gloriously Return to Homeland	HC-14-1	513
	星火燎原	Xinghuo Liaoyuan	Catching Fire from a Spark	HA-62-1	135
	星光闪闪	Xingguang Shanshan	Starlit Sparkle	HA-47-4	98

续表

笔画 Chinese Characters	品种名称 Name in Chinese	拼音名称 Name in Pinyin	英文含义 Meaning in English	所属组合 Combination	页码 Page
9	星光争辉	Xingguang Zhenghui	Starlit Contending for Brilliancy	HA-47-2	96
	星光灿烂	Xingguang Canlan	Starlit Splendid	HA-47-5	99
	星光高照	Xingguang Gaozhao	Starlit's Brightly Shining	HA-47-3	97
	星源云海	Xingyuan Yunhai	Xingyuan's Cloud Sea	HA-24-4	74
	星源红霞	Xingyuan Hongxia	Xingyuan's Red Glow	HA-24-3	73
	星源花歌	Xingyuan Huage	Xingyuan's Flowers Song	HA-84-1	210
	星源晚秋	Xingyuan Wanqiu	Xingyuan's Late Autumn	HA-83-1	207
	香纽扣	Xiangniukou	Scented Button	HD-39-1	562
	香菱紫衣	Xiangling Ziyi	Miss Xiangling's Purple Dress	HAR-37-1	438
	秋月闻莺	Qiuyue Wenying	Orioles Singing from the Autumn Moon	HAR-02-22	247
	秋风送霞	Qiufeng Songxia	Autumn Breeze Blowing Rosy Clouds	HAR-02-3	228
	秋冬桃红	Qiudong Taohong	Peach Red in Autumn & Winter	HA-34-9	81
	秋冬野美	Qiudong Yemei	Wild Beauty in Autumn & Winter	HA-11-3	52
	秋纹	Qiuwen	Miss Qiuwen	HC-02-7	502
	秋艳冬红	Qiuyan Donghong	Gorgeous Autumn & Red Winter	HA-34-8	80
	复色花容	Fuse Huarong	Flower Face Variegated	HC-02-5	500
	胖大姐	Pangdajie	Fat Eldest Sister	HB-37-1	492
	将军风度	Jiangjun Fengdu	General's Elegant Demeanor	HAR-09-1	295
	亭亭玉立	Tingting Yuli	Girls with Beautiful Standing Postures	HA-81-6	201
	姿容皆美	Zirong Jiemei	Pose & Face All Pretty	HAR-18-4	364
	美好向往	Meihao Xiangwang	Beautiful Yearning	HA-63-1	138
	美丽盛夏	Meili Shengxia	Beautiful Mid-Summer	HAR-14-4	330
	美秀映照	Meixiu Yingzhao	The Beauty from Reflecting	HA-76-1	180
	美秋	Meiqiu	Beautiful Autumn	HC-12-1	510
	美娇	Meijiao	Beautiful Charm	HB-30-1	477
	迷人红	Mirenhong	Charming Red	HC-02-4	499
	迷人红裙	Miren Hongqun	Attractive Red Dress	HAR-25-1	397
	迷你雅秀	Mini Yaxiu	Miniature Elegant Posture	HA-79-4	191
	迷你瑶姬	Mini Yaoji	Miniature Pretty Yaoji	HAR-12-1	313
	前程似锦	Qiancheng Sijin	Bright Future	HAR-18-15	375
	烂漫礼花	Lanman Lihua	Brilliant Fireworks	HAR-14-12	338
	洪福	Hongfu	Vast Happiness	HD-11-2	526
	恬静	Tianjing	Riposato	HD-23-5	539
	娇柔含羞	Jiaorou Hanxiu	Charming Soft with Pudency	HAR-14-17	343
	娇艳欲滴	Jiaoyan Yudi	Delicate & Charming	HAR-20-1	379
	盈盈笑口	Yingying Xiaokou	Smiling Mouths	HA-70-1	167
	勇士	Yongshi	Warrior	HC-17-1	516
	绚丽多彩	Xuanli Duocai	Bright & Colorful	HA-01-10	16
	绚丽夏秋	Xuanli Xiaqiu	Gorgeous Summer & Autumn	HAR-02-25	250
	绛唇映日	Jiangchun Yingri	Crimson Lips Reflected Sunlight	HA-56-2	114
10	艳红沐夏	Yanhong Muxia	Bright Red Bathed Summer	HAR-05-2	271
	艳紫妖红	Yanzi Yaohong	Brilliant Purple & Peculiar Red	HAR-29-1	405
	热恋季节	Relian Jijie	Lovestruck Season	HA-01-9	15

笔画 Chinese Characters	品种名称 Name in Chinese	拼音名称 Name in Pinyin	英文含义 Meaning in English	所属组合 Combination	页码 Page
	获奖喜悦	Huojiang Xiyue	The Joy of Winning an Award	HAR-06-9	281
	莺歌燕舞	Yingge Yanwu	Orioles Sing & Swallows Dancing	HA-67-7	155
	桃红如春	Taohong Ruchun	Peach Red as Spring	HAR-14-22	348
	桃红羞面	Taohong Xiumian	Peach Red & Bashful Face	HA-01-15	21
	桃红凝夏	Taohong Ningxia	Peach Red Coagulating Summer	HA-85-1	212
	夏云秋色	Xiayun Qiuse	Summer's Cloud & Autumn's Color	HAR-03-4	257
	夏日红妹	Xiari Hongmei	Summer's Red Sister	HA-68-4	163
	夏日红霞	Xiari Hongxia	Summer's Red Glow	HA-35-5	85
	夏日娇韵	Xiari Jiaoyun	Summer's Beautiful Rhymes	HAR-23-3	392
	夏日粉妹	Xiari Fenmei	Summer's Pink Sister	HA-68-3	162
	夏日海滩	Xiari Haitan	Summer's Beach	HAR-02-23	248
	夏令趣事	Xialing Qushi	Interesting Stories of Summer Camp	HA-67-9	157
	夏红秋丽	Xiahong Qiuli	Red Summer & Beautiful Autumn	HAR-18-7	367
	夏谷灵感	Xiagu Linggan	Summer Valley's Inspiration	HAR-18-8	368
	夏初巧遇	Xiachu Qiaoyu	Early Summer's Chance Encounter	HAR-08-1	286
	夏秋粉妞	Xiaqiu Fenniu	Pink Girl in Summer & Autumn	HA-05-5	33
	夏秋盛典	Xiaqiu Shengdian	Great Ceremony in Summer & Autumn	HA-10-20	50
	夏秋褶裙	Xiaqiu Zhequn	Wrinkled Skirt of Summer & Autumn	HA-06-2	35
	夏奥赛场	Xiaao Saichang	The Summer Olympic Games' Stadium	HAR-22-5	388
	烈焰红唇	Lieyan Hongchun	Flaming Lips	HA-67-5	153
	铁娘子	Tieniangzi	Iron Lady	HD-36-2	558
10	笑脸	Xiaolian	Smile Face	HD-20-10	537
	倚天望云	Yitian Wangyun	Look over Cloud Against Sky	HAR-35-3	419
	爱心	Aixin	Loving Heart	HB-02-6	448
	留芳亭	Liufangting	Gloriette with Fragrance	HC-16-1	515
	皱瓣金心	Zhouban Jinxin	Wrinkled Petals with Golden Heart	HAR-28-1	403
	恋人约会	Lianren Yuehui	Lovers Dating	HA-13-7	59
	高氏佳作	Gaoshi Jiazuo	Mr Gao's Excellent Masterpiece	HAR-23-2	391
	高朋满座	Gaopeng Manzuo	Distinguished Friends Party	HAR-14-8	334
	羞女	Xiunü	Shy Girl	HD-20-9	536
	羞涩桃腮	Xiuse Taosai	Peach Cheeks with Shyness	HAR-02-27	252
	粉元春	Fenyuanchun	Pink Early Spring	HB-08-5	452
	粉玉镯	Fenyuzhuo	Pink Jade Bracelet	HD-10-4	524
	粉玉露	Fenyulu	Pink Jade Dew	HB-26-4	466
	粉色田野	Fense Tianye	Pink Field	HAR-04-4	266
	粉色烛影	Fense Zhuying	Shadows of Pink Candle	HAR-08-7	292
	粉星	Fenxing	Pink Stars	HD-27-1	543
	粉娇醉秋	Fenjiao Zuiqiu	Pink Girls Making Autumn Intoxicate	HA-10-15	45
	粉柔舞裙	Fenrou Wuqun	Soft Pink Dance Dress	HA-59-2	127
	粉润心田	Fenrun Xintian	Pink Moistening Hearts	HA-54-1	109
	粉浪迎秋	Fenlang Yingqiu	Pink Waves Welcome Autumn	HAR-06-5	277
	粉萌妹	Fenmengmei	Pink Maidens	HB-29-3	476
	粉雅佳境	Fenya Jiajing	Pink Elegant & Wonderful Place	HAR-26-1	399

续表

笔画 Chinese Characters	品种名称 Name in Chinese	拼音名称 Name in Pinyin	英文含义 Meaning in English	所属组合 Combination	页码 Page
10	粉富美	Fenfumei	Pink, Rich & Beautiful Girl	HB-17-3	457
	粉增秀	Fenzengxiu	Adding Pink's Beauty	HB-12-2	454
	粉颜仙姿	Fenyan Xianzi	Pink Face & Fairy Posture	HAR-02-11	236
	浮翠流丹	Fucui Liudan	Floating Green & Flowing Red	HAR-04-5	267
	流光溢彩	Liuguang Yicai	Brilliant Color & Glitzy Light	HA-58-1	120
	浪漫夏景	Langman Xiajing	Romantic Summer Scenery	HAR-14-16	342
	窈窕淑女	Yaotiao Shunü	Fair Maiden	HAR-02-1	226
	陶醉	Taozui	Intoxication	HD-14-3	529
	娟好静秀	Juanhao Jingxiu	Beautiful in Appearance & Gentle in Disposition	HA-02-2	26
	娘子军	Niangzijun	Women Army	HC-17-2	517
	绣凤	Xiufeng	The Phoenix with Embroider	HD-30-3	551
11	堆金积玉	Duijin Jiyu	Store up Gold & Accumulate Jade	HAR-18-5	365
	梦中幻云	Mengzhong Huanyun	Magic Clouds in Dream	HAR-10-1	298
	梦幻世界	Menghuan Shijie	Dream-like World	HA-10-19	49
	梦归	Menggui	Realized Dream	HB-21-1	458
	梦境	Mengjing	Dream Scene	HB-27-4	470
	雪山	Xueshan	Snow Mountain	HD-13-2	527
	晨阳闪金	Chenyang Shanjin	Morning Sun Flashing Gold	HA-58-5	124
	野秋	Yeqiu	Wild Autumn	HC-12-2	511
	晚秋白云	Wanqiu Baiyun	Late-autumn's White Clouds	HA-83-2	208
	唯美	Weimei	Only Beauty	HC-10-7	508
	银边绣衣	Yinbian Xiuyi	Embroidered Clothes with Silver Borders	HA-81-4	199
	彩球发结	Caiqiu Fajie	Colorful Ball and Hair Knots	HAR-35-9	425
	清风胧月	Qingfeng Longyue	Gentle Breeze & Hazy Moon	HAR-05-1	270
	清丽少女	Qingli Shaonü	Charming & Young Girl	HAR-02-14	239
	淳朴	Chunpu	Unsophistication Simplicity	HB-02-8	450
	淡云阁雨	Danyun Geyu	Pale Cloud & Pavilion Rain	HAR-12-2	314
	淡雅柔粉	Danya Roufen	Quietly Elegant & Soft Pink	HAR-14-13	339
	深山九妹	Shenshan Jiumei	Ninth Sister in Remote Mountains	HAR-07-1	283
	惬意粉	Qieyifen	Pleasant Pink	HB-33-1	487
	婚庆元宝	Hunqing Yuanbao	Gold Ingot for Wedding	HA-41-5	89
	婚纱	Hunsha	Wedding Dress	HD-01-2	520
12	越秀	Yuexiu	Surpass Beauty	HB-30-5	481
	越松子	Yuesongzi	Surpass Pine Cone	HB-30-3	479
	超级粉冠	Chaoji Fenguan	Supreme Pink Crown	HAR-30-1	407
	超级叠粉	Chaoji Diefen	Overlapped Pink Supreme	HAR-13-1	322
	超丽	Chaoli	Surpassed Beauty	HD-20-6	533
	揽月阁	Lanyuege	Pavilion of Enjoying the Moon	HB-27-3	469
	喜迎曙光	Xiying Shuguang	Happily Welcome Dawn	HA-58-2	121
	酣悦	Hanyue	Drunken Delight	HB-25-1	462
	紫红苑	Zihongyuan	Purple Red Dwelling	HB-26-2	464
	紫花巷	Zihuaxiang	Purple Flower Alley	HD-30-1	549

续表

笔画 Chinese Characters	品种名称 Name in Chinese	拼音名称 Name in Pinyin	英文含义 Meaning in English	所属组合 Combination	页码 Page
12	紫胭脂	Ziyanzhi	Purple Rouge	HB-31-1	482
	紫粉舞秋	Zifen Wuqiu	Purple Pink Dancing Autumn	HA-43-5	92
	紫浪吻夏	Zilang Wenxia	Purple Waves Kissing Summer	HA-61-1	133
	紫绮	Ziqi	Purple Silk Cheongsam	HC-02-6	501
	紫绶金章	Zishou Jinzhang	Gold Seal with Purple Ribbon	HA-01-16	22
	紫绸	Zichou	Purple Silk	HD-20-8	535
	紫靓丽	Ziliangli	Purple Beauty	HC-11-1	509
	紫墙古宅	Ziqiang Guzhai	Purple Wall Ancient Messuage	HAR-37-3	440
	紫瓣融雪	Ziban Rongxue	Purple Petals Melting Snow	HAR-02-4	229
	暑期红艳	Shuqi Hongyan	Summer Holidays' Brilliant Red	HAR-14-7	333
	黑火喷金	Heihuo Penjin	Black Fire Spraying Gold	HAR-17-3	359
	黑红塔	Heihongta	Black-red Tower	HD-34-1	555
	黑红精灵	Heihong Jingling	Dark Red Elves	HAR-35-12	428
	黑亮哥	Heiliangge	Black-bright Brother	HD-33-1	554
	黑皱	Heizhou	Dark Crinkles	HD-38-1	561
	粤桂宝珠	Yuegui Baozhu	Guangdong & Guangxi's Pearl	HA-32-2	78
	童年回忆	Tongnian Huiyi	Childhood Memories	HAR-13-3	324
	温馨感觉	Wenxin Ganjue	Warm Feeling	HAR-14-5	331
	富丽堂皇	Fuli Tanghuang	Magnificence	HAR-14-15	341
13	瑞气祥云	Ruiqi Xiangyun	Happy Atmosphere & Auspicious Cloud	HA-68-2	161
	蒸蒸日上	Zhengzheng Rishang	Thriving upward	HA-59-1	126
	锦绣河山	Jinxiu Heshan	Land of Splendours	HAR-02-7	232
	微雕作品	Weidiao Zuopin	Miniature-sculpture Work	HA-81-2	197
	新发现	Xinfaxian	New Discovery	HB-40-1	495
	新作品	Xinzuopin	New Masterwork	HB-23-1	460
	新灵感	Xinlinggan	New Inspiration	HB-28-1	472
	新郎装	Xinlangzhuang	Bridegroom's Clothing	HB-14-10	455
	新面貌	Xinmianmao	New Look	HB-29-1	474
	新思路	Xinsilu	New Thoughts	HB-39-1	494
	新颖	Xinying	Novelty	HB-22-1	459
	新潮艳口	Xinchao Yankou	Trendy Pretty Mouth	HAR-35-4	420
	意外收获	Yiwai Shouhuo	Windfall	HAR-18-13	373
	满堂喝彩	Mantang Hecai	Universal Applause	HA-10-8	38
	福寿齐天	Fushou Qitian	Happiness & Longevity Comparable to the Universe	HA-59-4	129
	福寿宝塔	Fushou Baota	Happiness & Longevity Pagoda	HAR-23-1	390
	福娃新衣	Fuwa Xinyi	Mascots' New Clothes	HAR-35-15	431
	群蜂纷飞	Qunfeng Fenfei	A Swarm of Bees Flying	HAR-06-2	274
	叠瓣金蕊	Dieban Jinrui	Overlapped Petals with Golden Stamens	HA-35-4	84
14	碧血丹心	Bixue Danxin	Red Blood & Loyal Heart	HAR-24-2	395
	瑶林仙境	Yaolin Xianjing	Yaolin's Fairyland	HA-02-3	27
	暮鼓晨钟	Mugu Chenzhong	Evening Drum & Morning Bell	HA-34-10	82
	熙攘庙会	Xirang Miaohui	Bustling Fairs	HA-13-5	57

续表

笔画 Chinese Characters	品种名称 Name in Chinese	拼音名称 Name in Pinyin	英文含义 Meaning in English	所属组合 Combination	页码 Page
14	酷夏流金	Kuxia Liujin	Flowing Gold in Hot Summer	HAR-17-1	357
	貌美绝俗	Maomei Juesu	The Beautiful Looks that Never Seen	HAR-18-3	363
	端庄秀丽	Duanzhuang Xiuli	Dignified Pretty	HAR-03-2	255
	精致绣品	Jingzhi Xiupin	Delicate Embroidery	HAR-04-2	264
	精彩无限	Jingcai Wuxian	Splendid Infinite	HAR-18-1	361
	熊熊火焰	Xiongxiong Huoyan	Leaping Flames	HA-56-1	113
15	蕊珠头饰	Ruizhu Toushi	Miss Ruizhu's Headdress	HA-03-47	30
	醉春风	Zuichunfeng	Drunk Spring Breeze	HB-30-2	478
	墨皱锦	Mozhoujin	Black & Crinkled Brocade	HC-09-1	505
	黎明破晓	Liming Poxiao	Break of Dawn	HA-58-4	123
16	橘色晓霞	Juse Xiaoxia	Orange Color Foredawn	HAR-02-16	241
	橘红云路	Juhong Yunlu	Orange Red Cloud Road	HAR-12-6	318
	橘红淌金	Juhong Tangjin	Gold Trickling Down from Orange Red	HAR-18-12	372
	橘粉醉夏	Jufen Zuixia	Orange Pink Drunken Summer	HAR-02-5	230
17	繁华世界	Fanhua Shijie	Bustling World	HAR-12-8	320
	繁华街景	Fanhua Jiejing	Bustling Streetscape	HA-01-11	17
19	攀登	Pandeng	Climbing	HD-23-4	538
22	镶边彩扣	Xiangbian Caikou	Frilly Colorful Button	HA-67-4	152

2. 拼音名称索引（Index of the New Camellia Hybrids in Chinese Pinyin）

字母 English Letters	拼音名称 Name in Pinyin	品种名称 Name in Chinese	英文含义 Meaning in English	所属组合 Combination	页码 Page
A	Apoliu Nüshen	阿婆六女神	Apoliu's Goddess	HC-15-1	514
	Apoliu Shuixing	阿婆六水星	Apoliu's Mercury	HB-36-1	491
	Apoliu Zhichen	阿婆六之晨	Apoliu's Morning	HC-13-1	512
	Aixin	爱心	Loving Heart	HB-02-6	448
B	Bayue Honglang	八月红浪	August's Red Waves	HAR-14-23	349
	Bayue Piaoxue	八月飘雪	August's Snowing	HAR-03-8	261
	Bayue Talang	八月踏浪	Walking in Waves in August	HAR-10-6	303
	Baifumei	白富美	White, Rich & Beautiful Girl	HB-17-2	456
	Bainiangzi	白娘子	White Snake Lady	HD-20-4	531
	Baimei Qianjiao	百媚千娇	Enchanting Beauty	HAR-08-5	290
	Banjuan Honglian	半卷红帘	Semi-rolled Red Curtain	HA-86-1	214
	Baoshi Liuxia	宝石流霞	Gem Bathed in Flowing Rosy Clouds	HAR-18-9	369
	Beijibing	北极冰	Arctic Ice	HD-13-3	528
	Benfang Wuzhe	奔放舞者	Unrestrained Dancer	HAR-27-1	401
	Bixue Danxin	碧血丹心	Red Blood & Loyal Heart	HAR-24-2	395
	Bianzhuang Monü	变装魔女	Magic Girl in Changing Dresses	HA-01-17	23
	Bingyi	冰艺	Ice Art	HC-10-6	507
	Boguang Linlin	波光粼粼	Sparkling Ripples	HAR-14-9	335
	Bojite Aixin	波吉特爱心	Mrs Birgit Linthe's Loving Heart	HAR-03-1	254
	Bote Xinsheng	波特新生	Peter Pan's Rebirth	HA-57-1	117
	Bote Xinzi	波特新姿	Peter Pan's New Posture	HA-57-2	118
	Buye Hongcheng	不夜红城	Ever-bright Red City	HA-80-2	194
C	Caiqiu Fajie	彩球发结	Colorful Ball and Hair Knots	HAR-35-9	425
	Canlan Huihuang	灿烂辉煌	Splendid & Glorious	HA-10-12	42
	Canlan Yangguang	灿烂阳光	Brilliant Sunshine	HAR-02-6	231
	Cengceng Shiyi	层层诗意	Layers of Poetic Flavor	HAR-31-1	409
	Cengdie Meijing	层叠美景	Layering Beautiful Scenery	HAR-18-2	362
	Changbian Guniang	长辫姑娘	Long Braid Girl	HAR-04-3	265
	Changshoucun	长寿村	Longevity Village	HB-28-2	473
	Chaoji Diefen	超级叠粉	Overlapped Pink Supreme	HAR-13-1	322
	Chaoji Fenguan	超级粉冠	Supreme Pink Crown	HAR-30-1	407
	Chaoli	超丽	Surpassed Beauty	HD-20-6	533
	Chenyang Shanjin	晨阳闪金	Morning Sun Flashing Gold	HA-58-5	124
	Chenggong Zhixi	成功之喜	Exultation of Success	HAR-14-2	328
	Chicheng Hongxin	赤诚红心	Sincere Red Heart	HA-03-46	29
	Chicheng Zhixin	赤诚之心	Sincere Heart	HAR-18-11	371
	Chihong Huahai	赤红花海	Crimson Flower Sea	HA-75-1	178

续表

字母 English Letters	拼音名称 Name in Pinyin	品种名称 Name in Chinese	英文含义 Meaning in English	所属组合 Combination	页码 Page
C	Chiyan Luhuo	赤焰炉火	Red Flame Stove Fire	HA-77-3	184
	Chulian	初恋	First Love	HD-20-3	530
	Chutan	初探	First Exploration	HB-38-1	493
	Chuchu Huanteng	处处欢腾	Exult Everywhere	HA-14-5	61
	Chunhong	春红	Spring Red	HB-35-1	490
	Chunli	春丽	The Beauty of Spring	HB-30-4	480
	Chunmanyuan	春满园	Spring Fully in Gardens	HD-36-1	557
	Chunxi	春喜	Spring Delight	HB-27-2	468
	Chunyangui	春燕归	Spring Swallows Returning	HB-27-5	471
	Chunzhixiao	春之笑	Spring Laughter	HB-29-2	475
	Chunpu	淳朴	Unsophistication Simplicity	HB-02-8	450
D	Da Fensituan	大粉丝团	Big Group of Fans	HA-63-2	139
	Dafei Shengji	大菲升级	'Francis Eugene Phillis' Upgrade	HA-06-3	36
	Dafu Dagui	大福大贵	Big Happiness & Big Millionaire	HAR-02-20	245
	Dajia Guixiu	大家闺秀	Young Girl from Respectable Family	HAR-06-3	275
	Dakuayue	大跨越	Great Stride Leap	HB-32-1	484
	Daqipo	大气魄	Great Courage	HD-40-1	563
	Dashiye	大视野	Broad Field of Vision	HB-32-3	486
	Datang Gongdeng	大唐宫灯	Palace Lantern in Great Tang Dynasty	HAR-32-1	411
	Daxi Linmen	大喜临门	Great Happy Event is Coming	HAR-02-2	227
	Dayu	大玉	Large Jade	HD-25-1	540
	Dayuzhuo	大玉镯	Large Jade Bracelet	HD-10-5	525
	Danya Roufen	淡雅柔粉	Quietly Elegant & Soft Pink	HAR-14-13	339
	Danyun Geyu	淡云阁雨	Pale Cloud & Pavilion Rain	HAR-12-2	314
	Dangdai Qiaoli	当代俏丽	Contemporary Pretty Girl	HAR-02-13	238
	Dieban Jinrui	叠瓣金蕊	Overlapped Petals with Golden Stamens	HA-35-4	84
	Dongxia Chucai	冬夏出彩	Colorfull in Winter & Summer	HAR-06-8	280
	Duanzhuang Xiuli	端庄秀丽	Dignified Pretty	HAR-03-2	255
	Duijin Jiyu	堆金积玉	Store up Gold & Accumulate Jade	HAR-18-5	365
	Duoji Alan	多季阿兰	Multiseasonal Mark Alan	HA-62-2	136
	Duoji Hongguan	多季红冠	Multiseasonal Red Crown	HAR-18-10	370
	Duoji Juemei	多季绝美	Multiseasonal Absolute Beauty	HA-65-3	145
	Duoji Linglong	多季玲珑	Multiseasonal Exquisiteness	HA-72-1	172
	Duoji Naidong	多季耐冬	Multiseasonal Naidong	HA-60-1	131
	Duoji Runxiang	多季润香	Multiseasonal Moist Fragrance	HA-65-1	143
	Duoji Shengqing	多季盛情	Multiseasonal Great Kindness	HAR-02-10	235
	Duozi Liying	多姿丽影	Multi-Poses & Pretty Shadow	HAR-02-26	251
E	Ertong Leyuan	儿童乐园	Children's Playground	HAR-14-6	332
F	Fanhua Jiejing	繁华街景	Bustling Streetscape	HA-01-11	17
	Fanhua Shijie	繁华世界	Bustling World	HAR-12-8	320
	Feijian Huohua	飞溅火花	Flying Sparks	HAR-02-17	242

续表

字母 English Letters	拼音名称 Name in Pinyin	品种名称 Name in Chinese	英文含义 Meaning in English	所属组合 Combination	页码 Page
F	Feihou Erxi	肥后二喜	Higo's Second Pleasure	HA-44-2	94
	Fenfumei	粉富美	Pink, Rich & Beautiful Girl	HB-17-3	457
	Fenjiao Zuiqiu	粉娇醉秋	Pink Girls Making Autumn Intoxicate	HA-10-15	45
	Fenlang Yingqiu	粉浪迎秋	Pink Waves Welcome Autumn	HAR-06-5	277
	Fenmengmei	粉萌妹	Pink Maidens	HB-29-3	476
	Fenrou Wuqun	粉柔舞裙	Soft Pink Dance Dress	HA-59-2	127
	Fenrun Xintian	粉润心田	Pink Moistening Hearts	HA-54-1	109
	Fense Tianye	粉色田野	Pink Field	HAR-04-4	266
	Fense Zhuying	粉色烛影	Shadows of Pink Candle	HAR-08-7	292
	Fenxing	粉星	Pink Stars	HD-27-1	543
	Fenya Jiajing	粉雅佳境	Pink Elegant & Wonderful Place	HAR-26-1	399
	Fenyan Xianzi	粉颜仙姿	Pink Face & Fairy Posture	HAR-02-11	236
	Fenyuanchun	粉元春	Pink Early Spring	HB-08-5	452
	Fenyulu	粉玉露	Pink Jade Dew	HB-26-4	466
	Fenyuzhuo	粉玉镯	Pink Jade Bracelet	HD-10-4	524
	Fenzengxiu	粉增秀	Adding Pink's Beauty	HB-12-2	454
	Fenghe Rili	风和日丽	Breezy & Sunny	HAR-24-1	394
	Fenghua Zhengmao	风华正茂	Prime of Life	HAR-12-5	317
	Fengyunsao	风韵嫂	Charmed Sister-in Law	HB-26-1	463
	Fengguan Xiapei	凤冠霞帔	Phoenix Coronet & Robes	HAR-02-12	237
	Fenghuangting	凤凰亭	Phoenix Pavilion	HB-34-2	489
	Fozhi Huazhang	佛植华章	Foshan Botanical Garden's Brilliant Works	HA-01-18	24
	Fozhi Yingrui	佛植盈瑞	Foshan Botanical Garden's Full Auspiciousness	HA-78-1	186
	Fucui Liudan	浮翠流丹	Floating Green & Flowing Red	HAR-04-5	267
	Fushou Baota	福寿宝塔	Happiness & Longevity Pagoda	HAR-23-1	390
	Fushou Qitian	福寿齐天	Happiness & Longevity Comparable to the Universe	HA-59-4	129
	Fuwa Xinyi	福娃新衣	Mascots' New Clothes	HAR-35-15	431
	Fuse Huarong	复色花容	Flower Face Variegated	HC-02-5	500
	Fuli Tanghuang	富丽堂皇	Magnificence	HAR-14-15	341
G	Gaopeng Manzuo	高朋满座	Distinguished Friends Party	HAR-14-8	334
	Gaoshi Jiazuo	高氏佳作	Mr Gao's Excellent Masterpiece	HAR-23-2	391
	Gebi Tuoling	戈壁驼铃	Gobi Camel Bells	HAR-35-8	424
	Gudian	古殿	Ancient Temple	HB-24-1	461
	Gumiao Hongzhong	古庙红钟	Red Bell of Ancient Temple	HA-80-1	193
	Gupu Qingyou	古朴清幽	Ancient Beauty & Tranquility	HAR-37-2	439
	Guaiqiao Nühai	乖巧女孩	Well-behaved & Clever Girl	HA-81-3	198
	Guangchang Hongyi	广场红艺	Square Red Art	HA-53-2	106
	Guangchang Wubu	广场舞步	Dance Steps in Square	HA-10-10	40
	Guose Tianzi	国色天姿	Possess Surpassing Beauty	HA-13-6	58

续表

字母 English Letters	拼音名称 Name in Pinyin	品种名称 Name in Chinese	英文含义 Meaning in English	所属组合 Combination	页码 Page
H	Hanyue	酣悦	Drunken Delight	HB-25-1	462
	Hansen Zhiyue	汉森之悦	Mr Waldemar Max Hansen's Delight	HAR-08-3	288
	Hemu	和睦	Harmony	HB-26-3	465
	Heshun	和顺	Harmony & Smooth	HD-20-5	532
	Hexie Jiayuan	和谐家园	Harmonious Homeland	HA-79-1	188
	Heihong Jingling	黑红精灵	Dark Red Elves	HAR-35-12	428
	Heihongta	黑红塔	Black-red Tower	HD-34-1	555
	Heihuo Penjin	黑火喷金	Black Fire Spraying Gold	HAR-17-3	359
	Heiliangge	黑亮哥	Black-bright Brother	HD-33-1	554
	Heizhou	黑皱	Dark Crinkles	HD-38-1	561
	Hongban Jinxin	红瓣金心	Red Petals & Golden Heart	HAR-10-4	301
	Hongban Jingxiu	红瓣竞秀	Red Petals' Beauty Competitions	HAR-33-1	413
	Hongban Xiangxin	红瓣香心	Red Petals with Fragrant Heart	HA-77-2	183
	Hongcheng Jinku	红城金库	Red City's Gold Vault	HAR-35-17	433
	Hongcheng Manjin	红城漫金	Red City Overflowing Gold	HAR-35-13	429
	Hongcheng Yejing	红城夜景	Red City's Night Scenes	HAR-35-11	427
	Hongchou Wulang	红绸舞浪	Red Silk Dancing Waves	HA-10-14	44
	Hongerzhui	红耳坠	Red Eardrop	HD-27-3	545
	Hongfa Mote	红发模特	Red Hair Models	HAR-35-7	423
	Hongfen Wuhui	红粉舞会	Red-pink Dancing Party	HAR-06-4	276
	Honghong Huohuo	红红火火	Prosperity & Jollification	HAR-04-1	263
	Hongla Diaosu	红蜡雕塑	Sculpture with Red Wax	HAR-14-14	340
	Honglang Taotian	红浪滔天	Surge Red Waves	HAR-09-2	296
	Honglin Wazhai	红林瓦寨	Tile Village in Red Forest	HA-71-1	170
	Hongling Baoxi	红铃报喜	Happy News from Red Bells	HAR-06-6	278
	Honglong Wutian	红龙舞天	Red Dragon Dancing in the Sky	HA-53-1	105
	Hongman Jinshan	红漫金山	Red Covering Golden Mountains	HAR-14-11	337
	Hongpan Tuojin	红盘托金	Red Plate Holding up Gold	HAR-22-2	385
	Hongrong Gongzhu	红绒公主	Red Velvet Princess	HAR-35-16	432
	Hongse Changxiang	红色畅想	Red Imagination	HA-10-13	43
	Hongse Xingkong	红色星空	Red Starry Sky	HAR-10-3	300
	Hongse Xunzhang	红色勋章	Red Medal	HA-79-3	190
	Hongse Zhuti	红色主题	Red Theme	HA-59-3	128
	Hongshiwen	红诗文	Red Poem	HB-27-1	467
	Hongwutai	红舞台	Red Stage	HC-05-2	504
	Hongxing Jiangzhang	红星奖章	Red Star Medal	HA-82-1	205
	Hongyi Xiannü	红衣仙女	The Fairy Dressing Red	HAR-08-4	289
	Hongyuan Manfu	红院满福	Red Courtyard with Full Happiness	HAR-14-20	346
	Hongyun Shanguang	红云闪光	Lightening Red Clouds	HA-65-2	144
	Hongzhuang Yiren	红装伊人	Beloved Lady Dressed in Red	HA-67-8	156
	Hongfu	洪福	Vast Happiness	HD-11-2	526

续表

字母 English Letters	拼音名称 Name in Pinyin	品种名称 Name in Chinese	英文含义 Meaning in English	所属组合 Combination	页码 Page
H	Houlang Benyong	后浪奔涌	Back Wavy Surges	HAR-14-18	344
	Houqi Zhixiu	后起之秀	Up-rising Star	HA-64-1	141
	Huacheng Naoshi	花城闹市	Guangzhou's Busy Streets	HA-26-6	76
	Huadenglong	花灯笼	Colorful Lantern	HD-32-1	553
	Huaji Duige	花季对歌	Coupled Singing in Blooming Season	HA-56-3	115
	Huarong Yuemao	花容月貌	Beautiful Feature as Flower & Moon	HAR-02-18	243
	Huayang Nianhua	花样年华	Diversity of Life	HAR-06-1	273
	Huayi Xiaoxuan	花衣小旋	Miss Xiaoxuan Dressed Colorful Clothes	HA-14-6	62
	Huanqing	欢庆	Festivity	HB-02-9	451
	Huilong Chenxi	回龙晨曦	Huilong Town's Morning Light	HAR-01-1	224
	Huilong Songge	回龙颂歌	Huilong Town's Ode	HA-22-2	71
	Huilong Xinmao	回龙新貌	Huilong Town's New Look	HAR-10-9	306
	Huimou Zhili	回眸之丽	Beautiful Glancing Back	HAR-02-21	246
	Hunqing Yuanbao	婚庆元宝	Gold Ingot for Wedding	HA-41-5	89
	Hunsha	婚纱	Wedding Dress	HD-01-2	520
	Huohong Niandai	火红年代	Fire Red Era	HAR-03-3	256
	Huohong Shuangjie	火红双节	Fire Red Double Festivals	HA-11-4	53
	Huoshan Lijing	火山丽景	Beautiful Scene of Volcano	HAR-38-1	443
	Huojiang Xiyue	获奖喜悦	The Joy of Winning an Award	HAR-06-9	281
J	Jiri Liangchen	吉日良辰	Auspicious Day	HA-43-4	91
	Jixiang Ruyi	吉祥如意	Good Lucky as Desired	HA-67-1	149
	Jixing Gaozhao	吉星高照	Lucky Stars Shining Brightly	HAR-21-1	382
	Jijie Secai	季节色彩	Seasonal Colors	HA-01-13	19
	Jiamian Hongtan	加冕红毯	The Red Carpet for Crowning	HA-73-1	174
	Jiangjun Fengdu	将军风度	General's Elegant Demeanor	HAR-09-1	295
	Jiangchun Yingri	绛唇映日	Crimson Lips Reflected Sunlight	HA-56-2	114
	Jiaorou Hanxiu	娇柔含羞	Charming Soft with Pudency	HAR-14-17	343
	Jiaoyan Yudi	娇艳欲滴	Delicate & Charming	HAR-20-1	379
	Jinri Xier	今日喜儿	Today's Miss Xier	HAR-13-2	323
	Jincan Hongyun	金灿红云	Gold Shining in Red Clouds	HAR-11-1	309
	Jinheirong	金黑绒	Golden Black Charpie	HC-04-2	503
	Jinyan Liangying	金艳靓影	Jinyan's Pretty Shadow	HA-79-2	189
	Jinyuan	金源	Gold Fountainhead	HD-02-2	521
	Jinxiu Heshan	锦绣河山	Land of Splendours	HAR-02-7	232
	Jinzhaohui	尽朝晖	Morning Sunlight over the Sky	HD-37-1	559
	Jingcai Wuxian	精彩无限	Splendid Infinite	HAR-18-1	361
	Jingzhi Xiupin	精致绣品	Delicate Embroidery	HAR-04-2	264
	Jiujiu Tongxin	九九同心	Nine Persons with One Heart	HAR-35-6	422
	Jiujiu Yanyang	九九艳阳	The Bright Sun of Sept. Ninth	HA-01-14	20
	Jiutian Yaochi	九天瑶池	Jade Pool at the Ninth Heaven	HAR-14-1	327
	Jiuyue Jingyan	九月惊艳	September's Amazement	HAR-10-5	302

续表

字母 English Letters	拼音名称 Name in Pinyin	品种名称 Name in Chinese	英文含义 Meaning in English	所属组合 Combination	页码 Page
J	Jufen Zuixia	橘粉醉夏	Orange Pink Drunken Summer	HAR-02-5	230
	Juhong Tangjin	橘红淌金	Gold Trickling Down from Orange Red	HAR-18-12	372
	Juhong Yunlu	橘红云路	Orange Red Cloud Road	HAR-12-6	318
	Juse Xiaoxia	橘色晓霞	Orange Color Foredawn	HAR-02-16	241
	Juanhao Jingxiu	娟好静秀	Beautiful in Appearance & Gentle in Disposition	HA-02-2	26
K	Kongque Xuanli	孔雀炫丽	Peacock Flaunting Its Beauty	HAR-34-1	415
	Kuxia Liujin	酷夏流金	Flowing Gold in Hot Summer	HAR-17-1	357
	Kuanghuan Wuqu	狂欢舞曲	Orgiastic Dance Tune	HAR-19-1	377
L	Lanyuege	揽月阁	Pavilion of Enjoying the Moon	HB-27-3	469
	Lanman Lihua	烂漫礼花	Brilliant Fireworks	HAR-14-12	338
	Langman Xiajing	浪漫夏景	Romantic Summer Scenery	HAR-14-16	342
	Liming Poxiao	黎明破晓	Break of Dawn	HA-58-4	123
	Lianren Yuehui	恋人约会	Lovers Dating	HA-13-7	59
	Liangkou Hongqiu	良口红秋	Liangkou Town's Red Autumn	HA-77-1	182
	Lieyan Hongchun	烈焰红唇	Flaming Lips	HA-67-5	153
	Linjia Hongjie	邻家红姐	Neighbor's Red Sister	HAR-08-8	293
	Linglong Heimei	玲珑黑妹	Exquisite Black Sister	HA-66-1	147
	Liucun Caihong	刘村彩虹	Liu Village's Rainbow	HAR-10-8	305
	Liufangting	留芳亭	Gloriette with Fragrance	HC-16-1	515
	Liuguang Yicai	流光溢彩	Brilliant Color & Glitzy Light	HA-58-1	120
	Liuchao Zhifen	六朝脂粉	Six Dynasties' Face Powder	HAR-03-6	259
	Liujiaoting	六角亭	Hexagonal Pavilion	HD-29-1	548
	Liuliu Dashun	六六大顺	Double Sixs Making Everything Smooth	HAR-06-7	279
	Liumai Shenjian	六脉神剑	Six Pulses God Sword	HAR-04-6	268
	Longhuo Hexi	龙火贺喜	Dragon-fire Congratulations	HA-67-6	154
	Longzhang Fengzi	龙章凤姿	Dragon's Literate & Phoenix's Posture	HAR-37-4	441
M	Mantang Hecai	满堂喝彩	Universal Applause	HA-10-8	38
	Maomei Juesu	貌美绝俗	The Beautiful Looks that Never Seen	HAR-18-3	363
	Meihao Xiangwang	美好向往	Beautiful Yearning	HA-63-1	138
	Meijiao	美娇	Beautiful Charm	HB-30-1	477
	Meili Shengxia	美丽盛夏	Beautiful Mid-Summer	HAR-14-4	330
	Meiqiu	美秋	Beautiful Autumn	HC-12-1	510
	Meixiu Yingzhao	美秀映照	The Beauty from Reflecting	HA-76-1	180
	Menggui	梦归	Realized Dream	HB-21-1	458
	Menghuan Shijie	梦幻世界	Dream-like World	HA-10-19	49
	Mengjing	梦境	Dream Scene	HB-27-4	470
	Mengzhong Huanyun	梦中幻云	Magic Clouds in Dream	HAR-10-1	298
	Mini Yaoji	迷你瑶姬	Miniature Pretty Yaoji	HAR-12-1	313
	Mini Yaxiu	迷你雅秀	Miniature Elegant Posture	HA-79-4	191
	Mirenhong	迷人红	Charming Red	HC-02-4	499

续表

字母 English Letters	拼音名称 Name in Pinyin	品种名称 Name in Chinese	英文含义 Meaning in English	所属组合 Combination	页码 Page
M	Miren Hongqun	迷人红裙	Attractive Red Dress	HAR-25-1	397
	Minzu Fushi	民族服饰	National Dress	HA-81-8	203
	Mingxing Fengfan	明星风范	Star Manner	HAR-07-2	284
	Mozhoujin	墨皱锦	Black & Crinkled Brocade	HC-09-1	505
	Mugu Chenzhong	暮鼓晨钟	Evening Drum & Morning Bell	HA-34-10	82
N	Niannian Youyu	年年有余	Having Surplus Every Year	HA-74-1	176
	Niangzijun	娘子军	Women Army	HC-17-2	517
	Nümo Shizhuang	女模时装	Female Models' Fashionable Dress	HA-10-11	41
P	Pandeng	攀登	Climbing	HD-23-4	538
	Pangdajie	胖大姐	Fat Eldest Sister	HB-37-1	492
	Pihong Qiaonü	披红俏女	The Pretty Draped Red	HAR-22-4	387
Q	Qixi Lipin	七夕礼品	Double-Seventh Day's Gift	HAR-08-6	291
	Qixiannü	七仙女	The Seventh Fairy	HD-30-2	550
	Qiyue Kaimu	七月开幕	In July Curtain-up	HAR-10-7	304
	Qimiao Xingse	奇妙形色	Fantastic Shape & Color	HAR-18-14	374
	Qianqiuwang	千秋旺	Forever Prosperous	HD-20-7	534
	Qiancheng Sijin	前程似锦	Bright Future	HAR-18-15	375
	Qiaozhi Qianjin	乔之千金	Dr Georg Ziemes' Beloved Daughter	HAR-13-4	325
	Qieyifen	惬意粉	Pleasant Pink	HB-33-1	487
	Qingchun Zhige	青春之歌	Youth's Song	HAR-14-19	345
	Qingfeng Longyue	清风胧月	Gentle Breeze & Hazy Moon	HAR-05-1	270
	Qingli Shaonü	清丽少女	Charming & Young Girl	HAR-02-14	239
	Qiudong Taohong	秋冬桃红	Peach Red in Autumn & Winter	HA-34-9	81
	Qiudong Yemei	秋冬野美	Wild Beauty in Autumn & Winter	HA-11-3	52
	Qiufeng Songxia	秋风送霞	Autumn Breeze Blowing Rosy Clouds	HAR-02-3	228
	Qiuwen	秋纹	Miss Qiuwen	HC-02-7	502
	Qiuyan Donghong	秋艳冬红	Gorgeous Autumn & Red Winter	HA-34-8	80
	Qiuyue Wenying	秋月闻莺	Orioles Singing from the Autumn Moon	HAR-02-22	247
	Qunfeng Fenfei	群蜂纷飞	A Swarm of Bees Flying	HAR-06-2	274
R	Relian Jijie	热恋季节	Lovestruck Season	HA-01-9	15
	Richu	日出	Sunup	HD-28-1	547
	Riyue Chongguang	日月重光	The Sun & The Moon Appear Together	HAR-03-7	260
	Rizhao Xianglu	日照香炉	Sunshine Censer	HAR-15-2	353
	Ronggui	荣归	Gloriously Return to Homeland	HC-14-1	513
	Ruhua Mengjing	如画梦境	Picturesque Dream	HA-81-1	196
	Rumeng	如梦	Like a Dream	HD-37-2	560
	Ruizhu Toushi	蕊珠头饰	Miss Ruizhu's Headdress	HA-03-47	30
	Ruiqi Xiangyun	瑞气祥云	Happy Atmosphere & Auspicious Cloud	HA-68-2	161
S	Sanqian Moli	三千墨丽	Three Thousand Ink Works	HAR-35-10	426
	Shanhu Guniang	珊瑚姑娘	Coral Girl	HA-48-4	103
	Shanhu Tiange	珊瑚田歌	Coral's Field Songs	HA-48-3	102

续表

字母 English Letters	拼音名称 Name in Pinyin	品种名称 Name in Chinese	英文含义 Meaning in English	所属组合 Combination	页码 Page
S	Shanhu Xinmao	珊瑚新貌	Coral's New Look	HA-48-2	101
	Shanguang Fengche	闪光风车	Glittering Windmill	HAR-16-1	355
	Shaonüqun	少女裙	Maiden Skirt	HB-31-2	483
	Shenshan Jiumei	深山九妹	Ninth Sister in Remote Mountains	HAR-07-1	283
	Shengming Yangguang	生命阳光	Sunny Life	HA-01-12	18
	Shengri Yanhui	生日宴会	Birthday Party	HA-67-10	158
	Shiyue Hongtou	十月红透	Red Through in October	HAR-10-2	299
	Shiyue Yanhuo	十月烟火	October's Fireworks	HAR-14-21	347
	Shishang Xiaomei	时尚小妹	Stylish Little Sister	HAR-35-5	421
	Shuxiang Zhijia	书香之家	Literary Family	HAR-36-1	436
	Shuqi Hongyan	暑期红艳	Summer Holidays' Brilliant Red	HAR-14-7	333
	Shuanghong Licheng	双红荔城	Double Red Litchi City	HAR-02-15	240
	Shuangmian Jiaren	双面佳人	Beautiful Maiden with Double Faces	HAR-02-9	234
	Shuimo Danqing	水墨丹青	Chinese Ink Painting	HA-53-3	107
	Siji Fengqing	四季风情	Four-season Amorous Feelings	HA-21-8	68
	Siji Heyun	四季合韵	The Rhyme Conforming Four Seasons	HAR-22-3	386
	Siceng Xiangjian	似曾相见	Seems to Have Met Before	HAR-14-24	350
T	Tandang Xionghuai	坦荡胸怀	Magnanimous Mind	HAR-22-1	384
	Taohong Ningxia	桃红凝夏	Peach Red Coagulating Summer	HA-85-1	212
	Taohong Ruchun	桃红如春	Peach Red as Spring	HAR-14-22	348
	Taohong Xiumian	桃红羞面	Peach Red & Bashful Face	HA-01-15	21
	Taozui	陶醉	Intoxication	HD-14-3	529
	Tianlun Zhile	天伦之乐	Enjoying Family's Happiness	HA-10-18	48
	Tianyuan Fengguang	田园风光	Rural Scenery	HAR-35-14	430
	Tianjing	恬静	Riposato	HD-23-5	539
	Tieniangzi	铁娘子	Iron Lady	HD-36-2	558
	Tingting Yuli	亭亭玉立	Girls with Beautiful Standing Postures	HA-81-6	201
	Tonghai Yongqiu	彤海咏秋	Red Sea Singing Autumn	HA-12-2	55
	Tongnian Huiyi	童年回忆	Childhood Memories	HAR-13-3	324
	Tuanju	团聚	Reunion	HD-35-1	556
W	Waipo Tongyao	外婆童谣	Grandmother's Nursery Rhymes	HAR-35-2	418
	Wanqiu Baiyun	晚秋白云	Late-autumn's White Clouds	HA-83-2	208
	Wanjia Denghuo	万家灯火	Myriad Families' Lights	HA-10-16	46
	Wanshi Shunjing	万事顺景	Everything Good Fortune	HA-55-1	111
	Wangcai	旺财	Flourish Treasures	HB-02-7	449
	Wangdefu	旺德福	Wonderful	HB-09-4	453
	Weidiao Zuopin	微雕作品	Miniature-sculpture Work	HA-81-2	197
	Weimei	唯美	Only Beauty	HC-10-7	508
	Weida Fuxing	伟大复兴	Great Renaissance	HAR-15-1	352
	Wenxin Ganjue	温馨感觉	Warm Feeling	HAR-14-5	331
	Wufu Linmen	五福临门	Five Blessings Arriving at Home	HA-68-1	160

续表

字母 English Letters	拼音名称 Name in Pinyin	品种名称 Name in Chinese	英文含义 Meaning in English	所属组合 Combination	页码 Page
W	Wugu Fengdeng	五谷丰登	Abundant Harvest of All Crops	HA-81-5	200
	Wuye Denghuo	午夜灯火	Midnight Lights	HAR-10-10	307
	Wuye Jinguang	午夜金光	Midnight Golden Light	HA-67-2	150
X	Xiyang Yuhui	夕阳余晖	Afterglow of the Sunset	HA-10-17	47
	Xizi Wanxia	西子晚霞	West Lake's Sunset Glow	HA-69-1	165
	Xirang Miaohui	熙攘庙会	Bustling Fairs	HA-13-5	57
	Xiying Shuguang	喜迎曙光	Happily Welcome Dawn	HA-58-2	121
	Xiaao Saichang	夏奥赛场	The Summer Olympic Games' Stadium	HAR-22-5	388
	Xiachu Qiaoyu	夏初巧遇	Early Summer's Chance Encounter	HAR-08-1	286
	Xiagu Linggan	夏谷灵感	Summer Valley's Inspiration	HAR-18-8	368
	Xiahong Qiuli	夏红秋丽	Red Summer & Beautiful Autumn	HAR-18-7	367
	Xialing Qushi	夏令趣事	Interesting Stories of Summer Camp	HA-67-9	157
	Xiaqiu Fenniu	夏秋粉妞	Pink Girl in Summer & Autumn	HA-05-5	33
	Xiaqiu Shengdian	夏秋盛典	Great Ceremony in Summer & Autumn	HA-10-20	50
	Xiaqiu Zhequn	夏秋褶裙	Wrinkled Skirt of Summer & Autumn	HA-06-2	35
	Xiari Fenmei	夏日粉妹	Summer's Pink Sister	HA-68-3	162
	Xiari Haitan	夏日海滩	Summer's Beach	HAR-02-23	248
	Xiari Hongmei	夏日红妹	Summer's Red Sister	HA-68-4	163
	Xiari Hongxia	夏日红霞	Summer's Red Glow	HA-35-5	85
	Xiari Jiaoyun	夏日娇韵	Summer's Beautiful Rhymes	HAR-23-3	392
	Xiayun Qiuse	夏云秋色	Summer's Cloud & Autumn's Color	HAR-03-4	257
	Xiannü Xiafan	仙女下凡	Fairy from Sky Down to the Earth	HA-05-4	32
	Xianhua Yingchi	闲花映池	Wild Flowers Reflected from Pool	HAR-02-24	249
	Xiangcun Secai	乡村色彩	Rural Color	HAR-12-7	319
	Xiangling Ziyi	香菱紫衣	Miss Xiangling's Purple Dress	HAR-37-1	438
	Xiangniukou	香纽扣	Scented Button	HD-39-1	562
	Xiangbian Caikou	镶边彩扣	Frilly Colorful Button	HA-67-4	152
	Xiaofentong	小粉童	Little Pink Child	HD-27-4	546
	Xiaojingxi	小惊喜	Little Surprise	HD-10-2	522
	Xiaojiubei	小酒杯	Small Wine Cup	HB-34-1	488
	Xiaoliwu	小礼物	Small Gift	HD-26-1	542
	Xiaoluoxuan	小螺旋	Small Spiral	HB-32-2	485
	Xiaoniao Yiren	小鸟依人	Sweet & Helpless Birds	HAR-14-10	336
	Xiaoyu	小玉	Small Jade	HD-25-2	541
	Xiaoyuzhuo	小玉镯	Small Jade Bracelet	HD-10-3	523
	Xiaozitong	小紫童	Little Purple Child	HD-27-2	544
	Xiaolian	笑脸	Smile Face	HD-20-10	537
	Xinzui Shenmi	心醉神迷	Mental Intoxication	HA-01-8	14
	Xinxin Xiangrong	欣欣向荣	Flourishment	HA-10-9	39
	Xinchao Yankou	新潮艳口	Trendy Pretty Mouth	HAR-35-4	420
	Xinfaxian	新发现	New Discovery	HB-40-1	495

续表

字母 English Letters	拼音名称 Name in Pinyin	品种名称 Name in Chinese	英文含义 Meaning in English	所属组合 Combination	页码 Page
X	Xinlangzhuang	新郎装	Bridegroom's Clothing	HB-14-10	455
	Xinlinggan	新灵感	New Inspiration	HB-28-1	472
	Xinmianmao	新面貌	New Look	HB-29-1	474
	Xinsilu	新思路	New Thoughts	HB-39-1	494
	Xinying	新颖	Novelty	HB-22-1	459
	Xinzuopin	新作品	New Masterwork	HB-23-1	460
	Xingguang Canlan	星光灿烂	Starlit Splendid	HA-47-5	99
	Xingguang Gaozhao	星光高照	Starlit's Brightly Shining	HA-47-3	97
	Xingguang Shanshan	星光闪闪	Starlit Sparkle	HA-47-4	98
	Xingguang Zhenghui	星光争辉	Starlit Contending for Brilliancy	HA-47-2	96
	Xinghuo Liaoyuan	星火燎原	Catching Fire from a Spark	HA-62-1	135
	Xingyuan Hongxia	星源红霞	Xingyuan's Red Glow	HA-24-3	73
	Xingyuan Huage	星源花歌	Xingyuan's Flowers Song	HA-84-1	210
	Xingyuan Wanqiu	星源晚秋	Xingyuan's Late Autumn	HA-83-1	207
	Xingyuan Yunhai	星源云海	Xingyuan's Cloud Sea	HA-24-4	74
	Xingwei Yishu	行为艺术	Performance Art	HAR-35-18	434
	Xingfen Huayu	杏粉花雨	Apricot Pink & Flower Rain	HA-58-3	122
	Xingfu Madailin	幸福玛黛琳	Happy Madeline	HAR-02-8	233
	Xingfu Shidai	幸福时代	The Happy Era	HAR-11-3	311
	Xingfu Zhijia	幸福之家	Happy Family	HAR-18-6	366
	Xiongxiong Huoyan	熊熊火焰	Leaping Flames	HA-56-1	113
	Xiunü	羞女	Shy Girl	HD-20-9	536
	Xiuse Taosai	羞涩桃腮	Peach Cheeks with Shyness	HAR-02-27	252
	Xiufeng	绣凤	The Phoenix with Embroider	HD-30-3	551
	Xuri Dongsheng	旭日东升	The Sun Rising from East	HA-70-2	168
	Xuanli Duocai	绚丽多彩	Bright & Colorful	HA-01-10	16
	Xuanli Xiaqiu	绚丽夏秋	Gorgeous Summer & Autumn	HAR-02-25	250
	Xueshan	雪山	Snow Mountain	HD-13-2	527
Y	Yanhong Muxia	艳红沐夏	Bright Red Bathed Summer	HAR-05-2	271
	Yanxia Honglang	炎夏红浪	Red Waves in Hot Summer	HA-16-3	64
	Yanxia Hongsan	炎夏红伞	Red Umbrella in Hot Summer	HAR-17-2	358
	Yanxia Mohong	炎夏魔红	Hot Summer's Magic Red	HAR-08-2	287
	Yanzi Yaohong	艳紫妖红	Brilliant Purple & Peculiar Red	HAR-29-1	405
	Yaolin Xianjing	瑶林仙境	Yaolin's Fairyland	HA-02-3	27
	Yaotiao Shunü	窈窕淑女	Fair Maiden	HAR-02-1	226
	Yeqiu	野秋	Wild Autumn	HC-12-2	511
	Yeshi	夜市	Night Market	HB-41-1	496
	Yideng Yanzhi	一等颜值	First Class Pretty	HAR-03-5	258
	Yijian Zhongqing	一见钟情	Love at First Sight	HA-36-11	87
	Yitian Wangyun	倚天望云	Look over Cloud Against Sky	HAR-35-3	419
	Yishu Shijia	艺术世家	Artistic Family	HAR-20-2	380

续表

字母 English Letters	拼音名称 Name in Pinyin	品种名称 Name in Chinese	英文含义 Meaning in English	所属组合 Combination	页码 Page
Y	Yiwai Shouhuo	意外收获	Windfall	HAR-18-13	373
	Yinbian Xiuyi	银边绣衣	Embroidered Clothes with Silver Borders	HA-81-4	199
	Yingxiong Bense	英雄本色	Hero's True Quality	HAR-02-19	244
	Yingge Yanwu	莺歌燕舞	Orioles Sing & Swallows Dancing	HA-67-7	155
	Yingying Xiaokou	盈盈笑口	Smiling Mouths	HA-70-1	167
	Yongshi	勇士	Warrior	HC-17-1	516
	Yuxia Chengqi	余霞成绮	The Sunset Likely Beautiful Silk	HA-16-5	66
	Yuhou Mujing	雨后暮景	Evening Scene after Rain	HAR-12-4	316
	Yupu Xinxiu	玉浦新秀	Tama-no-ura's New Rookie	HA-81-7	202
	Yurun	玉润	Jade Moisten	HC-10-5	506
	Yuzhu	玉珠	Jade Bead	HD-41-1	564
	Yuexia Yaotai	月下瑶台	Superb Balcony under the Moon	HAR-11-2	310
	Yueyue Meihong	月月玫红	Monthly Rose Red	HAR-14-3	329
	Yuesongzi	越松子	Surpass Pine Cone	HB-30-3	479
	Yuexiu	越秀	Surpass Beauty	HB-30-5	481
	Yuegui Baozhu	粤桂宝珠	Guangdong & Guangxi's Pearl	HA-32-2	78
	Yunduopiao	云朵飘	Cloudlets Drifting	HD-31-1	552
Z	Zhangdeng Jiecai	张灯结彩	Decorated scenes with Lanterns & Streamers	HA-87-1	216
	Zhengqi Douyan	争奇斗艳	Contend in Beauty & Fascination	HA-21-9	69
	Zhengzheng Rishang	蒸蒸日上	Thriving upward	HA-59-1	126
	Zhizu Changle	知足常乐	Contentment Being Happiness	HA-67-3	151
	Zhongqiu Hongxi	中秋红喜	Mid-Autumn Festival's Red Delight	HA-16-4	65
	Zhouban Jinxin	皱瓣金心	Wrinkled Petals with Golden Heart	HAR-28-1	403
	Zhuoyue Fengzi	卓越风姿	Remarkable Posture	HAR-12-3	315
	Zirong Jiemei	姿容皆美	Pose & Face All Pretty	HAR-18-4	364
	Ziban Rongxue	紫瓣融雪	Purple Petals Melting Snow	HAR-02-4	229
	Zichou	紫绸	Purple Silk	HD-20-8	535
	Zifen Wuqiu	紫粉舞秋	Purple Pink Dancing Autumn	HA-43-5	92
	Zihongyuan	紫红苑	Purple Red Dwelling	HB-26-2	464
	Zihuaxiang	紫花巷	Purple Flower Alley	HD-30-1	549
	Zilang Wenxia	紫浪吻夏	Purple Waves Kissing Summer	HA-61-1	133
	Ziliangli	紫靓丽	Purple Beauty	HC-11-1	509
	Ziqi	紫绮	Purple Silk Cheongsam	HC-02-6	501
	Ziqiang Guzhai	紫墙古宅	Purple Wall Ancient Messuage	HAR-37-3	440
	Zishou Jinzhang	紫绶金章	Gold Seal with Purple Ribbon	HA-01-16	22
	Ziyanzhi	紫胭脂	Purple Rouge	HB-31-1	482
	Zuichunfeng	醉春风	Drunk Spring Breeze	HB-30-2	478
	Zuojia Jiaonü	左家娇女	Tender & Cute Girl	HAR-35-1	417